Studies in the History of Philosophy of Mind

Volume 32

The aim of the series *Studies in the History of Philosophy of Mind* is to foster historical research into the nature of thinking and the workings of the mind. The volumes address topics of intellectual history that would nowadays fall into different disciplines like philosophy of mind, philosophical psychology, artificial intelligence, cognitive science, etc. The monographs and collections of articles in the series are historically reliable as well as congenial to the contemporary reader. They provide original insights into central contemporary problems by looking at them in historical contexts, addressing issues like consciousness, representation and intentionality, mind and body, the self and the emotions. In this way, the books open up new perspectives for research on these topics.

This book series is indexed in SCOPUS.

Chiara Beneduce • Mattia Mantovani
Denise Vincenti
Editors

The Sense of Touch

Medieval and Modern Debates in Philosophy and Science

Springer

Editors
Chiara Beneduce
Institute of Philosophy, De Wulf-Mansion
Centre for Ancient, Medieval and
Renaissance Philosophy
KU Leuven
Leuven, Belgium

Mattia Mantovani
Institute of Philosophy, De Wulf-Mansion
Centre for Ancient, Medieval and
Renaissance Philosophy
KU Leuven
Leuven, Belgium

Denise Vincenti
Dipartimento di Filosofia, Scienze
Sociali, Umane e della Formazione
Università degli Studi di Perugia
Perugia, Italy

ISSN 1573-5834 ISSN 2542-9922 (electronic)
Studies in the History of Philosophy of Mind
ISBN 978-3-032-16307-3 ISBN 978-3-032-16308-0 (eBook)
https://doi.org/10.1007/978-3-032-16308-0

This Springer imprint is published by the registered company Springer Nature Switzerland AG
The registered company address is: Gewerbestrasse 11, 6330 Cham, Switzerland

Contents

1 **Introduction: The Sense of Touch. Medieval and Modern Debates in Philosophy and Science** 1
Chiara Beneduce, Mattia Mantovani, and Denise Vincenti

2 **The Aristotelians' Touch. Commenting *On Sense and What is Sensed* (*De sensu et sensato*) in Thirteenth-Century Oxford** 23
Monika Mansfeld

3 **Labour Pain and Touch in Late Medieval Science** 51
Chiara Beneduce

4 **Nicholas of Autrecourt on the Superiority of Touch and the Ontology of Sensible Qualities** 77
Aurélien Robert

5 **With Fingers and Tongue: The Primacy of Touch and Tactile Qualities in Renaissance Galenic Medicine** 103
Elisabeth Moreau

6 **Touch in Second Scholasticism: Francisco Suárez, John Poinsot, Bartolomeo Mastri and Bonaventura Belluto** 121
Daniel Heider

7 **The Sense of (Interoceptive) Touch in Descartes** 143
Alison Simmons

8 **From Fingertips to Papillae: Marcello Malpighi on the Sense of Touch** 173
Luca Tonetti

9 Mary Shepherd on Proprioception and Knowledge of the External World 209
Antonia LoLordo

10 Touch *A Priori*? Tactile Perception of Space in Nineteenth-Century Physiology and Psychology 229
Denise Vincenti

Index 267

Chapter 1
Introduction: The Sense of Touch. Medieval and Modern Debates in Philosophy and Science

Chiara Beneduce, Mattia Mantovani, and Denise Vincenti

Abstract This introductory chapter opens the volume by framing medieval and modern debates on the nature of touch. Spanning the period from the rise of Aristotelian university curricula to the flourishing of experimental psychology, the collection brings into focus the multifaceted discussions surrounding touch—its objects, limits, and operations. It highlights how, across changing intellectual contexts, touch emerged as a persistent and central concern in both philosophical and scientific inquiry, shaping broader reflections on sensation, embodiment, and knowledge.

Keywords Touch · Embodiment · Affective Experience and Sense Ecology · Classification of the Senses · Cognition and the Unity of Experience · Anatomy and Physiology

> One of us decided to ask our blind man whether he would like to have eyes. He replied, "If I was not so curious, I would just as well have long arms […]. Eyes stop seeing well before hands stop touching".
>
> Diderot, *Letter on the Blind for the Use of Those Who Can See* (1749).

C. Beneduce (✉) · M. Mantovani
KU Leuven, De Wulf-Mansion Centre for Ancient, Medieval and Renaissance Philosophy, Leuven, Belgium
e-mail: chiara.beneduce@kuleuven.be; mattia.mantovani@kuleuven.be

D. Vincenti
Università degli Studi di Perugia, Dipartimento di Filosofia, Scienze Sociali, Umane e della Formazione, Perugia, Italy
e-mail: denise.vincenti@unipg.it

C. Beneduce et al. (eds.), *The Sense of Touch*, Studies in the History of Philosophy of Mind 32, https://doi.org/10.1007/978-3-032-16308-0_1

1.1 In Touch with the World

In recent decades, perception has come to the forefront of research in fields as varied as the philosophy of mind and cognitive sciences, comparative ethology and neurophysiology, as well as in cultural and critical studies.[1] We are regularly confronted with new findings on perceptual processes in humans, and discovering novel sense organs and sense modalities in non-human animals and, some argue, even in plants.[2] These findings pose significant questions and invite a re-examination of the conceptual framework that grounds our understanding of sensory experience. This trend has also been reshaping our inquiry into the past, as attested by the growing number of publications on the history of the senses and of perception theories.[3] Indeed, it is no exaggeration to say that perception is one of the hottest topics in present-day research.[4]

While neuroscientists and ethologists unravel the deep structure of the brain and the wonders of the animals' perceptual systems—such as the magnetic compass embedded in the eyes of migratory birds, or the direct pathway between smell and emotional reactions—historians are still puzzled by the most easy-at-hand of our senses: touch. This sense still struggles to find its place in most histories of perception. Research on the long-term history of perception has mostly focused on vision and, more recently, on the so-called "internal senses"—such as memory—and on consciousness.[5] With something of a paradox, we seem to know how people of the past conceived of their inner states much better than of their own body and the workings of their hands. Recent publications have begun to explore the understandings of touch in Antiquity and from the late eighteenth century onwards.[6] By

---

[1] Limiting ourselves to the field of the humanities, see, for example, Howes (2022), showing how perception moved to the centre of research. On philosophy of perception in general and on the history of the philosophy of perception, see, for example, Hamlyn (2022, originally published in 1961); Fish (2010); and Matthen (2015). The original of the Diderot passage quoted in the epigraph above reads as follows: "Quelqu'un de nous s'avisa de demander à notre aveugle s'il serait bien content d'avoir des yeux: 'Si la curiosité ne me dominait pas', dit-il, 'j'aimerais bien autant avoir de longs bras […] les yeux cessent plutôt de voir que les mains de toucher'". Translated in Tunstall (2011, p. 176, punctuation modified).

[2] For a recent survey of the literature, see Yong (2022).

[3] See, for example, Jütte (2005); Knuuttila and Kärkkäinen (2008); Macpherson (2011); Knuuttila and Kärkkäinen (2013); Silva and Yrjönsuuri (2014); Bennett and Toivanen (2020); Heider (2021); Gensler et al. (2022); Toivanen (2022); Mantovani et al. (2023); Kettler and Tullett (2025). As for the early modern age, see also Buchenau and Lo Presti (2017) and Perler and Bender (2020).

[4] It came as no surprise when the Nobel Prize in Physiology or Medicine 2021 was jointly awarded to David Julius and Ardem Patapoutian for their studies of the receptors for temperature and touch: see www.nobelprize.org/prizes/medicine/2021/press-release (last accessed 15 October 2025).

[5] See, for example, Lagerlund (2007); Thörnqvist and Toivanen (2022). See also Mousavian and Fink (2020), which appeared in the same series in which our collection is also published (*Studies in the History of Philosophy of Mind*).

[6] Mark Bradley and Shane Butler lead an entire Routledge series titled "The Senses in Antiquity", which also hosts Purves (2018) on touch. While Pogliano (2015) is an important publication on the history of the sense of touch, it only cursorily addresses touch in the Middle Ages; see also Paterson (2016) and Vincenti (2021).

contrast, the medieval and early modern periods remain sorely underexplored, despite this being precisely when the sense of touch—as we intend to show—took centre stage in the philosophical and scientific debate.[7]

Written by nine leading experts in the field, the present collection aims to finally address this gap in the scholarship by examining the medieval, early modern, and modern debates on the sense of touch through the dual lenses of the history of philosophy and the history of science—two disciplines that were not well differentiated before the early modern period, and which even then, became distinct only problematically. The present collection complements the approach of cultural studies and of the history of comparative literature and the arts, which have of late enjoyed a great fortune, and have duly called attention to the most neglected and "forgotten" of all the senses.[8] We firmly believe that understanding the history of perception, and especially of touch, requires an interdisciplinary approach, one that captures its multifaceted nature and the shifting meanings it has assumed through cultures and centuries. True to this interdisciplinary approach, the present collection adopts the joint perspectives of the history of philosophy and the history of science—two disciplines that have, until now, remained largely absent from discussions on the nature and understanding of touch.

More specifically, the collection spans from the mid-thirteenth to the early twentieth century, namely, from the establishment of Aristotelian curricula in European universities to the heyday of experimental psychology. The long-term approach taken in this collection is intended to pinpoint some pivotal moments in the medieval, early modern, and modern transformation of the understanding of touch, most of which have rarely if ever been studied before. The nine in-depth studies contained in this volume showcase the far-reaching implications of a debate that has engaged philosophers and scientists for centuries, and will hopefully contribute to helping us better understand the framework of the contemporary research on these topics, and the many questions still waiting for an answer.[9] We are on the brink of a decisive shift in the understanding of the nature and of the long-term history of touch—or this, at least, is what the present collection seeks to advocate and promote. It is high time to finally start bringing back to light the hidden history of touch and to trace how our understanding of this sense was transformed throughout the Middle Ages, the Renaissance, the early modern and modern periods.

---

[7] The recently published Beneduce (2024) is an exception to the extent that it is entirely devoted to late medieval scientific views of touch.

[8] See, for example, O'Rourke Boyle (1998); Gubbini (2009); Classen (2012); Moshenska (2014); Maurette (2018); Kearney (2021); Subramanian (2021); Gagné (2021); and Pareti (2024); see also Crucianelli (2024) and Paterlini (2025), exploring the biology of touch and its social role.

[9] Good examples of how current philosophy of science tackles many of the issues on touch that appear in our historical reconstruction are Fulkerson (2014) and Fulkerson's useful survey in the *Stanford Encyclopedia of Philosophy* (first published in 2015 and last revised in 2024).

1.2 Topics under Debate

The present collection does not pretend to account for all the diverse theories of touch developed between the thirteenth and the nineteenth century. No single book could do full justice to the medieval, early modern and modern reflections on this sense, all the more so in light of the fact that we have only recently started to rediscover them. The preliminary research carried out within this collection has made it possible nonetheless to identify a few key topics that were for centuries at the forefront of the debate.

The historical actors' categories changed over time, and the nine chapters of this collection leave us in no doubt as to the sophistication of the concepts employed and of the distinctions at play—distinctions that require close readings of the texts, in their original languages (original-language quotations in our collection are always provided in the footnotes). It would be therefore a methodological error to project a single term or concept onto the entire period under examination. As the collection demonstrates, the very meaning and scope of "touch" (*tactus*, *toucher*, *Tastsinn*, etc.) were far from self-evident and constituted, in fact, one of the principal bones of contention. In order to chart so vast and shifting a landscape without falling into inadvertent anachronism, we believe it to be expedient in this Introduction for us to grasp the nettle and deliberately adopt English and present-day categories as a meta-language, leaving it to the individual chapters—and to the readers' wits—to articulate these topics in their concrete historical specificity and in their own conceptual terms. With these provisos in place, one cannot fail to notice that most of the debates on the sense of touch discussed in this collection revolve around the issues of (i) *embodiment*, (ii) *affective experience and sensory ecology*, (iii) *the classification of the senses*, (iv) *cognition, and the unity of experience* and, finally, (v) *the anatomy and physiology of touch.*

Embodiment Historical understandings of touch, as they emerge from our collection, are deeply connected to broader conceptions of the perceiver's embodied experience. "Touch […] is the sense of the body", and our perception depends on this body of ours being "touched", as only bodies can.[10] Lucretius (c. 98–55 BC), in his *On the Nature of Things*, argued that perception is a matter of touch, and touch alone. Since the rediscovery of this poem in the late fifteenth century, these lines have provided inspiration for physicists and have been a subject for much debate among philosophers.[11] Our collection demonstrates that, contrary to conventional belief, atomist-oriented interpretations of "the sense of the body" and its "sensibles" fol-

[10] "Tactus […] est corporis sensus". Lucretius, *De rerum natura*, II, ll. 434–435 (ed. and trans. Leonard and Smith, 1942, p. 352); "quae tamen omnia corporea constare necessest | natura, quoniam sensus inpellere possunt; | tangere enim et tangi, nisi corpus, nulla potest res". Idem, *De rerum natura*, I, ll. 302–304 (ed. and trans. Leonard and Smith, 1942, p. 235).

[11] See Maurette (2014). For an example of debates among philosophers, see Descartes, AT VI 142, Henry More to René Descartes, 11 December 1648. References for Descartes' works are to Descartes (1964–1974).

lowing in the footsteps of Democritus (c. 470/457 BC–c. 360/350 BC) were already present in philosophical and theological debates of earlier centuries (see ROBERT's chapter). But it was not only atomists and their supporters who credited touch with such an extraordinary role. Indeed, the greatest opponent of the atomists of old, Aristotle (384–322 BC), argued that "the primary form of sense is touch, which belongs to all animals".[12] Touch is not only the one sense necessary for an animal to survive, but its *essential* one, that is, the one sense an animal cannot lack lest it cease *being* an animal in the first place. In other words, an animal is an animal *insofar as* it possesses the sense of touch and, eventually, yet other senses superimposed—and this also holds true of rational animals, i.e., of human beings.[13] On this account, touch is therefore the most fundamental power of the animal's sensitive soul. As our collection shows, the thirteenth-century Aristotelians made these claims their own (see MANSFELD's chapter), and the debate that ensued among the so-called Scholastics was to continue well into the seventeenth century (see HEIDER's chapter). Even in the medical Renaissance, often cited as the clearest example of Lucretius' influence, Aristotle's conception of touch as a fundamental sense remained a key theoretical foundation (see MOREAU's chapter). Our collection illustrates, moreover, how the emergence of new philosophical understandings of the living body in the early modern period opened up new problems and conceptual possibilities. In particular, few thinkers have had so profound an influence as René Descartes (1596–1650) on our understanding of the senses and the philosophy of mind. This is the very domain to which the host series for our collection is devoted, as attested by its title, *Studies in the History of Philosophy of Mind*. The very concept of "mind", be it embodied or not—as opposed to the "soul" of the tradition—is one of the most contested legacies of Cartesian philosophy. Yet Descartes' own understanding of the sense of touch has, surprisingly, rarely been studied in depth, especially with regard to its twofold nature: its outward-facing "exteroceptive" side and its inward-facing "interoceptive" side (the perception of the body's internal states), a duality which sets it apart from all other sense modalities (see SIMMONS' chapter). Later thinkers, by contrast, and especially Mary Shepherd (1777–1847), contended that proprioception—i.e., perception of the body's position and motion in space—is distinct from touch (see LOLORDO's chapter). The rise of scientific physiology and psychology brought this debate to its apogee. By the late nineteenth century, the study of touch had become closely linked with experimental investigations into coenaesthesia, kinaesthesia, and the muscular sense, revealing the crucial role of touch in perceiving both the body itself and the external world (see VINCENTI's chapter).

Affective Experience and Sense Ecology All perceivers engage in touch by virtue of having a body—their own body. While this may seem self-evident, it is far from being a neutral claim, as it raises a range of theoretical questions, particularly con-

[12] Aristotle, *De anima*, II.2, 413b4–5. For Aristotle's works in their English translation, we refer the reader to Jonathan Barnes' edition: Aristotle (1984a and 1984b).

[13] Aristotle's account is not without tensions: see Aristotle, *De anima*, III.13, 435b20–26 and idem, *De sensu et sensato*, 1, 436b11–19.

cerning the inward dimension of touch. This fundamental character of the perception of touch opens up a discussion on *feeling* as an *affective* experience, a business of pleasure, pain, and passion. The nature of affective touch is currently being investigated in both the biomedical sciences and the philosophy of science, and is a privileged subject for phenomenological investigations. The present collection shows that quite a few medieval and early modern thinkers had long reflected on the pleasant and unpleasant characteristics of tactile experience. From Arabic philosophy—Ibn Sīnā (d. 1037) and Ibn Rushd (d. 1198), better known in the West as Avicenna and Averroes—to Latin medieval scholastic thought—e.g., Albert the Great (d. 1280), Peter John Olivi (d. 1298), Peter of Abano (d. 1316), and John Buridan (d. ca. 1361)—authors engaged in discussions on whether pleasure and pain are to be counted among the perceptual *objects* of touch. Among these pleasurable and painful sensations, our collection gives particular attention to *sexual pleasure* and the *pain of childbirth*, and makes a compelling case that the conceptual history of touch is greatly enriched by a diversity-sensitive approach—one that considers gender differences, highlighted through discussions of sex and generation, as well as experiences specific to women (see Beneduce's chapter). These investigations were further pursued by early modern anatomists, who dissected male and female genitalia in order to compare them and to account for their "exquisite sensitivity", *exquisitus sensus*, compared to the rest of the body. Marcello Malpighi (1628–1694) extended his research to the entire skin surface of non-human animals, observing under the microscope feathers and hairs in order to puzzle out their specific way of "getting in touch" with the world (see Tonetti's chapter).

The pleasurable and painful aspects of tactile experience led medieval and early modern thinkers to reflect on the broader "ecology" of the senses. In particular, considerations of sexual pleasure gave rise to a view of touch as not merely essential to *individual survival*, but also as vital to the *preservation* and *reproduction* of the *species*. Thirteenth-century Aristotelians already reflected at length, not without tensions, on the role of touch in securing the animal's *well-being*, by itself and in coordination to the other "external" senses, and to taste especially (see Mansfeld's chapter). Significantly enough, these insights were further developed by thinkers who, on all other matters, defended anti-Aristotelian accounts of the perceptual process. Suffice it here to notice that Descartes—who famously denied sense-perception and a "mind" to non-human animals—went even further in defending the "ecology" of the sense-perceptions, and singled out touch as the most effective of our "bodyguards" (see Simmons' chapter).[14] It was not just a philosophers' problem. For centuries in the wake of Galen (129–c. 216 AD), physicians had reflected on the hand as the medical organ and the most sophisticated diagnostic instrument at their disposal: health proves a touchy business, and medicine stood out as among the finer arts of the animal with hands (see Moreau's chapter).

Classification of the Senses The complexity of touch lies in both its "internal" and "external" side, each raising questions about the number and nature of perceptual

[14] See Mantovani (2022).

capacities ascribed to it, as well as its status within the hierarchy of the senses. In particular, the inward-facing—interoceptive—aspect of touch opened discussions on whether pleasure, pain, and bodily awareness are to be subsumed under this sense. Through an analysis of both edited and unedited sources, the collection identifies the medieval conceptual foundations for Renaissance and later discussions about the possibility of a "sixth sense", alternative to touch, in order to account for the perception of pleasure and pain (see BENEDUCE's chapter). A "sixth sense" was also called to the stage to account for proprioception, as notably argued by Shepherd against George Berkeley (1685–1753), Thomas Reid (1710–1796), and Étienne Bonnot de Condillac (1714–1780) (see LOLORDO's chapter).

The outward-facing side of touch—touch as an "external" sense—was no less problematic. Aristotle himself treated touch as a special case, one that raises difficulties with respect to his overall theory of perception. He held that each of the other "external" senses possesses its own specific class of objects—its proper sensible: colour for vision, sound for hearing, odour for smell, and flavour for taste. Our hands, by contrast, experience a much richer array of features in bodies: their being hot or cold, rough or smooth, dry or wet, soft or hard, light or heavy, and yet many others. Aristotle therefore debated whether touch was one unified sense or a cluster of separate senses, but ultimately affirmed it as a single sense modality. Yet the matter persisted as a point of contention among Scholastics from the thirteenth century to the seventeenth century (see MANSFELD's chapter and that of HEIDER).[15] This was not the only difficulty by which the Latin Aristotelians were beset. The sense of touch seemed to be at risk both of breaking apart and of broadening to incorporate other sense modalities. In particular, the strict similarities between touch and taste makes it difficult to draw a clear-cut distinction between the two. Touch and taste are in fact *contact* senses, namely, senses that operate by direct contact with the perceived object, contrary to the *distal* senses of sight, hearing, and smell.[16] Some thinkers went so far as to argue that touch is, in the end, the one and only sense: "Democritus and most of the natural philosophers […] make all sensibles to be tangibles".[17] Unwittingly, Aristotle's condemnation of the theory served to familiarize thirteenth-century philosophers with this fragment of theory (see ROBERT's chapter), and inspired yet bolder formulations in the early modern period (see TONETTI's chapter), most notably of all in Descartes and in Pierre Gassendi (1592–1655).[18] As our collection demonstrates, the purview of touch and its regard to the other senses was one of the most controversial issues of medieval and early modern theories of perception. Touch had facets almost as multiple as the thinkers who had tried to pin down this slippery sense for so many centuries.

---

[15] For Aristotle's own views, see Aristotle, *De anima*, II.11, 422b20 and III.1, 424b22–425a13. On the unity or plurality of touch in Aristotle, see Johnstone (2021). The question of whether touch is multisensory is still one of the most debated in philosophical discussions about touch; for an overview of such discussions, see Fulkerson (first published in 2015 and last revised in 2024).

[16] On contact senses, see, for example, Johansen (1998, pp. 178–188).

[17] Aristotle, *De sensu et sensato* 4, 442a29–b3.

[18] For Descartes, see Mantovani (2023). For Gassendi, see reference in TONETTI's chapter.

Cognition and the Unity of Experience "Nihil est in intellectu quod non fuerit in sensu". For centuries, university students all over Europe had been instructed that "nothing is in the intellect that was not first in the senses". A question thus arose as to whether some senses were more conductive than others to the acquisition of knowledge. Since ancient times, indeed, the investigation into their specific contribution to the perceiver's cognitive ability resulted in competing hierarchies of senses.[19] In particular, both Plato (ca. 428/423–348 BC) and Aristotle argued for the epistemic prominence of sight over all other sense modalities, and over touch especially.[20] Aristotle claimed that touch occupies the lowest rank in the hierarchy of the senses, insofar as it allows humans to experience only the lowest forms of brutish pleasures in contrast to the purer delights accessible through sight, and the insights it affords us into the heavenly spheres.[21] Elsewhere in his writings, however, Aristotle claimed that a softer flesh lends itself to a finer sense of touch and, therefore, to a more apt mind. This theory became one more point of reference and common lore in the Middle Age, encoded in the motto "molles carne, apti mente" (see BENEDUCE's chapter).[22] Touch, therefore, according to Aristotle and his followers, is, at once, the most "brute" of all senses, on the one hand, and the base of cognition, on the other. As such, it is the mark of humans' animality *and* of their cognitive preeminence: the hierarchy of the senses resulted in a hierarchy of perceivers (see MANSFELD's chapter). By the thirteenth century, some thinkers were already pushing this point further still, and argued that touch is the sense modality least prone to errors, and the only sense that permits perceivers to ascertain the real existence of an object, as opposed to the "appearances" that fool the eyes (see ROBERT's chapter). These long-standing debates about the value and ranking of the senses led to ongoing re-evaluations of how sensory perception contributes to knowledge. Thus, for example, in order to cast doubts onto the reliability of perceptions, Descartes set out to prove that vision and touch are *both* prone to errors, and to analogous ones, whether in the form of double vision or in the experience of one marble as if it were two upon touching it with crossed fingers (see Fig. 1.1)—an illusion first described by Aristotle, and that still goes under his name.[23]

With regard to cognition, the asymmetry between the senses—not to say their open opposition—posited a challenge to the possibility of integrating them into *one* unified experience. Aristotelians introduced a "common sense" for this task, amongst its many other functions, but it was not until the early modern period that

[19] On the *querelle* over sight versus touch, see, for example, Kambaskovic-Sawers and Wolfe (2014).

[20] There is a famous passage in Plato: *Rep.* VI, 508a–b. For Plato's text in English, we refer the reader to Paul Shorey's translation: see Plato (1969). Famous passages in Aristotle include Aristotle, *Metaphysica*, I.1, 980a21–28 and *De anima*, III.3, 429a3–4.

[21] See Aristotle, *Ethica Nicomachea*, X.5, 1175b24–1176a3; III.10, 1118a2–b7.

[22] The *Auctoritates Aristotelis* list: "Molles carne mente aptos dicimus, duros vero ineptos". Anonymous, *Auctoritates Aristotelis*, AA 6:99 (ed. Hamesse, 1974, p. 182). Cf. Aristotle, *De anima*, II.9, 421a7–26.

[23] Descartes, AT VI 142, *Dioptrique* VI. On "problems, perceptual errors, and misperceptions" in the history of perception theories, see Glenney and Silva (2019).

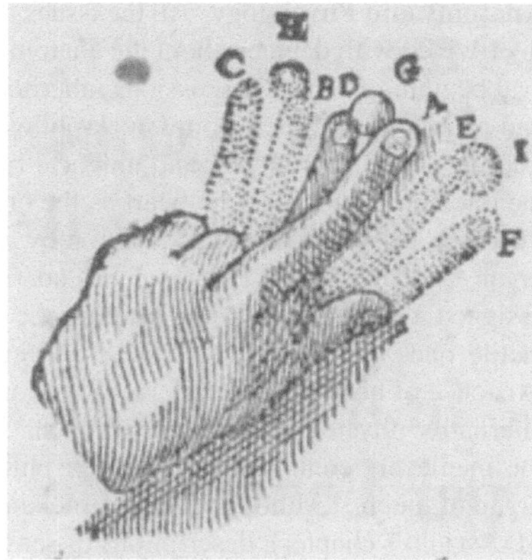

Fig. 1.1 Descartes' depiction of Aristotle's Illusion in the *Dioptrique*, VI. Open domain

this debate reached its peak.[24] In 1688, William Molyneux (1656–1698) famously posed a question to John Locke (1632–1704), asking whether, according to his theory, a blind person who suddenly regained their sight would be able to immediately distinguish between a cube and a sphere, as could be done with the hands.[25] Arguably, Molyneux's Problem is *the* debate concerning the sense of touch that had attracted the greatest deal of attention in scholarship, including a recent collection of essays entirely devoted to its long-term history and attempts at new answers in light of up-to-date research in the cognitive sciences.[26] Our collection builds upon these studies and further pursues their approach and their invitation to reconsider the early modern discourse on the senses from a more inclusive viewpoint. The case of Mary Shepherd, indeed, leaves no doubt as to the need to revisit lesser-studied thinkers and to bring back their voices in the debate (see LoLordo's chapter). Shepherd's reflection shows that, from the first half of the nineteenth century onwards, the sense of touch was addressed in the larger framework of the muscular sense(s) and of bodily feeling more generally, as exemplified by *The Hand* by Charles Bell (1774–1842). The late–nineteenth century debate opposing "empiricist" and "nativist" thinkers—to employ the terms used by Hermann von Helmholtz (1821–1894)—further complicated the issues at stake, including Molyneux's problem. Immanuel Kant (1724–1804)'s notion of the *a priori* provided a powerful tool for rethinking the role of touch in the perceptual process, as well as in bodily organization, self-affection, coenaesthesia and kinaesthetic movements (see Vincenti's chapter).[27]

[24] On the common sense, see, among others, Gregoric (2007).

[25] See Locke, *An Essay Concerning Human Understanding* (1786, vol. 1, pp. 107–108); Locke. *Letters 849–1241* (ed. E.S. de Beer, 1978); and Molyneux, *Dioptrica nova* (1692).

[26] See, respectively, Ferretti and Glenney (2020) and Glenney (2025); see also Morgan (1977), Degenaar (1996), and Chottin (2014).

[27] On innatism and nativism in eighteenth-century theories of perception, see Hatfield (1991).

Anatomy and Physiology All the issues discussed in this Introduction were closely intertwined with debates about the anatomy and physiology of touch. At their medieval beginnings, these discussions inherited reflections on the anatomy, physiology, and pathology of touch from Greek philosophical and medical traditions. Yet these traditions had left no clear and univocal biological accounts of touch. Aristotle, on the one hand, identified the heart as the organ of touch and flesh as its medium; on the other, he wavered on this position by ascribing to flesh, at times, the role of the organ of touch itself. Drawing instead from the Greek medical tradition, Galen assigned a central role to the brain in the processes of sense perception, including tactile ones. In contrast to Aristotle's cardiocentrism, Galen also recognized the existence of nerves and their crucial role in sensory perception, thereby offering an alternative physiological model of touch.[28] These two views were widely debated in the medieval "controversies between philosophers and physicians" regarding the organ of touch.[29] Although the Aristotelian position enjoyed a lasting influence (see MANSFELD's chapter), descriptions of sense perception based on the function of the nerves also proved to be influential—thanks in part to the mediation of Avicenna, whom medieval Latin university masters considered to be the main authority for the idea that nerves, rather than the heart, serve as the organs of touch.[30] From Galen, premodern and early modern science also inherited the emphasis on the hand as a heuristic and diagnostic instrument, particularly in medicine.[31] Renaissance physicians regarded the hand as essential for assessing a patient's bodily constitution and state of health (see MOREAU's chapter). In the early modern period, with the advent of the microscope and increasingly refined techniques of dissection, discussions of touch and of the role of the nerves in tactile perception became more sophisticated. These culminated in theories and anatomical drawings such as those of Malpighi, who proposed the existence of "papillae"—the extremities of nerve branches, and thus the external "organs" of both taste and touch (see TONETTI's chapter).

Aristotle's treatment of the so-called "contact senses" was no more definitive than his view of the organ of touch. He acknowledged that touch and taste differ from the other senses, as they perceive their objects through direct contact; yet he insisted that perception in all senses requires a *medium*, serving as a conduit between the object

[28] For Aristotle, see the following passages: Aristotle, *De anima*, II.11, 423b18–27; *De sensu et sensato*, 2, 439a2–3; *De partibus animalium*, II.1, 647a19–21; II.8, 653b23–27; and II.10, 656b33–37. In Galenic thought, bodily functions are mainly distributed between the brain, the heart, and the liver, but it is the brain that is specifically responsible for sensation. On Galen's principles of the body, see Galen, *De Placitis Hippocratis et Platonis*, VII, 3 (ed. and transl. De Lacy, 1978–1984, pp. 439–449).

[29] On the medieval "controversy between philosophers and physicians" regarding the organ of touch, see, for example, Hasse (2000), Lagerlund (2008), Beneduce (2019) and (2023), and Knuuttila and Kärkkäinen (2013).

[30] On Avicenna's position regarding the organ of touch, see Avicenna Latinus, *Liber de anima seu sextus de naturalibus*, pars II, cap. 3 (ed. Van Riet, 1972, p. 138).

[31] See Von Hoffmann (2023).

and the sensory organ.[32] The distinction between the contact senses and those operating at a distance, and the question of whether a medium must be posited to account for tactile and gustatory perception, remained matters of debate long after Aristotle (see MANSFELD's chapter). These discussions had far-reaching implications, such as the question, for example, of whether *sensible species* ought to be assumed not only for the distance senses but also for the contact senses (see HEIDER's chapter).

The analysis of the anatomy and physiology of touch underwent further elaboration in the modern period. The question of the relationship between touch and the muscular system—whether touch and the muscular sense are in fact one and the same or whether they are two distinct senses—marks a crucial stage in this development. The problem—first explicitly formulated by Bell in 1833—can be seen as a further elaboration of ideas already found in earlier philosophers such as Berkeley (see LOLORDO's chapter), and, more distantly, in much older debates (see BENEDUCE's chapter). Similarly, the analysis of inner touch as a means of apprehending the intimate feeling of existence—the body's awareness of itself as alive—which famously occupies a central place in Condillac's *Treatise on Sensations* (1754), in fact underpins much of the modern discourse on the senses (see SIMMONS' chapter). It was above all in the late–nineteenth century reflections arising from scientific physiology and psychology that the anatomy and physiology of touch found new methodological and instrumental foundations. In particular, the compass of Ernst Weber (1795–1878) made it possible to trace, with experimental precision, the sensory circles composing the body and determining the tactile sensitivity of each region. The mapping of the body according to its zones of sensitivity paved the way for a new tactile anatomo-physiology, one capable of assigning to this sense a fundamental role in the perception of one's own body and of the external world. Similarly, the study of the relation between touch and movement, by the end of the nineteenth century, came to rely on new technical, instrumental, and conceptual apparatuses that both clarified the nature of this sense and underscored its importance in perceptual and proprioceptive processes (see VINCENTI's chapter).

Embodiment, *affective experience and sense ecology*, *the classification of the senses*, *cognition and the unity of experience*, and *anatomy and physiology*: this list is by no means exhaustive. As this overview suggests, and as the contributions to this collection make clear, many of these themes were addressed simultaneously by the thinkers involved, often blurring the boundaries between them and revealing their deep interconnections. The complex, fascinating field of the history of the conceptualization of the sense of touch undoubtedly warrants further investigation. Indeed, this collection aims not only to offer readers a sense of the diversity and relevance of the many of the positions at play and of the questions at stake but also to serve as a call to arms. We look forward to future studies in the field, hoping that our collection will help to pave the way for new lines of research.

[32] On the medium of perception, see Johansen (1998, pp. 116–147, pp. 178–193, pp. 199–212), and Massie (2013).

1.3 Overview of the Chapters

In the first chapter of the collection, Monika Mansfeld's "The Aristotelians' Touch. Commenting *On Sense and What is Sensed* (*De sensu et sensato*) in Thirteenth-Century Oxford", the main theoretical issues of the Aristotelian theory of touch are presented alongside an examination of how the complex set of problems concerning this sense, as found in Aristotle, was developed in some of the earliest surviving Latin commentaries on his psychological writings. While recent scholarship has investigated conceptualizations of touch in fourteenth-century Parisian natural philosophy, no study prior to Mansfeld's research had focused on how touch was understood in Oxford natural philosophy a century earlier.[33] The sources discussed by Mansfeld primarily consist of unedited commentaries on Aristotle's *De sensu et sensato*, in various forms (literal expositions, question-commentaries, and doubts), which were written by both anonymous and prominent authors from the thirteenth-century Oxford milieu, e.g., Adam of Bockenfield (d. before 1294), and Geoffrey of Aspall (d. 1287). Mansfeld identifies and describes the main themes related to touch that emerged in thirteenth-century Oxford commentaries on the *De sensu*. In her reconstruction, the commentaries outline the concepts of "spiritual touch" and "material touch" as two different types of touch, or rather, as two different modalities by which the senses work: "spiritual touch" is proper to senses that work at a distance—namely, sight, hearing, and smell –, while "material touch" is proper to senses that work by contact—namely, touch and taste. Mansfeld explores the possible sources for this Oxford discussion and frames these concepts in the broader theories of touch developed by the Oxford commentators. In addition to the concepts of "spiritual" and "material" touch, the commentaries studied by the author present a set of reflections on touch largely influenced by Aristotelian thought, especially related to the role of touch in animal life, its epistemic significance within the sensory system, its perceptual objects, and its anatomo-physiological functioning. She emphasizes the importance of these texts for mapping medieval interpretations of Aristotle's theory of touch and for the establishment of a Latin Aristotelianism, as they were composed at the very outset of the Latin reception of his psychological works. While these commentaries generally remain faithful to Aristotle's teachings, Mansfeld demonstrates that they also exhibit originality—particularly in their reflections on the role of touch in distinguishing animals from other forms of life, and in their exploration of the concepts of "perfect" and "spiritual touch".

The second chapter, Chiara Beneduce's "Labour Pain and Touch in Late Medieval Science", shows how the Aristotelian problem of touch's multiple perceptual objects was extended in the late Middle Ages to include pleasure and pain as potential objects of touch. Aristotle himself had already reflected on the experience of pleasure through touch, mostly from an ethical viewpoint.[34] Beneduce singles out four main medieval frameworks of discussions of pleasure and pain as potential perceptual objects of touch. The first two are (i) discussions over whether touch is a

[33] On theories of touch in fourteenth-century Parisian natural philosophy, see Beneduce (2024).

[34] See footnote 21 in this Introduction.

single sense or a plurality of senses, along with its objects, and (ii) Aristotle's claim that corporeal pleasures are objects of touch. With regard to (i) and (ii), the chapter browses a range of sources: from Avicenna and Averroes—authorities for later medieval Latin discussions of pleasure and pain as the object of touch—to prominent philosophical and medical authors of late-medieval philosophy and science, including Albert the Great, Peter John Olivi, Peter of Abano, and John Buridan; and lesser known or anonymous authors, such as John Vath (active between the end of the thirteenth century and beginning of the fourteenth century) and an anonymous commentary on *On Animals* (*De animalibus*), preserved in the Vatican Library (ms. Vat. Lat. 2164) and probably written in the early fourteenth-century. By concentrating on pleasure, particularly in its sexual dimension, Beneduce highlights the medieval roots underpinning Renaissance and subsequent debates on pleasure and pain as tactile experiences, the notion of sexual pleasure as a potential "sixth sense", and the differentiation between the muscular sense and ordinary touch. The other medieval theoretical frameworks highlighted by Beneduce for conceiving the relationship between touch and the perception of pleasure and pain are (iii) debates on the nature of pleasure and pain (including, indeed, which senses are involved in their perception), and (iv) investigations into specific types of pleasure and pain. By drawing attention to gender differences in reconstructing the conceptual history of touch, Beneduce analyses, in the last part of her chapter, late-medieval scientific views of labour pain, as a case study for exploring to what extent touch was involved in medieval explanations of pain. For this analysis, she draws on commentaries on Aristotle's *On Animals*, on Avicenna's *Canon*, and on Hippocratic works produced in the university milieu—including edited and unedited works, both by famous and lesser known authors, i.e., Peter of Spain (thirteenth century), Albert the Great, Pedro Gallego (d. ca. 1267), Master Vericus (end of the thirteenth century), Dino del Garbo (d. 1327), Gentile da Foligno (d. 1348), and Jacques Despars (d. 1458). Though touch is very frequently involved in late-medieval explanations of the perception of pleasure and pain, Beneduce concludes that this does not seem to be the case for most medieval scientific explanations of labour pain.

The third chapter in the collection, AURÉLIEN ROBERT's "Nicholas of Autrecourt on the Superiority of Touch and the Ontology of Sensible Qualities", reconstructs the distinctive—and sorely neglected—position of Nicholas of Autrecourt (d. 1369) in the longstanding *querelle* over the epistemic primacy of sight versus touch. For Robert, the debate about hierarchy of the senses provides an occasion to consider Autrecourt's view on the nature of sensible phenomena (the so-called "apparentiae"), and to reassess interpretations of Autrecourt's epistemology and ontology. Autrecourt approaches the hierarchy of the senses via the question of their reliability, reflecting, in other words, on which sense should be held most accountable in our perceptions of reality. This reflection on the reliability of the senses is found in Autrecourt's work *A Useful Treatise to See Whether the Arguments of the Aristotelians Were Demonstrative*, which is most commonly known as *Exigit ordo*. By carefully reading the relevant passages of the *Exigit ordo* (in a chapter tellingly titled "Does everything that appears exist?"), Robert shows the prominence that Autrecourt ascribed to touch in the system of the senses. In Autrecourt's construction, touch provides the most reliable access to reality, insofar as it guarantees the least doubtful assessment of the existence and nature of sensible objects. According to Autrecourt, touch is the only sense that can achieve certainty regarding the object's real existence, since it is the only sense that apprehends something that exists in a subject and does not continually vary through time. Robert underscores that this epistemic primacy that Autrecourt ascribed to touch sets him apart from Aristotle and most medieval authors. In Robert's interpretation, the epistemic re-evaluation of touch proposed by Autrecourt situates him instead within another philosophical tradition: Democritus' theory of sensible qualities. Robert suggests that Autrecourt, possibly aligning with the interpretation of Democritus' explanation of colours given by Giles of Rome (d. 1316), claims, first, that touch assures us of the existence of the material bodies, and secondly, that sensible appearances are fluctuating properties arising from material bodies. Robert's chapter compellingly shows that Aristotle was far from being the only authority for theories of touch developed at medieval universities: Democritus was as a key figure in medieval discussions of the epistemic role of the senses.

The humanist rediscovery of yet other ancient sources—like the Stoics, Lucretius and Sextus Empiricus (c. 160 AD–c. 220 AD)—presented early modern thinkers with a vast array of alternative conceptual possibilities concerning the nature of touch: atomism, for example, permitted a convenient explanation of touch's working as a contact sense, whereas the arguments of the sceptics called into question touch's reliability. In this historical reconstruction, scholarship has particularly identified *On the Nature of Things* (*De rerum natura*) by Lucretius as the primary source for the Renaissance's elevation of touch.[35] The fourth chapter in our collection, Elisabeth Moreau's "With Fingers and Tongue: The Primacy of Touch and Tactile Qualities in Renaissance Galenic Medicine", aims to problematize this well-established scholarly interpretation of Renaissance theories of touch. By retracing the medical theories of Jean Fernel (1497–1558) and Jean Riolan the Elder (1539–1605), Moreau argues that Renaissance physicians, when promoting touch and taste as *the* epistemological tools for understanding the constitution of material bodies, were mostly following the Galenic tradition. Galen and his view of touch were taken up by both Fernel and Riolan as the point of departure for further elaborating their physiological and pharmacological theories. In the case of Renaissance physiology, moreover, the prominence of touch was based not only on Galen's authority; Galen himself had based some of his views about touch on Aristotle, and the theory of touch advanced by Renaissance physicians was indebted to both Galen and Aristotle. Deriving from these ancient authorities, Renaissance physiology was itself grounded upon the following ideas: the superiority of human touch compared to touch in other animals; the hand as the instrument for assessing the patient's bodily constitution (the so-called "complexion"); the tactile nature of primary and secondary qualities; and the similarity between touch and taste as contact senses. This last-mentioned aspect, inspired by Aristotle, became particularly relevant to Renaissance pharmacology. Because of the similarity—spelled out by Aristotle—between touch and taste as contact senses, Renaissance physicians involved in drug analysis considered taste and texture tactile

[35] See footnote 11 in this Introduction.

qualities. Based on Galen's writings, moreover, taste was granted an especially prominent role in assessing the active powers of drugs. Taste was considered, in fact, to offer the advantage of detecting qualities potentially present within bodies, since those qualities were activated when a drug was placed in the mouth. Thus, it was not just the hand but also the tongue that became a key tool in Renaissance medicine. In short, building upon Aristotle's characterization of touch as a contact sense and its connection to taste, and incorporating the importance of taste emphasized in Galenic medicine, Renaissance physicians placed touch and, especially, taste at the core of their discipline and practice. Despite the many references found in Renaissance works to *De rerum natura*—as, for example, in Fernel's masterpiece the *Universa medicina*—more "traditional" authorities (namely, Aristotle and Galen) are likely to have played a more foundational role than Lucretius in the importance ascribed to touch and taste in Renaissance medicine.

Moreau's chapter ultimately shows that Aristotle's views of touch had a substantial impact on Renaissance theories of touch, even in a medical context. This proves even truer for the so-called "Second Scholasticism", an influential wave of late-sixteenth and mid-seventeenth century thinkers, mostly active in universities, who professed strict adherence to Aristotle's views. The fifth chapter in our collection, Daniel Heider's "Touch in Second Scholasticism: Francisco Suárez, John Poinsot, Bartolomeo Mastri, and Bonaventura Belluto", finally does justice to debates about the sense of touch in the Second Scholasticism. Heider illustrates the considerable diversity of positions at play in these discussions by focusing on a few exemplary thinkers, some of whom are rarely considered in the literature: the Spanish Jesuit Francisco Suárez (1548–1617), the Portuguese Dominican and Thomist John Poinsot, also known as John of St. Thomas (1589–1644), and two Italian Conventual Franciscans and Scotists, Bonaventura Belluto (1600–1676) and Bartolomeo Mastri (1602–1673). Heider reconstructs the subtle points of controversy that distinguished these thinkers, particularly in response to Suarez's account. His study reveals a broader trend and parting of the ways among these Aristotelian commentators, with Poinsot criticizing Suarez, and Mastri and Belluto, in turn, challenging Poinsot in attempts to reconcile Suarez and Duns Scotus (1265/66–1308). Heider considers their divergences to pivot upon four issues: (i) the unity or plurality of touch; (ii) whether sensible *species* should be considered not only for distance senses but also for contact senses; (iii) the organ of touch; and (iv) the necessity or dispensability of the tactile medium. These questions all had clear origins in the Aristotle's puzzling theory of touch, and especially his account of the contact senses. Heider shows that all four authors agreed on the answer to the first two questions, arguing that touch is a *single* sense and, like all other external senses, cannot function without *species*—in this case, tangible ones. They significantly disagreed, however, on the latter two points. In particular, Poinsot stood apart from the others by identifying the nerves as the organ of touch and fine-tuning this theory to argue for the necessity of a medium, even for the paradigmatic contact sense. Heider highlights the sophistication of these diverging theories and the strategies they employed to stay consistent with Aristotle's teachings, even if not through literal readings.

Close study of historical debates surrounding the nature of touch—its objects, boundaries, and workings—calls into question any overly simplistic division between an "Aristotelian" Middle Age and an anti-Aristotelian Renaissance. Over the course of the seventeenth century, issues became further complicated by the emergence of novel philosophical paradigms, which led historical actors to speak of a "new" philosophy. One thinker was soon to stand out among of these *novatores* and to modify the intellectual landscape of the time: Descartes. The sixth chapter in our collection, "The Sense of (Interoceptive) Touch in Descartes", written by ALISON SIMMONS, addresses the complexity of touch in Descartes' thought. Within a broader interpretation of Descartes' theory of the senses—which presents the senses as primarily aimed at ensuring the perceiver's survival and well-being while navigating a world of bodies poised to benefit or harm us—Simmons shows that touch operates as an exceptional "bodyguard." For Descartes, it is through the sense of touch that we perceive our own body "from the inside", as in the experiences of pain and tickling. Simmons demonstrates this through detailed analysis of Descartes' peculiar taxonomy of the "external" and "internal" senses and its evolution across different works, foregrounding both its sophistication and its unresolved issues (the chapter concludes with a few open questions for future research). In Simmons' reading, "the neglected side of touch" is not only an effective "bodyguard" but also proves crucial to Descartes' very account of the human being as an embodied mind. It is because of interoception that each of us is able to single out *their own* body from others: the distinction between interoceptive and exteroceptive sensations therefore precedes—and grounds—the distinction between *my* body and *external* ones. Analogously, Simmons highlights the significance of interoception for health, and its role in relation to the passions and human agency. She substantiates her reading by engaging with various alternative interpretations of Descartes' theory of perception and contemporary accounts, effectively undermining Antonio Damasio's influential claim that Descartes failed to recognize the mind's embodiment within an organism.[36] The mention of Descartes in the 2021 Nobel Prize for Medicine announcement, which recognized the research by Julius and Patapoutian on temperature and touch receptors, emphasizes the ongoing relevance of Descartes' reflections, and of Simmons' chapter, for contemporary debates.[37]

In the latter half of the seventeenth century, the senses became a topic of widespread debate across all of Europe, partly in response to the distinction between primary and secondary qualities canonized by Locke in his *Essay Concerning Human Understanding* (1690).[38] Locke's claim that the non-geometrical qualities of bodies aren't quite as we perceive them called into question the ontology of sensible qualities and of "the proper sensibles" underpinning Aristotelian accounts. An opponent of this new philosophical perspective complained that "it was no longer in fashion to believe that fire is hot, that marble is hard: [...] these truths were too ancient for those who loved novelty".[39] These intense debates relied extensively on new observations,

[36] See Damasio (2008, pp. 245–252).

[37] See footnote 4 in this Introduction.

[38] Locke, *An Essay Concerning Human Understanding* (1786, vol. 1, p. 102).

[39] "Ce n'estoit plus la mode de croire que le feu estoit chaud, que le marbre estoit dur [...] ces veritez estoient trop anciennes pour ceux qui aimoient la nouveauté". Babin, *Journal ou relation fidele* (1679, p. 1).

including those made using the microscope, as exemplified by the well-known case of Robert Boyle (1627–1691).[40] It was not only the "sensibles", however, that were subject to the scrutiny of the "new" science: the sense organs, including those for touch, also assumed a new look when placed under the magnifying lens. The seventh chapter in the collection, LUCA TONETTI's "From Fingertips to Papillae: Marcello Malpighi Exploring the Sense of Touch", examines this important line of empirical research by considering one of the most celebrated anatomists of the early modern period: Marcello Malpighi. Suffice it to note that Malpighi's international reputation led to his election in 1669 to the Royal Society, which went on to publish all of his subsequent works. Tonetti's chapter focuses on Malpighi's published and unpublished writing from the 1660s onwards, bringing to light fresh textual and iconographical evidence regarding the evolution of his ideas about the anatomy and physiology of touch, which evolved partially in response to the critiques of his research methods and to the results advanced by Giovanni Girolamo Sbaraglia (1641–1710). In two works published in 1665—*On the Tongue* (*De lingua*) and *On the External Organ of Touch* (*De externo tactus organo*)—Malpighi caused a stir with his discovery of tiny structures in the tongue, and later in the rest of the skin, which he observed using his microscope. Malpighi maintained that these "papillae" represented the terminal points of nerve branches and thus functioned as a sort of external organ responsible for taste and touch. Tonetti painstakingly analyses Malpighi's lecture notes to reconstruct how he developed these views and the many challenges he faced, starting with an identification of which preparation and dissection techniques to employ. Tonetti shows that Malpighi's work on the tongue drew inspiration from culinary recipes—a remarkable exemplification of Malpighi's scientific "practice". Drawing was also essential to Malpighi's work, and Tonetti closely examines the various representations of these "papillae" to emphasize the subtle evolution of Malpighi's views, along with his intense focus on observing and representing everything from "monstrous" feet to the surface of the penis glans. Malpighi maintained that the numerous "papillae" of the lips and glans were responsible for the "exquisite sensitivity" (*exquisitus sensus*) of these body parts compared to others. Indeed, Tonetti shows that comparative anatomy played an important role in Malpighi's observations, by permitting him to test his theory that "papillae" are not components of the skin but of nerve endings. Malpighi also drew on metaphors from the animal world to explain the transmission of sensory stimuli along the nerves, comparing them to the tentacles of an octopus—an analogy which explicitly indexed the Stoic idea of a centralized soul that extends to the animal's peripheral sense organs. Tonetti documents Malpighi's continuous engagements with other anatomists and thinkers, from Antiquity through to his own time, which emerge in his remarks on the phantom limb illusion, his reference to Gassendi, and in implicit dialogue with Descartes. Tonetti's compelling argument regarding the fundamental role of illustrations in the development of Malpighi's views on the "papillae" also offers a fresh perspective on the interplay of touch and the other senses, particularly sight.

In her chapter "Mary Shepherd on Proprioception and Knowledge of the External World", the eighth of our collection, ANTONIA LOLORDO analyses the *Essays on the*

[40] Boyle, *Certain Physiological Essays* (1661).

Perception of an External Universe (1827) by Shepherd. Shepherd's theory of perception asserted that proprioception—perception of the body's position and motion in space—was distinct from touch, and thereby called into question a century-long debate surrounding the classification of human perceptual abilities. Shepherd argued that motion constitutes a "sixth sense" and that the "power of motion" functions as a "sixth organ of sense". LoLordo skilfully positions Shepherd's philosophical views within the context of earlier discussions, most notably those of Berkeley, Reid, and Condillac. In Shepherd's account, these earlier thinkers granted importance to bodily awareness in perception but tended to absorb it into the sense of touch, thus overlooking proprioception as an independent sensory faculty. As LoLordo notes, Shepherd broke with this tradition by arguing that bodily awareness, arising from proprioception, yields distinctive sensations and conceptual content not reducible to touch. Shepherd thus attributed an essential epistemic role to proprioception in the formation of concepts crucial to understanding the external world, including "extension", "externality", and "the self". LoLordo shows how Shepherd's account of proprioception informed her broader epistemology, which sought to reconcile reason and sensory experience in explicit response to the scepticism of David Hume (1711–1776). Shepherd redefined the concept of the self, moreover, by blending proprioceptive and interoceptive awareness. This nuanced view allowed her to advance a coherent account of how humans acquire knowledge of their own bodies, of external objects, and of the spatial relationship between them. In recovering Shepherd's important contributions to these debates, LoLordo also emphasizes her originality and relevance to the history of philosophy. By recognizing proprioception as a distinct sensory modality, indeed, Shepherd prefigured twentieth-century phenomenological accounts of embodiment, particularly those defended by Merleau-Ponty (1908–1961).

Shepherd's work is an interesting example of how certain conceptualizations of touch within the "empiricist" tradition offered insights for later theories and positions, even theories and positions contradicting the views initially put forward by the empiricists themselves. In the concluding chapter of our collection, "Touch *a priori*? Tactile Perception of Space in Nineteenth-century Physiology and Psychology", Denise Vincenti investigates the redefinition of the empiricist/rationalist debate in nineteenth-century German physiology and psychology, highlighting how it expanded the analysis of touch. As the title suggests, the chapter explores the (sometimes problematic or elusive) link between tactile perception of space and certain *a priori* elements of knowledge. In nineteenth-century German discussions on space perception, indeed, both nativist and empiricist thinkers often resorted to the notion of *a priori* to characterize perception of space. In this context, one of the main legacies of the philosophy of Kant was the idea that the innate cause of necessary knowledge *differs from* that knowledge itself.[41] The *a priori*, in other words,

[41] Kant's philosophy, indeed, made it clear that the "innate cause of necessary knowledge does not *resemble* that knowledge itself, and the great influence of this [Kantian] theory in this respect finally puts an end to the doctrine of 'innate ideas' as such". See Moore (1957, p. 131). As a consequence, the question of what is innate and what is *a priori* in knowledge became connected to the problem of biological inheritance, or rather, to some congenital, inborn aspects of the knowing subject.

ceases to be a structure of knowledge and acquires new features. Illustrative examples can be found in the reflections of Johannes Peter Müller (1801–1858) and of Ernst Weber, where the *a priori* was both embodied and physiologically defined. That which, for these thinkers, precedes and spatially organises sensory experience in space perception is the structure of the sense organs. Indeed, the spatial arrangement of the nerves sensitizing the sense organs of both touch and sight are innate, congenital, or inborn, and thus immediately provide a spatial quality to their sensations. Touch, in particular, constitutes the standard by which the perceiver assesses the spatiality of objects. Vincenti shows how, despite some important differences, the nativism of Carl Stumpf (1848–1936) similarly asserts that space is immediately given in sensation. Having abandoned a strictly physiological viewpoint, however, Stumpf connects the immediacy of spatiality to the idea that some structural forms (first and foremost *space*) are immanent to sensory materials. For Stumpf, space is therefore conceived as a sort of *common sensibile*, a cross-sensory notion, indissolubly linked to sensations that are grasped by different senses (primarily, by touch and by sight). Vincenti also emphasizes the connection between apriorism and tactile experience in the empiricist discourse of the time. Although Hermann Lotze (1817–1881) argued, for instance, that sensations are not immediately endowed with spatial qualities, his theory of space perception recognized that the perception of space is inherently and *a priori* part of the nature of the mind. It is the mind, therefore, that spatially organizes sensations (which, in themselves, lack such order), with the help of the *local signs* provided by sight and touch. In harmony with this idea, Wilhelm Wundt (1832–1920) emphasises the mind's role in imposing spatial order on sensations, introducing the notion of psychological "fusion": space perception is the result of a *fusion* of data coming from touch and from sight. Vincenti highlights, however, that Wundt's reflection lacks reference to the notion of *a priori*. What is instead preserved is the indisputable role of touch: Wundt privileges tactile representations and representations of movement over other sensory representations, as the foundation for the perceiver's apprehension of spatial form.

Acknowledgements This collection is the outcome of the conference *The Sense of Touch. Medieval and Modern Debates in Philosophy and Science* held at Radboud University in Nijmegen on 19–21 June 2023. We sincerely thank our speakers—and, now, authors—for their splendid contributions and their enthusiasm throughout the entire project. Our special thanks go also to the directors of the series *Studies in the History of Philosophy of Mind* for hosting this collection, and to the anonymous referees for their most precious advice, especially as regards the present Introduction. The research for this Introduction was made possible thanks to the financial support of the "ANR Access ERC Stg – édition 2023" grant, awarded by the *Agence nationale de la recherche* (ANR) for the project "Female Bodies, Female Sensations: A History of Female Corporeity and Sense-Perception in Medieval Science" (project number ANR-23-AERC-0017) and thanks to the generous funding of the *Research Foundation—Flanders* (FWO) for the project "From Truth to Ecology: Reassessing the History of the Senses, 1250–1750" (12ZY922N).

References[42]

Primary Literature

Anonymous. (1974). Auctoritates Aristotelis. In J. Hamesse (Ed.), *Les* Auctoritates Aristotelis: *Un florilège médiéval. Étude historique et édition critique*. Publications Universitaires.

Aristotle. (1984a). In J. Barnes (Ed.), *The Complete Works of Aristotle: The Revised Oxford Translation* (Vol. 1). Princeton University Press.

Aristotle. (1984b). In J. Barnes (Ed.), *The Complete Works of Aristotle: The Revised Oxford Translation* (Vol. 2). Princeton University Press.

Avicenna Latinus. (1972). Liber de anima seu sextus de naturalibus II. In S. Van Riet (Ed.), *Avicenna Latinus. Liber de anima seu sextus de naturalibus, I–II–III*. Peeters; Brill.

Babin, F. (1679). *Journal ou relation fidele de tout ce qui s'est passé dans l'université d'Angers au sujet de la philosophie de Des Carthes en l'execution des ordres du Roy pendant les années 1675, 1676, 1677, et 1678.*

Boyle, R. (1661). *Certain Physiological Essays and Other Tracts Written at Distant Times, and on Several Occasions by the Honourable Robert Boyle*. Henry Herringman.

Descartes, R. (1964–1974). *Œuvres*, In C. Adam & P. Tannery (Eds.), nouvelle présentation par J. Beaude, P. Costabel, A. Gabbey and B. Rochot, 11 vols. Vrin.

Diderot, D. (1749). *Lettre sur les aveugles à l'usage de ceux qui voient*.

Galen (1978–1984). De Placitis Hippocratis et Platonis. In P. De Lacy (Ed. and Trans.), *Galeni De Placitis Hippocratis et Platonis*. Akademie-Verlag (CMG V 4, 1, 2).

Locke, J. (1786). In W. Sleater, H. Chamberlaine, & J. Potts (Eds.), *An Essay Concerning Human Understanding* (Vol. 1–2) (Original work published 1690).

Locke, J. (1978). Letters 849-1241. In E. S. de Beer (Ed.), *The Correspondence of John Locke* (Vol. 3). Clarendon Press.

Lucretius. (1942). De rerum natura. In W. E. Leonard & S. B. Smith (Ed. and Trans.), *T. Lucreti Cari De rerum natura libri sex*. The University of Wisconsin Press.

Molyneux, W. (1692). Dioptrica nova: *A Treatise of Dioptricks, in Two Parts*. Tooke.

Plato (1969). Republic. In P. Shorey (Trans.), *Plato in Twelve Volumes* (Vols. 5 and 6). Harvard University Press; William Heinemann Ltd.

Secondary Literature

Beneduce, C. (2019). La fisiologia del tatto nel XIV secolo: Il caso di Giovanni Buridano. In O. Grassi & G. Catapano (Eds.), *Rappresentazioni della natura nel Medioevo* (pp. 207–220). SISMEL–Edizioni del Galluzzo.

Beneduce, C. (2023). John Buridan: The Human Body at the Intersection of Natural Philosophy and Medicine. In J.A. Zupko (guest ed.). *The Philosophy of John Buridan, American Catholic Philosophical Quarterly, 97*(2), 161–182.

Beneduce, C. (2024). *La scienza del tatto: Un percorso tra filosofia naturale e medicina*. Editrice Bibliografica.

Bennett, D., & Toivanen, J. (Eds.). (2020). *Philosophical Problems in Sense Perception: Testing the Limits of Aristotelianism*. Springer.

Buchenau, S., & Lo Presti, R. (Eds.). (2017). *Human and Animal Cognition in Early Modern Philosophy and Medicine*. University of Pittsburgh Press.

[42] This Introduction displays only selected references. The reader will find a much more extended apparatus of relevant literature on the topic of the history of philosophical and scientific conceptualizations of the sense of touch (both primary and secondary) in the individual chapters.

Chottin, M. (2014). *Le Partage de l'empirisme: Une histoire du problème de Molyneux aux XVIIe et XVIIIe siècles*. Champion.

Classen, C. (2012). *The Deepest Sense: A Cultural History of Touch*. University of Illinois Press.

Crucianelli, L. (2024). *Storia naturale del tatto*. UTET.

Damasio, A. (2008). *Descartes' Error: Emotion, Reason, and the Human Brain*. Vintage Books. (Original work published 1994).

Degenaar, M. (1996). *Molyneux's Problem: Three Centuries of Discussion on the Perception of Forms*. Kluwer.

Ferretti, G., & Glenney, B. (Eds.). (2020). *Molyneux's Question and the History of Philosophy*. Routledge.

Fish, W. (2010). *Philosophy of Perception: A Contemporary Introduction*. Routledge.

Fulkerson, M. (2014). *The First Sense: A Philosophical Study of Human Touch*. MIT Press.

Fulkerson, M. (2015, last modified 2024). Touch. In Edward N. Zalta & Uri Nodelman (Eds.), *The Stanford Encyclopedia of Philosophy* (Summer 2024 Edition). Accessed October 15, 2025, from https://plato.stanford.edu/archives/sum2024/entries/touch/

Gagné, A. (2021). *Embodying the Tactile in Victorian Literature: Touching Bodies/Bodies Touching*. Routledge.

Gensler, M., Mansfeld, M., & Michałowska, M. (Eds.). (2022). *The Embodied Soul: Aristotelian Psychology and Physiology in Medieval Europe between 1200 and 1420*. Springer.

Glenney, B. (2025). *A Pluralist's Guide to Solving Molyneux's Problem*. Routledge.

Glenney, B., & Silva, J. F. (Eds.). (2019). *The Senses and the History of Philosophy*. Routledge.

Gregoric, P. (2007). *Aristotle on the Common Sense*. Oxford University Press.

Gubbini, G. (2009). Tactus, osculum, factum. *Il senso del tatto e il desiderio nella lirica trobadorica*. Nuova Cultura.

Hamlyn, D. W. (2022). *Sensation and Perception: A History of the Philosophy of Perception*. Routledge. (Original work published 1961).

Hasse, D. N. (2000). *Avicenna's* De anima *in the Latin West: The Formation of a Peripatetic Philosophy of the Soul 1160–1300*. The Warburg Institute; N. Aragno Editore.

Hatfield, G. (1991). *The Natural and the Normative: Theories of Spatial Perception from Kant to Helmholtz*. The MIT Press.

Heider, D. (2021). *Aristotelian Subjectivism: Francisco Suárez's Philosophy of Perception*. Springer.

Howes, D. (2022). *The Sensory Studies Manifesto: Tracking the Sensorial Revolution in the Arts and Human Sciences*. University of Toronto Press.

Johansen, T. K. (1998). *Aristotle on the Sense-Organs*. Cambridge University Press.

Johnstone, M. A. (2021). Aristotle on the Unity of Touch. *Journal of the History of Philosophy, 59*(1), 23–43.

Jütte, R. (2005). *A History of the Senses: From Antiquity to Cyberspace*. Polity.

Kambaskovic-Sawers, D., & Wolfe, C. T. (2014). The Senses in Philosophy and Science: From the Nobility of Sight to the Materialism of Touch. In H. Roodenburg (Ed.), *A Cultural History of the Senses in the Renaissance* (pp. 107–126). Bloomsbury Publishing.

Kearney, R. (2021). *Touch: Recovering our Most Vital Sense*. Columbia University Press.

Kettler, A., & Tullett, W. (Eds.). (2025). *The Routledge History of the Senses* (1st ed.). Routledge.

Knuuttila, S., & Kärkkäinen, P. (2013). Sense Perception. Medieval Theories. In S. Knuuttila & J. Sihvola (Eds.), *Sourcebook for the History of the Philosophy of Mind: Philosophical Psychology from Plato to Kant* (pp. 61–81). Springer.

Knuuttila, S., & Kärkkäinen, P. (Eds.). (2008). *Theories of Perception in Medieval and Early Modern Philosophy*. Springer.

Lagerlund, H. (2007). *Forming the Mind: Essays on the Internal Senses and the Mind/Body Problem from Avicenna to the Medical Enlightenment*. Springer.

Lagerlund, H. (2008). Pietro d'Abano and the Anatomy of Perception. In S. Knuuttila & P. Kärkkäinen (Eds.), *Theories of Perception in Medieval and Early Modern Philosophy* (pp. 117–130). Springer.

Macpherson, F. (Ed.). (2011). *The Senses: Classic and Contemporary Philosophical Perspectives*. Oxford University Press.

Mantovani, M. (2023). *Renatus Democritus*. Descartes on Atoms and the Senses. In M. Mantovani, R. Zambiasi, and G. Zuccolin (guest eds.), *Aristotle's* De sensu *in the Latin Tradition, 1250–1650. Micrologus. Nature, Science and Medieval Societies, 31*(2), 331–357.

Mantovani, M. (2022). The Institution of Nature: Descartes on Human and Animal Perception. In D. Rutherford (Ed.), *Oxford Studies in Early Modern Philosophy* (Vol. XI). Oxford University Press.

Mantovani, M., Zambiasi, R., & Zuccolin, G. (guest eds.). (2023). *Aristotle's* De sensu *in the Latin Tradition, 1250–1650. Micrologus. Nature, Sciences and Medieval Societies, 31.*

Massie, P. (2013). Touching, Thinking, Being: The Sense of Touch in Aristotle's *De anima*. *Minerva. An Internet Journal of Philosophy, 17*, 74–101.

Matthen, M. (Ed.). (2015). *The Oxford Handbook of the Philosophy of Perception*. Oxford University Press.

Maurette, P. (2014). *De rerum textura*: Lucretius, Fracastoro, and the Sense of Touch. *The Sixteenth Century Journal, 45*(2), 309–330.

Maurette, P. (2018). *The Forgotten Sense: Meditations on Touch*. The University of Chicago Press.

Moore, G. E. (1957). Nativism. In J. M. Baldwin (Ed.), *Dictionary of Philosophy and Psychology* (Vol. 2). Peter Smith. (Original work published 1902).

Morgan, M. J. (1977). *Molyneux's Question: Vision, Touch and the Philosophy of Perception*. Cambridge University Press.

Moshenska, J. (2014). *Feeling Pleasures: The Sense of Touch in Renaissance England*. Oxford University Press.

Mousavian, S. N., & Fink, J. L. (Eds.). (2020). *The Internal Senses in the Aristotelian Tradition*. Springer.

O'Rourke Boyle, M. (1998). *Senses of Touch: Human Dignity and Deformity, from Michelangelo to Calvin*. Brill.

Pareti, G. (2024). *Il tatto e la profondità: La rivincita dei sensi oscuri*. Rosenberg & Sellier.

Paterlini, M. (2025). *La pelle che pensa*. Codice edizioni.

Paterson, M. (2016). *Seeing with the Hands: Blindness, Vision and Touch after Descartes*. Edinburgh University Press.

Perler, D., & Bender, S. (Eds.). (2020). *Causation and Cognition in Early Modern Philosophy*. Routledge.

Pogliano, C. (2015). *Senso lato: Il tatto e la cultura occidentale*. Carocci.

Purves, A. (Ed.). (2018). *Touch and the Ancient Senses*. Routledge.

Silva, J. F., & Yrjönsuuri, M. (Eds.). (2014). *Active Perception in the History of Philosophy: From Plato to Modern Philosophy*. Springer.

Subramanian, S. (2021). *How to Feel: The Science and Meaning of Touch*. Columbia University Press.

Thörnqvist C.T. and Toivanen J. (eds.) (2022). *Forms of Representation in the Aristotelian Tradition*. Volume 2: Dreaming. Brill.

Toivanen J. (ed.) (2022). *Forms of Representation in the Aristotelian Tradition*. Volume 1: Sense Perception. Brill.

Tunstall, K. E. (2011). *Blindness and Enlightenment: An Essay. With a New Translation of Diderot's* Letter on the Blind *(1749) and a Translation of La Mothe Le Vayer's* Of a Man Born Blind *(1653)*. Continuum.

Vincenti, D. (2021). Espace tactile et espace visuel: L'origine de la notion d'étendue chez Jules Lachelier. *Revue de Métaphysique et de Morale, 111*(3), 385–404.

Von Hoffmann, V. (2023). Can Mixtures be Identified by Touch? The Reception of Galen's *De complexionibus* in Italian Renaissance Medicine. In C. Beneduce (guest ed.). Complexio. *Across Disciplines, Early Science and Medicine, 28*(3–5), 498–525.

Yong, E. (2022). *An Immense World: How Animal Senses Reveal the Hidden Realms Around Us*. Random House Publishing Group.

Chapter 2
The Aristotelians' Touch. Commenting *On Sense and What is Sensed* (*De sensu et sensato*) in Thirteenth-Century Oxford

Monika Mansfeld

Abstract The purpose of this chapter is to present opinions on the sense of touch found in medieval Aristotelian tradition, especially in the Oxford commentaries on *De sensu et sensato* written in the thirteenth century, and demonstrate how they eventually made it into a coherent doctrine of sense perception. The chapter's main focus is on establishing whether there are any specific concepts concerning the sense of touch that are characteristic for Oxford scholars alone or whether there are only topics and issues derived directly from Aristotle's treatise they were commenting on. More specifically, it brings to light the similarities in their argumentative techniques and in their doctrinal solutions. By doing so, it offers the first throughout examination of the Oxford milieu of the thirteenth century—a decisive period for the formation and development of Latin Aristotelianism—and sheds light on these scholars' views on touch and its role in sense perception, as well as their influence—or its lack—on later academic generations.

The chapter presents the main tenets on the sense of touch from a variety of commentaries written at the University of Oxford in the thirteenth century, in different formats: literal expositions (*expositiones*), the so-called "doubts" (*dubitationes*), and question-commentaries (*quaestiones*). The introductory part gives a brief overview of Aristotle's views on sense perception in general and on the sense of touch specifically. The second part of the chapter provides an overview of a representative set of commentaries selected from a dozen Oxonian sources, rarely, if ever, studied before. Six texts, anonymous and with confirmed authorship, are selected, which serve as a basis for the subsequent doctrinal analysis.

This analysis focuses on four problems related to the sense of touch as found in the Oxford commentaries considered in this chapter: (1) the necessity of touch; (2) the organ of touch, the medium, and the object of touch; (3) the spirituality of touch; and (4) touch in the hierarchy of the senses. Conclusions and future research possibilities are presented in the final part of the chapter.

M. Mansfeld (✉)
University of Łódź, Łódź, Poland
e-mail: monika.mansfeld@uni.lodz.pl

C. Beneduce et al. (eds.), *The Sense of Touch*, Studies in the History of Philosophy of Mind 32, https://doi.org/10.1007/978-3-032-16308-0_2

Keywords Oxford · *De sensu et sensato* · Aristotle · Adam of Bockenfield · Geoffrey of Aspall · Thirteenth century natural philosophy

2.1 Touch in Aristotle

The translation into Latin of Aristotle's (384–322 BC) works represented a watershed for Western philosophy. This also proves true of his psychological writings, *On the Soul* (*De anima*) and *On Sense and What is Sensed* (*De sensu et sensato*). Knowledge of these treatises deeply and widely influenced theories of the senses in the medieval West, including the understanding of touch.[1] Indeed, the Oxford debates about touch of the thirteenth century often revolved a few difficulties already singled out for special consideration by Aristotle himself, who insisted time and again on the anomalous status of touch as compared to the other senses. Significantly enough, touch is the last sense considered by Aristotle in this treatise *On the Soul*, and the one that posited the greatest challenges to his perception theory.

Throughout his works on natural philosophy, Aristotle repeatedly expresses the opinion that an animal is a being endowed with a sensitive soul, which enables sense perception.[2] In his view, sensations are attributed to all and only animals, because they serve to distinguish between what is and what is not an animal.[3] Animals are, therefore, by definition, perceivers. This does not mean, however, that all animals must possess *all* five senses to be found in humans (sight, hearing, smell, taste, and touch). On the contrary, only certain senses are deemed essential for an animal. In particular, Aristotle ascribes a central role to touch in the system of the senses with respect to what we would nowadays label as biology and ecology. In his *De anima*, he claims that touch is the only sense possessed by all animals and that touch is the most necessary of the senses, especially in view of animal's survival: without touch,

---

[1] On the reception of Aristotle in the Middle Ages, see, for example, Marenbon (2020). For an example of how Aristotle's views on touch were received in the late Middle Ages, see Beneduce (2023, pp. 161–182) and Beneduce (2022, pp. 253–275, especially pp. 253–255).

[2] For instance, see Aristotle, *De anima*, III.12, 434b18–24, 435b22–24; *De sensu et sensato*, 2, 438b30–439a3; *De iuventute et senectute*, 469a10ff; *De partibus animalium*, 656a27–31. The passages of Aristotle's *De sensu et sensato* used in this chapter come from an unpublished critical edition, access to which I owe to Griet Galle, namely Aristotle, *De sensu et sensato* (eds. Galle and Peeters, in preparation). The edition itself is a profound reworking of Peeters (1996). The numbering of chapters within *De sensu* is based on this edition. Due to the lack of page numbers in this edition, references are restricted to Bekker's verse numbers.

[3] Aristotle, *De sensu et sensato*, 1, 436b10–11. For more on the relationship between being an animal and having sense perception, see Schriefl and Seger (2023, pp. 587–619) and Hasse (2010, p. 306).

an animal dies.[4] Aristotle also highlights the fundamental role played by touch in animal's survival in his *On Sense and What is Sensed* (*De sensu et sensato*).[5]

Although Aristotle regarded touch as a fundamental sense in his account of living organisms, he nonetheless singled it out as challenging his general theory of perception.[6] He argued that all other "external" senses have *one* specific object of perception—one "proper sensible", as he called it: colour is the perceptual object of sight, sound of hearing, odour of smell, and flavour of taste. This perceptual object, according to Aristotle, is a domain ranging between contraries—in the case of colour, for example, between light (white) and darkness (black). By contrast, touch appeared to him as linked to perceive several contrary qualities (most fundamentally, hot-cold and wet-dry) and, therefore, to have more than one perceptual object. Based on this anomaly of touch with respect to the other four senses, a question was thereby aroused by Aristotle on whether touch is one sense or a cluster thereof.[7]

No less challenging was Aristotle's attempt to identify substantial differences between touch and the other "external" senses. He particularly acknowledged the difficulty in drawing a clear distinction between touch and taste, as he found them to be remarkably similar in nature. This led him to describe taste as a form of touch, or as a derivative of it, so to speak.[8]

The subtle boundary between touch and taste brought Aristotle to some hesitations. Most notably, Aristotle at times considered touch as the only sense needed for an animal to *exist* and the other senses (including taste) as only instrumental to animal's *well-being*; in other cases, he ascribed both touch and taste with the peculiarity of being the only senses necessary for animals to survive.[9] Nonetheless, Aristotle was clear about the other three senses (sight, hearing, and smell), the "distant" ones: contrary to touch and taste, these are said to be useful but not necessary for animals.[10] His identification of touch as the most necessary and basic sense for animals is, however, paired with another opinion—already notably expressed by Plato (ca. 428–423 to 348 BC)—about the superiority of sight over touch as concerning their specific contribution to the perceiver's cognitive life.[11] This claim,

[4] Aristotle, *De anima*, II.2, 413b4–5; II.2, 414a3–5; III.13, 435b3–8.

[5] Aristotle, *De sensu et sensato*, 1, 436b12–13.

[6] For a general introduction to Aristotelian sense perception theories, see, for instance Knuuttila (2008, pp. 1–22). On Aristotle on the sense of touch, see especially Freeland (1992); Johansen (1998); Polansky (2007, pp. 321–337); Johnstone (2021); Sorabji (1971); and Steiner Goldner (2018).

[7] Aristotle, *De anima*, II.11, 422b20 (ed. Barnes, 1984, vol. 1, p. 672) and III.1, 424b22–425a13 (*Ibidem*, p. 675).

[8] The claim that taste is a form of touch is found in Aristotle, *De anima*, II.9, 421a18.

[9] The reader can compare Aristotle, *De anima*, III.13, 435b20–26 to Aristotle, *De sensu et sensato*, 1, 436b11–19.

[10] Aristotle, *De sensu et sensato*, 1, 436b18–22.

[11] With respect to this topic, Plato's opinion is most explicitly expressed in: Plato, *Republic*.VI, 508a–b (transl. Shorey, 1969). The same claims can be found in Aristotle: Aristotle, *Metaphysica*, I.1, 980a21–28 and Aristotle, *De anima*, III.3, 429a3–4.

well-known and influential for later philosophical reflections, grounded the long-lasting *querelle* over the primacy of either sight or touch in the system of the senses.[12]

The similarity between touch and taste led Aristotle to think that they share the same *organ*, namely the heart.[13] Yet, in his system, this statement also was not clear-cut: although Aristotle usually adhered to his cardiocentrism, namely the identification of the heart as the central organ for all vital operations of the body, he sometimes hesitated whether to ascribe flesh to the role of touch's organ instead.[14] This last identification is, however, complicated by Aristotle's treating of touch and taste as "contact senses". For Aristotle believed that both touch and taste work by direct contact with the perceived object, contrary to the case of sight, hearing, and smell, which instead work by distance with respect to the object of perception. Although he sharply distinguished touch and taste from the other senses, because they perceive by contact with the object, he still wished to underline the need of positing a medium, i.e., something that acts as a conduit between the object and the sense's organ, to explain sense perception for all five senses.[15] As a result, he decided to treat flesh as the proper medium for touch and not its organ *sensu stricto*.[16]

When identifying heart as touch's organ and flesh as its medium, Aristotle clearly analysed touch on the physiological level. In his theory of the senses, he further ascribed properties of the four elements to the senses and he noted that touch—as well as taste—was predominantly earthy in nature.[17] This points to the fact that earth constitutes a kind of material composition of the organ of taste that is necessary for realizing its activities and ends.[18]

For Aristotle, touch was a puzzling sense indeed and so it was for scholars commenting on his works, especially on his *De sensu et sensato*. It was so, because in the treatise itself, apart from repeated remarks pointing to the significance of this sense, Aristotle does not expand on touch, but only makes three claims: 1) touch is necessary to animal's survival; 2) while touch and taste are similar in nature, touch, however, dominates this pair; 3) touch is predominantly earthy in nature.

---

[12] On this, see, for example, Kambaskovic-Sawers and Wolfe (2014).

[13] Aristotle, *De sensu et sensato*, 3, 439a1–3.

[14] See the following passages: Aristotle, *De anima*, II.11, 423b18–27, *De sensu et sensato*, 2, 439a2–3, *De partibus animalium*, II.1, 647a19–21, II.8, 653b23–27, II.10, 656b33–37. On interpretative problems towards Aristotle's position on the organ and medium of touch, see, for example, Johansen (1998, pp. 193–199 and pp. 212–215) and Redmond (2019).

[15] On touch and taste as contact senses in Aristotle's philosophy, see, for example, Johansen (1998, pp. 178–188). On the medium of perception, on the medium in contact senses, and on interpretative problems arising from this Aristotelian theory, see, for example, Johansen (1998, pp. 116–147, pp. 178–193, pp. 199–212).

[16] On this position, see again: Aristotele, *De anima*, II.11, 423b18–27.

[17] Aristotle, *De sensu et sensato*, 2, 438b30–31.

[18] For more information of the earthy nature of the organ of touch, see Johansen (1998, pp. 40–44).

2.2 Oxford Commentaries on *De sensu et sensato* in the Thirteenth Century

The doctrinal analysis of theories of touch in the thirteenth century Oxonian commentaries on the *De sensu et sensato* should be preceded by a general presentation of the commentaries themselves. Together with the early commentary tradition on Aristotle's *De anima*, these constitute the first attempts to introduce the Aristotelian theory of sense perception into Latin scholarship. An introduction is therefore needed, especially because these works have been understudied so far. There still remain several questions surrounding some of them, including their authorship and authenticity, as well as their times and places of origin.[19]

It can be reasonably assumed that the tradition of Latin commentaries on *De sensu* started only at the beginning of the thirteenth century, since there is no evidence for a Latin translation of the treatise before that time period.[20] Averroes's *Epitome of Aristotle's On Sense and What is Sensed* (*Epitome in De sensu et sensato*), along with those on other treatises from the *Parva naturalia* set, was probably translated into Latin around 1230. By the middle of the thirteenth century, Latin scholars were supplementing Aristotle's *De sensu* with Averroes's commentary to such an extent that we can find very few commentaries on the former with no references to the latter.[21] Oxford, where the tradition of academic commentary on the *De sensu* began, is no exception to this rule. In this chapter, I analyse six out of about a dozen commentaries written in Oxford between ca. 1230 to the turn of the thirteenth and fourteenth centuries, by choosing the ones in which discussions of touch are longer and more substantial compared to the ones I have not selected.[22]

*2.2.1 Expositions (*Expositiones*)*

2.2.1.1 Anonymous, *Finito libro De anima* Commentary

To the best of my knowledge, the earliest commentary on *De sensu* is an anonymous literal exposition called, after its incipit, *Finito libro De anima*. It consists of a dozen chapters (*capitula*), each of which has a part devoted to the text division and

[19] I will refrain, however, from discussing the authorship of particular works in this chapter to avoid disrupting my line of reasoning. My purpose is neither to determine whether a particular commentary was inspired by another one or that these similarities between them are mere coincidences. This chapter is intended to present a comprehensive overview of the opinions expressed on touch in Oxford *De sensu* commentaries in the thirteenth century as such, regardless of who presented them and who reused them.

[20] For more information about the earliest translation of Aristotle's *De sensu et sensato*, see Galle (2008b, pp. 105–150) and Galle (2008c, pp. 1–90). Cf. Bloch (2008, pp. 83–103). For more on the Latin reception of Aristotle's *De sensu et sensato*, see Mantovani et al. (2023).

[21] Nelis (2022, p. 6). For a critical edition of this text see Averroes, *In De sensu et sensato* (ed. Shields and Blumberg, 1949a, pp. 1–46).

[22] For more information on the Latin reception of the *Parva Naturalia* treatises and the place of the *De sensu et sensato* in the collection, see, for instance, De Leemans (2020, pp. 1385–1397).

explanatory remarks, sentence-for-sentence (less frequently word-for-word), on the commented text; such a structure is most common in literal expositions of its times. It was initially attributed to Adam of Bockenfield (d. before 1260) and believed to be the second recension (or redaction) of his commentary.[23] Yet, when preparing its critical edition it became evident to me that this attribution was false.[24]

The commentary was composed before 1230, when the Latin translation of *De sensu et sensato* started circulating alongside the Latin translation of Averroes's epitome of this treatise, because *Finito libro De anima* lacks in fact any references to Averroes. Since Adam of Bockenfield was active in Oxford in the late 1230s and early 1240s, he could not have authored this commentary. Judging from its contents, its author may have been an Oxford University academic a generation older than Bockenfield, perhaps a Franciscan from the circle of Robert Grosseteste.

This anonymous work appears both innovative and original. It is innovative, because it is the first attempt to comment upon *De sensu* that we know of. The commentator visibly begins from scratch and struggles with both the only translation of *De sensu* that was available to him, known as the "old translation" (*translatio vetus*), and the lack of secondary literature. Unsurprisingly, since the sources available at his time were scant, he had almost no clue of the Greek and Arabic reception of *De sensu*. It can be also inferred from the commentary that its author had limited exposure to the Aristotelian literature: he refers to only three of Aristotle's works, namely *De anima*, *De generatione et corruptione*, and *De caelo*. It is evident that he reconstructs the views of Plato, Empedocles, and Democritus from *De sensu* alone; sometimes the poor translation he uses makes him misinterpret them anyway. Interestingly, Euclid is the only author quoted by the commentator whose opinions could not have been taken from Aristotle.

Ironically, the originality of the commentator's approach is a side-effect of his faithfulness to Aristotle: he believes in the importance of interpreting Aristotle's texts as accurately as possible. He tries to avoid arbitrary reinterpretations of Aristotle, instead preferring to create a comprehensive and consistent presentation of the original text for his students. Nevertheless, the poor quality of the *De sensu* Latin translation sometimes forces him to think outside the text—and the box. For example, when he encounters Greek terms left untranslated, blurred aporias, or seemingly contradictory arguments, he has to look for the logic behind them. He is creative in his approach to understand the text by all means. In the process, he sometimes presents analyses and interpretations not found elsewhere. As we shall see, with respect to the sense of touch, the results of such a procedure can be surprising.

2.2.1.2 Adam of Whitby, *Sententia libri De sensu et sensato*

The closest commentary, in both form and doctrine, to *Finito libro De anima* was written by Adam of Whitby (active in mid thirteenth c.). The information about his authorship is confirmed by the colophon of one of the manuscripts preserving the text. Unfortunately, the colophon is not helpful in dating the text. Nor does it

[23] Cf. Lohr (1967, p. 321), Galle (2010, pp. 48–51).

[24] The critical edition is still in preparation.

provide any information about who the author was; a learned guess is that he may have been related to the London sheriff in 1210. No other works by him are preserved.[25]

What can be learned from Adam of Whitby's commentary on *De sensu* is that he was better versed in Aristotelian natural philosophy than the author of *Finito libro De anima*. He evidently had more books at his disposal, but, most importantly, he had access to other treatises from the *Parva Naturalia* set. With a similar doctrinal background as the author of *Finito libro De anima*, he could build on the latter's ideas to coin more complex arguments. As a result, his commentary reveals a deeper understanding of Aristotle's works and his ability to use them to make meaningful connections.

2.2.1.3 Adam of Bockenfield, *Cum intentio physici quod physicus*

The most heated debates over the thirteenth century Oxonian commentaries on Aristotle concern these by Adam of Bockenfield (also known as Buckfield).[26] Modern scholars have attributed a number of commentaries to him; a dozen manuscripts are supposed to contain his writings or their later shortcuts, paraphrases, and reworkings. However, only a single colophon of the so-called first redaction (or recension) of the commentary on *De sensu et sensato* attests his authorship, thus it is the only genuine commentary on this treatise penned by Adam of Bockenfield.[27]

Since much more is known of this work than the two above, it can be used as a historical beacon. It was written between 1238 and the late 1240s by a young scholar who had just completed his studies. From the manner he handled the topic, it appears that he was well-versed not only in the whole Aristotelian corpus in Latin translation but also in Averroes's commentaries, at least to natural philosophical works. A particularly interesting feature of this commentary[28] is that, apart from literal explanations, it also contains *dubitationes*, a kind of proto-questions discussing topics less directly related to the commented text. In them, Bockenfield problematizes sense perception issues not found in *De sensu*. He strives to link them to other Aristotelian philosophical works. Thus, his work provides a valuable insight into the period's intellectual climate and explains the process of problematizing certain issues not directly discussed by Aristotle himself.

---

[25] Cf. Lohr (1967, p. 324), Weijers (1994, p. 31), Köhler (2017, p. 52), and Mansfeld (2021, p. 135).

[26] This will emerge from my discussion below in this section of the chapter. However, only a small part of the general controversy over Adam of Bockenfield's philosophical works pertains to the commentaries on *De sensu et sensato*. For more information about the complexity of establishing his genuine works, see, for example, Nelis (2022, pp. 1–56), Galle (2023, pp. 15–44), and Brumberg-Chaumont and Poirel (2022).

[27] Galle (2010, pp. 48–49).

[28] Lohr (1967, pp. 317–323), Callus (1943, pp. 29–30). For other commentaries on the *Parva naturalia* set, see: Adam of Bockenfield (2022).

2.2.2 *Doubts* (Dubitationes)

If the presentation of commentaries on *De sensu et sensato* were to be strictly chronological, the next section would have to be devoted to the so-called "Oxford Gloss", a collection of glosses dispersed throughout numerous copies of early Latin translations of Aristotle's natural philosophical works.[29] As a collective effort by Oxford scholars around the middle of the thirteenth century, it shows the state of the art in natural philosophy developed and accumulated through added notes and remarks to those left by their predecessors. The Oxford Gloss both resulted from teaching and served as teaching aids for the next generation of students. Nonetheless, as a source of knowledge concerning Oxford scholars' theories of touch it is too limited. There are only a few notes on this topic, short and lacking a doctrinal context, which makes their analysis difficult. Thus, to gain a better understanding of Oxford scholars' thoughts on touch, we need to look elsewhere. Fortunately, there is a set of so-called *dubitationes*, another type of proto-questions, in Ms. 80 of the Prague Metropolitan Chapter, that were directly inspired by the Oxford Gloss.[30] Therein, common views on touch are explained in a more comprehensible manner.

The commentary consists of a series of short discursive pieces called *dubitationes*, which are divided into two thematically distinct blocks beginning with lemmas. The first block deals with the relationship between an organism viewed as a psychophysical unity and its various operations, such as sensations and emotions. The other block contains four short *dubitationes* on animal senses. This is where three *dubitationes* on touch can be found: "Do all animals have touch and taste" (*Utrum tactus et gustus insint omni animali*), "Does perceiving sensible qualities of touch and taste require the medium" (*Utrum tactus et gustus immutentur per medium*), and "Does touch make up an animal so that without it an animal would not exist" (*Utrum tactus det esse animali sine quo animal esse non possit*). What is most important for our current considerations is that, by and large, the commentary is a reworking of glosses coming from two manuscripts making up the Oxford Gloss, namely: London, British Library, Ms. Royal 12.G.II and Cambridge, Gonville and Caius, Ms. 506 (384). These glosses serve as determinations for particular *dubitationes*. The resulting commentary is then an interplay between the incidental notes derived from Oxford Gloss and the need to give its contents a more structured and philosophically coherent form. This can be seen in both choosing subject-matters for particular *dubitationes* and the phrasing of their solutions. This is a significant aspect of the work and a testimony to the author's skill and creativity.

[29] For more information about the so-called "Oxford Gloss", see Galle (2008a, pp. 197–281).

[30] For more information on this work along with its critical edition, see Mansfeld (2022a, pp. 329–348).

2.2.3 *Questions (*Quaestiones*)*

2.2.3.1 Geoffrey of Aspall, Quaestiones super *De sensu et sensato*

Geoffrey of Aspall (d. 1287) was an English cleric, master of Arts at Oxford, active there from about 1250 to 1263. He was a colleague of John Peckham—probably in Oxford—who reproached him for his innumerable benefices in 1286.[31] Aspall was the Canon of St. Paul in London, Archdeacon of Dublin, and worked as a clerk of the Chancellery of the Queen of England.[32] He wrote numerous commentaries on Aristotle, for instance on *Metaphysica*, *De generatione et corruptione*, *Physica*, and on several works from the *Parva Naturalia* set, some of which were most likely known and used in Paris as well.[33]

His *Quaestiones super De sensu et sensato*, with as many as ninety four questions, is beyond comparison the longest, most developed, and detailed commentary on this treatise in the Latin West written in the thirteenth century. It may be somewhat misleading to call it a question-commentary, since it actually consists of questions and intermittent expository parts. These expositions usually precede a set of several questions on a similar topic, serving as a doctrinal introduction and terminological base for them; for this reason, I call such sets *lectiones*. There are twenty five such *lectiones*, the largest of which contains fifteen questions, the smallest one—a single one. Aspall uses a vast number of Aristotle's works and he must have been familiar with some of the contemporary commentaries on *De sensu* too, as he frequently, yet tacitly, refers to them.[34]

---

[31] For more information about Aspall's academic career, see Geoffrey of Aspall, *Questions on Aristotle's* Physics (ed. Donati and Trifogli, 2017, p. xi).

[32] Russell (1936, pp. 28–31).

[33] Cf. Lohr (1968, pp. 150–152).

[34] As for examples relating to the sense of touch, see the discussion of spiritual touch below. However, Aspall's respect for his predecessors and colleagues is more apparent in his treatment of problems not fully explored by Aristotle. There is a discussion, for instance, regarding whether brute animals can discriminate between good and evil. Aspall's predecessors would unanimously agree that a) animals that are prudent use sight and hearing for utility reasons, b) in contemplation these senses serve to discriminate between true and false and in practice they distinguish good and evil. In most cases, this opinion is expressed as in the *Finito libro De anima* commentary, cf. Anonymus, *Finito libro De anima*, Erfurt, Universitätsbibliothek, Dep. Amplon. Ms. 2° 318, f. 151va; London, Wellcome Collection, Ms. 3, f. 54ra; Città del Vaticano, Biblioteca Apostolica Vaticana, Vat. lat. 5988, f. 34rb: "Consequenter ponit specialem causam quare visus et auditus insunt animali. Dicit ergo quod isti sensus insunt omnibus animalibus disciplinabilibus sive habentibus prudentiam non propter esse sed causa utilitatis, quia multas differentias rerum manifestant isti sensus animalibus. Et mediantibus etiam istis contemplativi discernunt verum et falsum, practici bonum et malum. Et hoc est: *habentibus autem*". By contrast, Aspall would see the template theme and disagree, since combining claims a) and b) contributes to misunderstanding the natures of humans and animals respectively. Cf. Geoffrey of Aspall, *Quaestiones super De sensu et sensato Aristotelis*, lectio VIII, q. 24: *Utrum sensitiva possit super discretionem boni a malo*, corpus quaestionis, Cambridge, Gonville and Caius College 509/386, f. 292ra–b; Oxford, New College 285, f. 169vb; Oxford, Merton College Library 272, f. 261va; Todi, Biblioteca Comunale Lorenzo Leoni 15, f. 104va–b: "Dico quod non est bonum morale nisi in habentibus rationem. Est ergo bonum

Only a single *lectio* (VII) is devoted to the sense of touch and, to some extent, to the sense of taste.[35] It is, unfortunately, one of the shortest and least developed ones. Its author does not seem interested in touch and its functions at all; he restricts himself to analysing the "obligatory" issue on whether all animals must possess touch (q. 19).

2.2.3.2 Anonymous, Oriel 33, *Quaestiones in De sensu et sensato*

Oriel 33 is different in this respect. It is the second most detailed Oxonian question-commentary on *De sensu*. It contains twenty six questions, similar in size (each of them being approximately a column, or slightly more than a column, long) and execution (a standard format with a couple of arguments against the main thesis, one or two in favour of it, a short presentation of the author's position, and responses to the arguments). Judging from its handwriting, the commentary was written at the turn of the thirteenth and fourteenth centuries.

Its author devotes three questions to the sense of touch, namely q. 3: "Is it necessary for an animal to have senses of touch and taste" (*Utrum tactus et gustus sint sensus necessarii animali*), q. 16: "Is the organ of touch predominantly earthy" (*Utrum organum tactus sit terrae a dominio*), and q. 17: "Is the organ of touch situated close to the heart" (*Utrum organum tactus situetur iuxta cor*). They are good examples of what was considered controversial for scholars of the late thirteenth century as well as what views were considered common knowledge in that academic circle. For these reasons they are crucial for our considerations in this chapter.

2.3 The Sense of Touch in Thirteenth-Century Oxford Commentaries on *De sensu et sensato*

Thirteenth century Oxonian commentators on the *De sensu et sensato* reacted to Aristotle's breviloquence and sketchiness of Aristotle's description of touch in his work employing several techniques. Their standard approach was to analyse the aforementioned remarks on the nature of touch and to use them as a springboard for

naturae et est bonum consuetudinale, bonum naturale aut solum in essendo aut solum propter naturalem motum rei. Primo modo reperitur bonum naturale in quolibet ente, quia quantum habet unumquodque de entitate tantum de bonitate. Secundo modo dicitur quod quaedam bruta ex naturali instinctu et motu sunt bene disposita ad exercendum bonas operationes vel malas, et hoc in eadem specie. Bonum consuetudinale autem est ex proprio motu proprii. Proprius consensus <est> scilicet deliberando quoniam facere sic est bonum vel malum, et hoc solum est per intellectivam, et tale solum est in hominibus, aut ex coactione, et tunc aut sit consensus per coactionem aut non. Si sit consensus, aut ergo sit consensus cum deliberatione, et tale solum reperitur in hominibus, aut non cum deliberatione, immo solum ex assuetudine, videlicet per assuentiam prosumpat in opus, et tale reperitur in brutis". In this chapter, all transcriptions and editions are mine, if not indicated otherwise.

[35] For a provisional list of questions see Ebbesen et al. (2016, pp. 66–70).

raising broader questions about the nature and workings of touch—for example, whether an animal might possess only a single sense, whether it is possible to find an animal that can touch but not taste, and whether plants, which are affected by tangible objects and nourish themselves, might also possess some form of sensation. These issues, it is to be noted, are not addressed in the *De sensu et sensato* directly, and feature only faintly, if ever, in the rest of Aristotle's corpus.[36]

In this chapter, I will present four topics discussed by the thirteenth century Oxford scholars in their commentaries on the *De sensu*. They will be as follows: the necessity of touch; the organ of touch, the medium, and the object of touch; the spirituality of touch; touch in the hierarchy of the senses.

2.3.1 The Necessity of Touch

The issue of the necessity of touch is the one which drew the most interest of the early commentators of *De sensu* in Oxford. In various forms, it is present in all the commentaries. The problem is dealt with either in separate questions or in longer passages of literal expositions. In practice, however, it consists of two interrelated yet different discussions: whether touch is that which makes an animal an animal and whether living creatures other than animals can have touch. Interestingly, questions with titles directly referring to the first problem usually discuss the other instead.

2.3.1.1 Does Touch Make an Animal an Animal?

The *Finito libro De anima* simply states that there are two kinds of senses: the ones necessary for an animal, viz. touch and taste, and the ones that are useful but not necessary, viz. sight, hearing, and smell. Most commentaries on the *De sensu et sensato* in Oxford in the thirteenth century glide past this theme, which is rooted in Aristotle's text. The author of the *Finito libro De anima* commentary does not provide any details on this division other than a reference to the third book of *De anima* where, in his opinion, it is sufficiently described.[37] The same information can be

[36] For more information about the reception of the views on the sense of touch derived from Aristotle's *De sensu et sensato* as well as for their analyses in the Second Scholaticism, see Daniel Heider's chapter in this volume.

[37] Anonymous, *Finito libro De anima*, Erfurt, Universitätsbibliothek, Dep. Amplon. Ms. 2° 318, f. 151va; London, Wellcome Collection, Ms. 3, f. 54ra; Città del Vaticano, Biblioteca Apostolica Vaticana, Vat. lat. 5988, f. 34rb: "Consequenter ponit causam specialem quare gustus et tactus insunt animali dicens quod isti sensus sequuntur omni animali necessario et convertibiliter insunt unicuique secundum quod animal est. Et tamen quare tactus sic inest omni animali dictum est in fine tertii *De anima*. Gustus autem inest omni animali propter nutrimentum ut discernat sapidum ab insipido, et hoc fugiat et illud prosequatur, quia sapor universaliter, qui est proprium obiectum gustus, est propria passio nutrimenti. Et hoc est: *proprie autem*".

found in Adam of Whitby, who simply states that an animal must have both touch and taste to exist.[38]

Adam of Bockenfield, in turn, has more to say why touch and taste are necessary; he does so by referring to Aristotle's *De anima* too. He presents the following syllogism to prove the necessity of touch: "Touch is necessary for an animal as an animal [...], because touch is the sense of nutrition; nutrition is necessary for an animal as an animal; therefore, touch is necessary for an animal as an animal".[39] One may think that this is the function of taste rather than touch, but Bockenfield uses the argument given by Averroes to make his point: touch is the only sense causing nutrition of the organism in its purely physiological form.[40] Taste enhances the experience of food by demonstrating its various aspects. In other words, the author argues that touch is responsible for nutrition as such, while taste is responsible for feelings associated with eating, pleasant or not. Consequently, touch is necessary for animals' survival and taste is not.[41]

As Bockenfield develops his argument, he touches upon another topic: whether there are species of animals that are deprived of all senses except touch. Aristotle's opinion is that there are no such animals. In contrast, Averroes points to some sea creatures, like sponges, that do have touch but lack other senses, including taste.[42] According to him, animals that feed through their roots and do not have mouths have only the sense of touch. Adam of Bockenfield favours Averroes's position over Aristotle's. Nevertheless, he is convinced that these two do not directly contradict

---

[38] Adam of Whitby, *Sententia libri De sensu et sensato Aristotelis*, Firenze, Biblioteca Nazionale Centrale, Conv. Soppr. G 3 464, f. 74ra: "Secundo dat causam propter quam unusquisque sensus inest animalibus et procedit sic: primo dat causam propter quam tactus et gustus, et est propter esse: *Sensus autem qui*".

[39] Adam of Bockenfield, *Cum intentio physici quod physicus*, Philadelphia, Free Library, Lewis European 53, f. 52rb; Madrid, Biblioteca Nacional de España 3314, f. 100vb; Milano, Biblioteca Ambrosiana H.105 inf., f. 2rb; Oxford, Balliol College 313, f. 133rb: "[...] Tactus est necessarius animali secundum quod animal propter causam praedictam in libro *De anima*, scilicet quia tactus est sensus nutrimenti; et nutrimentum est necessarium animali secundum quod animal; quare sensus nutrimenti, scilicet tactus, necessarius est animali secundum quod animal".

[40] Cf. Averroes, *Commentarium magnum in De anima* (ed. Crawford, 1953, p. 156): "Et prima virtus sensus, que est prior naturaliter in esse aliis virtutibus sensus, est sensus tactus; quoniam, sicut virtus nutritiva potest separari in plantis a tactu et ab omni virtute sensus, ita tactus potest separari ab aliis sensibus; idest, cum fuerit inventus, non est necesse ut inveniantur alii sensus, et cum alii sensus fuerint inventi, necesse est ut ille inveniatur. Est igitur prior naturaliter aliis sensibus, sicut nutrimentum est prius naturaliter sensu tactus".

[41] Adam of Bockenfield, *Cum intentio physici quod physicus*, Philadelphia, Free Library, Lewis European 53, f. 52va; Madrid, Biblioteca Nacional de España 3314, f. 101ra; Milano, Biblioteca Ambrosiana H.105 inf., f. 2va; Oxford, Balliol College 313, f. 133rb: "Notandum est etiam ibidem secundum Commentatorem quod nullum sensibile praeter sensibile sensus tactus est in cibo secundum quod est cibus, sed solum faciunt alia sensibilia ad delectationem et non ad necessitatem, sicut est dictum, non appetitur alimentum quia dulce aut amarum, sed per ista discernunt animalia sua alimenta. Et hoc est quod dicit in littera: *sapidum et insipidum*".

[42] Cf. Averroes, *Commentarium magnum in De anima* (ed. Crawford, 1953, p. 163): "[...] Quedam enim animalia habent quinque virtutes sensus et quedam quasdam tantum, ut talpa, et quedam unum, scilicet tactum, ut spongia maris".

one another; the scope of Aristotle's explanation is narrower, since he does not seem to consider any animals with such features as animals in the strict sense.[43]

The necessity of touch and taste is also discussed in Prague *Dubitationes* but the solution is slightly different. According to the author of this text, animals cannot exist without touch but they cannot survive without taste either. Even primitive animals, such as worms, have both of these senses. He claims that, unlike perfect animals (i.e., animals that have a head), primitive animals do not have specific organs for either touch or taste. They can feel, however, both types of sensations everywhere in their bodies.[44]

2.3.1.2 Can Anything Other than an Animal Have the Sense of Touch?

Adam of Bockenfield and the author of Oriel 33 testify how a question whether touch and taste are necessary for an animal is transformed into the discussion of whether touch and, consequently, taste can be found in plants or inanimate objects, such as stones. They both agree that touch is crucial for survival, which is achieved by nutrition. According to many commentators, an animal uses touch to approach what is beneficial to it and avoid danger. What then could be the use of touch for plants and animals that do not move? Does receiving and digesting food by an organism imply that it has touch and/or taste?[45]

According to Adam of Bockenfield, the answer to both these questions is negative: although both plants and animals are nourished, their requirements for food differ significantly. Since a plant's nutrients are dispersed throughout its surroundings, it does not have to choose what to consume: as soon as something edible is present, a plant feeds on it spontaneously. Unlike plants, animals have to distinguish between what is harmful and what is beneficial. Bockenfield believes that touch

---

[43] Adam of Bockenfield, *Cum intentio physici quod physicus*, Philadelphia, Free Library, Lewis European 53, f. 52rb; Madrid, Biblioteca Nacional de España 3314, ff. 100vb–101ra; Milano, Biblioteca Ambrosiana H.105 inf., f. 2rb; Oxford, Balliol College 313, f. 133rb: "Et ita relinquitur quod gustus est necessarius omni animali sicut et tactus. Cuius tamen contrarium videtur dicere in secundo *De anima*, ubi dicit quod nullus aliorum sensuum posset esse sine tactu, tactus autem sine aliis esse posset. Plura enim animalia, ut dicit ibi Commentator, carent visu et auditu, odoratu et gustu. Ad quod respondet ibi Commentator quod, si aliquid habeat sensum tactus et nutriatur suis radicibus, sicut plantae nutriuntur, ut dicit de spongia maris, universaliter opinandum est quod istud habet sensum tactus sine gustu. Et forte innuit huiusmodi animalia, cum dixit hoc ibi in secundo *De anima*, omne tamen animal habens os habet gustum aliquem, ut dicit Commentator, sed hic non intelligit de talibus, unde non est contrarietas".

[44] Anonymous, *Dubitationes super De sensu et sensato Aristotelis*, dubitatio 2: *Utrum tactus et gustus insint omni animali* (ed. Mansfeld, 2022, p. 345): "Dicendum quod animalibus perfectis, cuiusmodi sunt capita habentia, inest organum gustus determinatum et deputatum ad gustum. Animalibus vero imperfectis inest organum gustus indeterminate, unde fortassis in vermibus est organum gustus per totum corpus indistinctum sicut et organum tactus est per totum corpus. In animalibus vero perfectis tactus est ubique in corpore, gustus vero in parte determinata".

[45] For more information on the solutions of such questions in Aristotle and medieval commentators, see Murphy (2005, pp. 295–339) and Thomsen Thörnqvist (2022, pp. 183–212).

enables them to avoid the former and approach the latter. Putting it another way, plants do not have touch or taste, because they do not need them, while animals cannot survive without them.[46]

The author of Oriel 33 reaches the same conclusion even though his premises are different. He says that animals are higher than plants in the hierarchy of animated creatures due to the fact that their forms subsume those of plants as well. Their advantage lies in their sensory perception as they receive information about their surroundings in the form of sensible species. This is considered to be the highest form of perception in the animal kingdom. At the same time, the complexity of sensory perception makes it difficult for the animal to sustain it. For an animal to survive, there must be such a balance between the elements which make up its organism that enables it to perform its actions properly. In short, the commentator believes that nature provided animals with the sense of touch to avoid corruptive factors and approach beneficent ones with respect to their elemental composition. This mechanism works particularly well in animals that move. He sees it as the efficient cause of touch. This kind of argument flows from the conviction that nature acts so as to promote higher forms of existence. If an animal did not have the sense of touch, it could be easily destroyed by external factors, such as fire. With touch, its chances of survival are much higher. Thus, we can observe the interplay between what the author of Oriel 33 believes are the efficient and the final causes of touch.[47]

---

[46] Adam of Bockenfield, *Cum intentio physici quod physicus*, Philadelphia, Free Library, Lewis European 53, f. 52rb–va; Madrid, Biblioteca Nacional de España 3314, f. 101ra; Milano, Biblioteca Ambrosiana H.105 inf., f. 2rb–va; Oxford, Balliol College 313, f. 133rb: "Notandum est etiam secundum Commentatorem ibi super secundum *De anima* propter quid omne animal indiget sensu et plantae non, cum tamen sensus primus, scilicet tactus, sit sensus nutrimenti, et plantae indigeant nutrimento sicut et animalia. Et est causa secundum ipsum ibi, quia nutrimentum animalis non est impositum in pluribus rebus sicut est in plantis, et ideo plantae non indigent sensu quo distinguant cibum. Animal tamen indiget sensu per quem comprehendat conveniens et inconveniens nutrimentis, ut expellat nocivum et attrahat iuvamentum".

[47] Anonymous, *Quaestiones in De sensu et sensato Aristotelis*, q. 3: *Utrum tactus et gustus sint sensus necessarii animali*, corpus quaestionis, Oxford, Bodleian Library, Oriel Ms. 33, f. 195ra: "Dicendum est quod isti sensus cuilibet animali sunt necessarii. Cuius ratio est: animalia excedunt res inanimatas in hoc quod non solum continent in se suas formas, sed continent in se formas aliorum, ut plantarum, et ideo attingunt infimum gradum cognitionis. Recipiunt enim species sensibilium sine materia, ut dicit Philosophus in tertio *De anima*. Sed corpus animalis est corpus mixtum ex simplicibus, est etiam unum et nutribile sive alimentale. Si ergo consideretur inquantum est mixtum, sibi competit debita commixtio calidi et frigidi, humidi et sicci, et quia medium corrumpitur extremis, excellentia calidi corrumpit tepidum, et ita de aliis. Et natura magis solicita est circa salvationem rerum nobilium quam aliarum, ideo natura attribuit animali sensum tactus per quem declinaret vel vitaret corruptiva sive corrumpentia et prosequetur convenientia, et hoc maxime necessarium est animalibus progressivis. Nisi enim animal progressivum tactum haberet, mitteret se in ignem, similiter in aerem. Si autem consideratur inquantum corpus, sic est unum et multum. Quidlibet non sit natum quodlibet nutrimentum, necessarius est ei gustus per quem discernat alimentum conveniens a disconvenienti. Unde tactus est necessarius animali propter suum esse et gustus est necessarius ei ut in esse continuetur".

2.3.2 *The Organ of Touch, the Medium, and the Object of Touch*

Fundamentally, all early Oxford commentators on *De sensu et sensato* agree that touch, along with taste, uses the so-called internal medium. The term "internal medium" is used by them to underline that touch is a contact sense.[48] Consequently, they consider flesh to be the medium rather than a proper organ of touch.[49] The latter is located inside the body, near the heart. This default opinion[50] is interestingly modified by Adam of Whitby who seems to suggest that our body (*corpus nostrum*) may be "[...] nearly an instrument of touch [...]", namely the organ of touch.[51]

The physiological reason why the organ of touch is located near the heart is given by almost all commentators in a very similar manner. They claim that it is predominantly earthy in nature, mainly because of the firmness, stability, and fundamental character of its functions.[52] The earthy nature means that the two dominant primary

---

[48] Anonymous, *Finito libro De anima*, Erfurt, Universitätsbibliothek, Dep. Amplon. Ms. 2° 318, ff. 157vb–158ra; London, Wellcome Collection, Ms. 3, f. 57va; Città del Vaticano, Biblioteca Apostolica Vaticana, Vat. lat. 5988, f. 38va: "Ex hoc arguit quod, cum sensus sint numero impares, quoniam quinque, sensus olfactus est medius illorum. Sunt enim duo sensus posteriores, scilicet tactuales, id est per contactum sentientes, cuiusmodi sunt tactus et gustus, et duo priores, sentientes per media extrinseca, scilicet visus et auditus, quorum medius est olfactus. Et hoc est: *videtur autem*".

[49] Anonymous, *Dubitationes super De sensu et sensato Aristotelis*, Ms. 80, dubitatio 3*: Utrum tactus et gustus immutentur per medium* (ed. Mansfeld, 2022a, p. 345): "Dicendum quod hoc est verum, sed dissimiliter, quia tactus et gustus immutantur per medium intrinsecum, ut per carnem, vel si quid aliud fuerit medium in tactu et gustu, de quo medio dubium sit quid sit illud. Sensus vero visus et auditus et olfactus immutantur per medium extrinsecum, scilicet aerem, quod medium non est pars corporis".

[50] See, for example, Adam of Bockenfield, *Cum intentio physici quod physicus*, Philadelphia, Free Library, Lewis European 53, f. 53rb; Madrid, Biblioteca Nacional de España 3314, f. 103vb; Milano, Biblioteca Ambrosiana H.105 inf., f. 3rb; Oxford, Balliol College 313, f. 135rb: "Consequenter, cum dicit: *Tactuale*, determinat cuiusmodi instrumentum dominio attribuenda sunt instrumenta sentientia per media intrinseca, scilicet instrumenta gustus et tactus, intendens quod instrumentum tactus et gustus attribuendum est terrae secundum dominium".

[51] Adam of Whitby, *Sententia in librum De sensu et sensato Aristotelis*, Firenze, Biblioteca Nazionale Centrale, Conv. Soppr. G 3 464, f. 75va: "Nos autem habemus tactum certissimum, et causa est, quia corpus nostrum quod est quasi instrumentum tactus est compositum ex differentiis tangibilibus in maxima aequalitate, et ideo subtilissime discernit de illis differentiis. Gustus autem est quidam tactus, quare gustum habemus certiorem olfactu, et ideo manifestior est nobis natura saporis quam odoris: *Natura autem aquae* etc.".

[52] For example, Adam of Whitby, *Sententia in librum De sensu et sensato Aristotelis*, Firenze, Biblioteca Nazionale Centrale, Conv. Soppr. G 3 464, f. 74vb: "Consequenter determinat quod instrumentum tactus et instrumentum gustus attribuenda sunt terrae. Instrumentum autem tactus attribuendum est terrae hac de causa, ut firmum sit quod quidem necessarium est, quia tactus est essentialis animali. Instrumentum autem gustus attribuendum est terrae, quia gustus est quidam tactus. Numquam enim fit gustus sine tactu sibi coniuncto. Et probat sic: primo determinat quod sic est (*Et ideo iuxta* etc.), secundo infert quoddam corrolarium (*Inde manifestum*). Et est quod instrumenta horum sensuum sita sunt iuxta cor, quia in eis abundat terra, ideo in eis abundat frigiditas. Et ideo necesse habent vigorari per caliditatem quae maxime viget circa cor: *De sensibus ita* etc.".

qualities of touch are coldness and dryness. To perform its actions, the organ of touch must be activated by the opposite qualities, namely of heat and humidity. Because the heart is considered the hottest organ in an animal's body, the organ of touch is located close to it so that the heart can enhance the operations of touch.[53]

For the thirteenth century Oxford commentators of the *De sensu*, the object of touch does not seem to be worthy of attention. By possibly following the Aristotelian tautology in *De anima*, III.1, 424b22–425a13, where Aristotle states that everything tangible can be perceived by touch, they claim that object of touch are simply tangible qualities.[54] At times, they try to specify what a tangible quality is and, usually,

[53] Anonymous, *Finito libro De anima*, Erfurt, Universitätsbibliothek, Dep. Amplon. Ms. 2° 318, f. 153ra; London, Wellcome Collection, Ms. 3, f. 54vb; Città del Vaticano, Biblioteca Apostolica Vaticana, Vat. lat. 5988, f. 35rb: "Consequenter dicit quod instrumentum tactus attribuendum est terrae et etiam instrumentum gustus, quia gustus est species tactus. Et hoc est: *et tactuale*. Consequenter infert ex hoc corollarie quod instrumentum gustus et tactus situm est iuxta cor. Cor enim opponitur cerebro in qualitate, eo quod est calidissima pars animalis, et ita, cum huiusmodi instrumentum sit terreum, per calorem cordis eius frigiditas relevatur. Et hoc est: *et ideo*". Adam of Bockenfield and Oriel 33 present this argument in a more elaborate form. Cf. Adam of Bockenfield, *Cum intentio physici quod physicus*, Philadelphia, Free Library, Lewis European 53, f. 53rb; Madrid, Biblioteca Nacional de España 3314, f. 102vb; Milano, Biblioteca Ambrosiana H.105 inf., 3rb–va; Oxford, Balliol College 313, f. 135rb: "Consequenter, cum dicit: *Tactuale*, determinat cuiusmodi instrumentum dominio attribuenda sunt instrumenta sentientia per media intrinseca, scilicet instrumenta gustus et tactus, intendens quod instrumentum tactus et gustus attribuendum est terrae secundum dominium. Et primo dicit quod instrumentum tactus attribuendum est terrae secundum dominium, et hoc hac de causa: ut firmum sit quod quidem necessarium est ad esse animalis, et tactus est sensus primus dans esse animali. Secundo, cum dicit: *Gustabile*, determinat quod instrumentum gustus attribuendum est terrae, et hoc quia gustus quidam tactus est, numquam enim sit gustus sine tactu. Tertio, cum dicit: *Et ideo quattuor*, ex his infert collorarium quod instrumentum istorum sensuum, scilicet tactus et gustus, sita sunt iuxta cor. Quia enim in eis abundat terra, ideo in eis abundat frigiditas propter quod necesse habent vigorari per caliditatem quae maxime viget circa cor. Et quod cor sit tale demonstrat, cum dicit: *Cor enim*, dicens quod cor est oppositum cerebro et in situ et in qualitate, et ita est calidissimum membrum animalis, cum cerebrum sit frigidissimum". Cf. Anonymous, *Quaestiones in De sensu et sensato Aristotelis*, q. 17: *Utrum organum tactus situetur iuxta cor*, corpus quaestionis, Oxford, Bodleian Library, Ms. Oriel 33, f. 195ra: "Dicendum est quod organa tactus et gustus sunt interius iuxta cor, quantum ad ipsam radicem sui et originem. Et huius ratio est, quia nihil sensitivum est sine calore (per Philosophum in secundo *De anima*), unde Commentator dicit ibidem quod, cum sensus infrigidantur, in operatione deficiunt; sed organum tactus et gustus est terrae per dominium et per consequens est siccum et frigidum de se; oportet ergo quod sit in loco calido, ita quod per participationem caloris exerceat suam operationem; sed cor membrum calidissimum est, cerebro enim opponitur et secundum situm et secundum qualitatem, ut per frigiditatem cerebri mitigetur calor cordis et ut per calorem cordis temperetur frigiditas cerebri. Et ideo homines qui habent magnum cor et parvum cerebrum sunt impetuosi, tamquam caliditate cordis per frigiditatem cerebri non sufficit mitigatio. Et e converso: qui habent magnum cerebrum et parvum cor sunt pingui et molli, tamquam frigiditate cerebri per calorem cordis non sufficit temperantia".

[54] Cf. Adam of Bockenfield, *Cum intentio physici quod physicus*, Adam of Bockenfield, *Cum intentio physici quod physicus*, Philadelphia, Free Library, Lewis European 53, f. 56ra; Madrid, Biblioteca Nacional de España 3314, f. 107rb; Milano, Biblioteca Ambrosiana H.105 inf., f. 12va; Oxford, Balliol College 313, f. 141ra: "Sapor enim est passio qualitatum tangibilium quae sunt quinque in numero, quare odor convenientiam habet cum sapore qui est sensibile gustus et etiam cum qualitatibus tangibilibus quae sunt sensibilia tactus".

identify them with the four primary qualities. Geoffrey of Aspall, however, discovers that this opinion is not as unproblematic as it may seem. He notes that from the fact that touch can feel primary qualities, one could infer that there are actually two distinct senses in the sense of touch: one would be responsible for sensing hotness and coldness and the other, for sensing humidity and dryness.[55] He cannot accept such a view: according to him, the object of touch, called "a tactile" (*tactibile*), is a composition of four elements in an appropriate proportion; if the proportion is not sound, nothing can be sensed (*non tactibile*).[56] This may also be the reason why animals cannot perceive elements in their pure form: no sentient subject cannot experience anything but the mixture.

2.3.3 Spiritual Touch

The views on touch presented in Oxford commentaries of the thirteenth century typically derive directly from Aristotle. Deviations from this trend are rare. There is, however, a feature found in several commentaries that is unique to the Oxford tradition of *De sensu et sensato* commentaries and is not found in Aristotle as such. In the earliest Oxonian commentaries, i.e., the anonymous *Finito libro De anima* and Adam of Bockenfield's *Cum intentio physici secundum quod physicus* commentaries, this feature appears in a "raw" form. It also reappears later, slightly altered, in the anonymous Oriel 33 and Geoffrey of Aspall's question-commentary.

According to the view, there are two kinds of touch: the *material* one (*tactus materialis*) and the *spiritual* one (*tactus spiritualis*), each of which gathers

[55] Geoffrey of Aspall, *Quaestiones super De sensu et sensato Aristotelis*, lectio VII, q. 19: *Utrum tactus sit necessarius animali*, Cambridge, Gonville and Caius College 509/386, f. 290vb; Oxford, New College 285, f. 168rb; Oxford, Merton College Library 272, f. 260va; Todi, Biblioteca Comunale Lorenzo Leoni 15, f. 103rb–va: "Item, si necesse est omne animal habere tactum, tunc esset aliquod animal habens sex sensus, quod videtur impossibile primo (?) commune, quia sunt quattuor sensus alii a tactu; sed sensus tactus est duo sensus; quare etc. Minor istius patet, quia unus sensus est unius contrarietatis, sed tactus est quattuor qualitatum contrariarum quae efficiunt duas contrarietates quarum neutrum ad aliam reducitur, quare tactus est duo sensus [...]".

[56] Geoffrey of Aspall, *Quaestiones super De sensu et sensato Aristotelis*, lectio VII, q. 19: *Utrum tactus sit necessarius animali*, Cambridge, Gonville and Caius College 509/386, f. 291ra; Oxford, New College 285, f. 168va; Oxford, Merton College Library 272, f. 260va–vb; Todi, Biblioteca Comunale Lorenzo Leoni 15, f. 103va: "[...] Dico quod, sicut auditus est audibilis et non audibilis et visus visibilis et non visibilis, sic tactus est tangibilis et non tangibilis; et ad istam contrarietatem reducuntur duae primae contrarietates. Vel dicerem quod Aristoteles innuit in quarto *Meteororum* quod omnia sensibilia calore participant, ista igitur contrarietas: siccum et humidum reducitur ad istam contrarietatem: calidum et frigidum, sicut contrarietas magis materialis ad contrarietatem magis formalem, unde secundum ordinem in extendendo per naturam prius est calidum et frigidum quam humidum et siccum".

information from different senses.[57] Material touch is characteristic of the contact senses, namely touch and taste, while spiritual touch works with sight, hearing, and smell. The spiritual touch is considered less material than the material one but not immaterial in an absolute sense. Generally, it is attributed to senses that require an extrinsic medium to receive sensory information. Material touch, in turn, is attributed to the ones requiring internal media only.

In thirteenth century Oxonian commentaries on *De sensu*, the idea of spiritual touch as opposed to the material one is discussed in the context of Aristotle's critique of Democritus's atomism. According to Aristotle, Democritus believed that sense perception is nothing more than just a variety of tactile impulses coming through our sensory system. As was typical of him, Aristotle did not elaborate on the details of his opponents' theories. For this reason, one cannot tell how *technically* he understood Democritus' claim that every sense is in some way touch. It could mean that he was treating all the senses as contact senses or, perhaps, that sensing *physically* involved athoms of particular shapes hitting the sensory organs. Anyway, Aristotle and his medieval Oxford commentators criticised this conviction saying that there is a significant difference between touch and other senses. Were it not for that difference, animals would perceive both common sensible qualities, such as shape, number, or magnitude, and qualities proper to particular senses, such as colour, sound, and flavour, by touch alone.[58] The idea behind the two kinds of touch was introduced to maintain the integrity of Aristotle's critique without losing the advantage of treating sensory perception as a contact between the object and subject of perception.[59]

[57] For instance in Anonymous, *Finito libro De anima*, Erfurt, Universitätsbibliothek, Dep. Amplon. Ms. 2° 318, f. 162ra; London, Wellcome Collection, Ms. 3, f. 59vb; Città del Vaticano, Biblioteca Apostolica Vaticana, Vat. lat. 5988, f. 41va: "Et nota quod dicit: *non tactualia tangi*, id est illa sensibilia quae non tanguntur materialiter, scilicet quae neque sunt tangibilia nec gustabilia, tangi spiritualiter. Et forsan excipit tangibilia et gustabilia, quia manifestum est nullum tale infinitum esse".

[58] Proper sensibles are characteristics of a unique sense only. The concept of common sensibles, in turn, refers to qualities that can be felt by more than one sense at the same time. For more information on proper and common sensibles, see, for example, Mansfeld (2022b, pp. 239–240).

[59] Anonymus, *Finito libro De anima*, Erfurt, Universitätsbibliothek, Dep. Amplon. Ms. 2° 318, f. 156ra; London, Wellcome Collection, Ms. 3, f. 56va; Città del Vaticano, Biblioteca Apostolica Vaticana, Vat. lat. 5988, f. 37rb: "Consequenter destruit hanc opinionem dicens quod, si omnia sensibilia propria essent tangibilia, tunc quilibet sensus proprius tactus esset, quod est impossibile. [...] Et nota quod accipiendo tactum pro tactu spirituali est haec opinio vera, de naturali autem tactu sive materiali falsa est. Consequenter ponit opinionem aliorum circa sensibilia communia. Et est quod omnia sensibilia communia sunt propria <sensibilia tactus>, sicut magnitudo, figura, asperum, lene, quia haec sensibilia communia sunt et non propria <alicui sensui>. Et etiam acutum et obtusum, quae posuit esse Democritus in suis athomis, quae sunt principia rerum, secundum ipsum communia sensibilia sunt". Cf. Adam of Bockenfield, *Cum intentio physici secundum quod physicus*, Philadelphia, Free Library, Lewis European 53, ff. 55vb–56ra; Madrid, Biblioteca Nacional de España 3314, f. 107ra; Milano, Biblioteca Ambrosiana H.105 inf., f. 12rb; Oxford, Balliol College 313, f. 141ra: "Consequenter, cum dicit: *Noster autem sensus quidam*, nunc ultimo dictum est quod quaedam animalia fugiunt odorabile inquantum gustabile et tangibile, id est inquantum gustui et tactui est nocivum, et in praecedentibus habetur quod odorabile odoratur mediante aere et aqua et posset alicui videri quod haec duo non sunt comprehensibilia in odore. Ad

In general, however, the idea that not only touch but also senses in general work in a *spiritual* way, unintuitive as it may seem to modern readers, was hardly surprising in the late Middle Ages. The idea of spirituality of sensory perception was first introduced into Latin philosophy through the translation of Averroes' (Ibn Rushd, 1126–1198) *Compendium libri Aristotelis de memoria et reminiscentia* in ca. 1230,[60] where he presented how sensory perception was, in fact, a process of gradual purification of sensory data from the object, through the medium, to the subject. As a starting point, "the Commentator" takes the crudest, corporeal, and, therefore, the least spiritual form of a physical object's surface. As it enters the medium, such as air or water, it becomes less corporeal and more spiritual. It further loses corporeality and gradually gains spirituality when it is received by one of the external sense organs and passes through the subsequent internal senses—common sense, imagination, the cogitative faculty—and, finally, actualises its maximum degree of spirituality in memory.[61]

By introducing this idea, Averroes and the medieval thinkers who follow in his footsteps, such as Richard Rufus of Cornwall (d. ca. 1260), Geoffrey of Aspall (d. 1287) and, later, Walter Burley (c. 1275–1344/45),[62] tried to bridge the gap between the material (physical objects) and the abstract (forms/images/intentions of these objects).[63] They frequently employed this concept in their theories of cognition and sensory perception on the occasion of explaining how sensory knowledge was possible in general and how sentient beings could know physical objects that they do not have access to physically (such as through sight) in particular.

It is in this context that remarks on spiritual touch usually appear in *De sensu* commentaries. If one sees this through the lens of Averroes, it means that touch, although operating on a very limited scale, filters sensory data from materiality into something more subtle and spiritual, which is later reworked by the internal sensorium. In the *De sensu* commentaries, there is not any clear distinction between spiritual and material touch, because they do not oppose but supplement one another. For they point to different features of touch: the materiality of touch is caused by it being a contact sense, and its spirituality enables sensory data to enter the animal's sensorium.

Besides the Averroistic tradition, other sources that could have been a source for the idea of spiritual touch are scarce; they originated in the contemplative theology

---

auferendum istam dubitationem determinat naturam et proprietatem circa instrumentum olfactus declarans quod olfactus est quasi mediae naturae inter visum et auditum ex una parte, quae sunt virtutes spirituales et immateriales ab eis receptivae, et inter gustum et tactum ex alia parte, quae sunt virtutes materiales".

[60] Wood (2019, pp. 157–158).

[61] Averroes (1949b, pp. 58–59).

[62] For more information on the medieval reception of this concept, see Wood (2019). A more detailed description of Burley's views on the spirituality of the senses can be found in Mansfeld (2022b).

[63] Wood (2019, p. 157). On how the Greek concept of *λόγος* ultimately become the Latin *intentio*, see Black (2011, pp. 159–161).

of the twelfth century. For instance, Hugh of Saint Victor (d. 1141) re-evaluates the tactile experience for spirituality as the most appropriate for the human encounter with the divine.[64] Baldwin of Canterbury (d. 1190), in turn, compares the soul touched by God to a musical instrument.[65] Within the same tradition, Bonaventure (d. 1274) calls touch "the most perfect spiritual sense and the most spiritual, superior even to sight, because it proceeds from charity".[66]

Whichever the source, the concept of spiritual touch in the *De sensu* commentaries was probably a well-known idea ubiquitous to such an extent that it did not require further explanation. It served as a handy reason why touch was regarded as the least spiritual sense. Yet, introducing the twofold division of touch did not make it possible to reconcile Democritus's and Aristotle's theories of sensory perception in the way it was intended by medieval scholars. It seems simply to move the problem to another level: if all external senses are basically touch, then why do particular senses differ so much?

The idea of spiritual touch as opposed to the material one was apparently at its peak in the middle of the thirteenth century and gradually became less popular in later authors. For example, without referring to any names, Geoffrey of Aspall brings back this concept but he does so dismissively.[67] By the 1270s, it was transformed into a concept of two different types of spirit rather than two different types of touch. It is the more material spirit that works for touch and taste, and the lighter, less material and more *spiritual* one that works for sight, hearing, and smell. The idea goes well with the conviction that, from a physiological point of view, spirits (*spiritus*) responsible for sensory perception are activated by either the brain (the coolest organ in an animal body) or the heart (the hottest organ in an animal body). The more material and, therefore, passive the spirits initially are, the more heat they need to perform their actions. Consequently, their organ must be situated closer to the strongest source of heat, i.e. the heart. As a result, the heart is considered the proper organ of both touch and taste.[68] In turn, the subtler and more spiritual the

---

[64] Cf. Hugh of Saint Victor, *Soliloquium de arrha animae* (ed. Migne, 1854, p. 970).

[65] Cf. Baldwin of Canterbury, *De duplici resurrectione, quae per oboedientiae perseverantiam obtinetur* (ed. Migne, 1855, p. 438).

[66] The translation is mine. The source of quotation is Adnès (1991, p. 1077). For more information on the concept of "spiritual touch", see Largier (2004, pp. 233–249).

[67] Cf. Geoffrey of Aspall, *Quaestiones super De sensu et sensato Aristotelis*, lectio V, q. 16: *An hoc sit verum <quod sensus sit ipsius coniuncti>*, Cambridge, Gonville and Caius College 509/386, f. 289vb; Oxford, Merton College Library 272, f. 259va–b; Todi, Biblioteca Comunale Lorenzo Leoni 15, f. 102va: "Quidam dicunt quod est sensus spiritualis, et est animae sensitivae in se et non fit per instrumentum corporeum, et est sensus corporalis, et iste est ipsius coniuncti, et iste fit per instrumentum corporeum, et de tali intendit hic. Mihi tamen videtur quod omnis perceptio quae fit per iudicium hominis est ipsius coniuncti et non animae, et hoc sive sit intellectualis sive sensibilis".

[68] This concept is characteristic, for instance, for Walter Burley. See Mansfeld (2022b, p. 234) and Gensler (2020).

spirits are, the more they require cooling from the brain: this is why sensory organs using the subtle spirit, viz. sight, hearing, and smell, are situated near the brain.[69]

2.3.4 Hierarchy of the Senses

The other context in which the spirituality of touch is discussed is the hierarchy of the senses as presented in the *Dubitationes* from Prague, Ms. 80. Spirituality of touch reappears in an independent note on the order of the powers of the soul; this is peculiar, because the note has little in common with the text of *De sensu et sensato* itself.

The commentator states that the capacities of the soul in every living being can be arranged in such a way that their positions reflect their level of perfection. According to him, perfection of a capacity of the soul is inversely related with its materiality: the more material and organic a power is, the lower is its position on the scale of perfection; conversely, the less material and the subtler it is, the higher is its position. Consequently, the vegetative powers occupy the lowest position on the scale, since they are entirely contained in one's body. The sensitive power is next in line, because it is a little less material and organic. The sensitive power is subdivided into five senses, each of which is ranked according to the degree of its materiality. Consequently, the sense of touch is situated at the bottom of the scale, followed by taste, smell, hearing, and sight. Sight is the highest external sense, because it is the least material. Still higher positions are held by subsequent internal senses: the common sense, imagination, and estimative power, which he calls the "distinctive power" (*potentia distinctiva*), in animals, which is equivalent to the "rational power" (*potentia rationalis*) in humans.[70]

---

[69] Anonymous, *Quaestiones in De sensu et sensato Aristotelis*, Oxford, Bodleian Library, Ms. Oriel 33, q. 17: *Utrum organum tactus situetur iuxta cor*, f. 195rb: "Ratio autem quare visus, auditus et olfactus magis manifestantur circa cerebrum, et gustus et tactus circa cor est haec, quia omnis sensus operatur mediante spiritu, sed spiritus deservientes gustui et tactui sunt materiales, et ideo indigunt magno calore (colore *ms.*); sed spiritus ministrantes aliis tribus sensibus sunt minus materiales et magis spirituales, et ideo indigunt quadam remissione per frigiditatem, quae frigiditas abundat in cerebro".

[70] Anonymous, *Dubitationes super De sensu et sensato Aristotelis, Nota de ordine inter virtutes animae* (ed. Mansfeld, 2022, pp. 342–343): "Propter hoc hic nota diligenter ordinem inter virtutes animae. Quaedam enim <sunt> grossiores et materialiores, et hae magis dicuntur organicae, quia sitae sunt in aliqua parte determinata ipsius corporis deputata ad tale opus. Quaedam enim sunt subtiliores et spiritualiores, et hae minus sunt organicae. Est ergo ordo talis in primo gradu et inter grossiores virtutes est virtus vegetativa, quae indifferenter est in animalibus et plantis, cuius organa fere omnes partes sunt corporis et maxime quattuor partes principales, ut cor etc. Secundo in ordinatione inter virtutes grossiores est sensus particularis, in quo etiam sunt gradus in grossitie et subtilitate. Tactus enim est sensuum particularium grossissimus, deinde gustus, tertio olfactus, quarto auditus, quinto visus qui sensuum particularium magis spiritualis et subtilis est, quia inter ceteros sensus particulares virtus visiva maxime habet de luce quae est forma cuiuslibet sensus et minime de materia; alii sensus e converso, quia tactus maxime de materia et minime de luce habet".

Touch, seen as the most material sense, is thus considered the least perfect of the senses. It deals with the most basic, crudest form of sensory data. Although he does not praise it, the commentator sees it as fundamental to sensory perception. Without it, sensory information cannot be further refined either by the higher external or internal senses.

This view on the role of touch is shared by Geoffrey of Aspall, who discusses whether touch or sight should be considered as the primary sense (in q. 19: *Utrum tactus sit necessarius animali*). His premises are the supposed opposition between sight and touch and the common belief that sight is of primary importance, whereas touch is merely secondary. Aspall explains that sight is primary because of its dignity and nobility. In turn, touch is primary when it comes to its source and operation. This means that Aspall considers touch as the most basic sense, i.e., the one from which all sensory operation starts. For this reason, touch does not provide as nuanced and adequate information to the perceiver as other, more developed senses. He argues, however, that touch is the sense on which all other senses are dependent, as it provides the foundation from which sensory operations begin. Therefore, he seems to argue that touch is the key in understanding sensory operations as such.[71]

The same point is also made in the Prague *Dubitationes* mentioned above, in the *dubitatio* "Whether touch makes an animal an animal so that without it an animal would not exist" (*Utrum tactus det esse animali sine quo animal esse non possit*). Its author admits that touch is the first sense to perform the fundamental function of a sentient being, which is grasping information. Nonetheless, he does not consider it as noble as other senses; still, touch is a prerequisite to all other sense perceptions for him as well.[72]

As I have just pointed out, thirteenth century Oxford commentators consider touch less subtle than the other senses but still important, since it is claimed to provide the basis for further and higher sensory operations.[73] The commentators also remarks a distinctive trait of touch: because of its earthiness, which makes it very

[71] Geoffrey of Aspall, *Quaestiones super De sensu et sensato Aristotelis*, lectio VII, q. 19: *Utrum tactus sit necessarius animali*, ad 1, Cambridge, Gonville and Caius College 509/386, ff. 290vb–291ra; Oxford, New College 285, f. 168va; Oxford, Merton College Library 272, f. 260va; Todi, Biblioteca Comunale Lorenzo Leoni 15, f. 103va: "Ad primam quaestionem dico ad primam rationem quod est prius secundum nobilitatem et intentionem naturae, et sic visus est prior tactu et est prius secundum originem et operationem, et sic tactus est prior. Natura enim semper incipit operari ab incompletiori".

[72] Anonymous, *Dubitationes super De sensu et sensato Aristotelis*, dubitatio 4: *Utrum tactus det esse animali sine quo animal esse non possit* (ed. Mansfeld, 2022, p. 346): "Dicendum probabiliter quod respiciendo ad operationes sensuum et ad effectus illorum erit tactus sensus nobilissimus, respiciendo vero ad causam visibilis et visus, scilicet ad lucem, dicendum quod visus est nobilior tactu. Praeterea, respiciendo ad ordinem sensuum dicendum quod alii sensus sunt nobiliores tactu. Tactus enim naturaliter est primus sensuum et in potentia ad omnes alios sensus. Alii vero sensus, cum sint naturaliter posteriores tactu, sunt tamquam actus respectu ipsius tactus; et actus est nobilior potentia".

[73] See *supra*, 39–44.

resistant, touch is considered capable of preserving impressions to a degree unattainable for other senses.[74]

Other interesting reflections are found in the Oxford commentators with respect to Aristotle's claim that human sense of touch is significantly better than touch in other animals.[75] This is because, in their opinion, the complexion of human body (*corpus humanum*) is based on the most balanced mixture of elements achievable and, for this reason, we can distinguish between various touch qualities with ease.[76] In short, since the most balanced mixture is the most receptive, even the slightest interference would result in humans detecting the change in touch quality; in less balanced mixtures, the interference must be bigger to detect anything. Surprisingly, only Adam of Bockenfield observes that the human sense of touch is not unrivalled in the animal kingdom. Although he is convinced that touch provides us with certain data, humans do not rank among the best in terms of detecting the subtlest touch qualities. Here, spiders hold the crown, since they are able to detect significantly weaker, more subtler, as Bockenfield would call them, "qualities" than humans.[77]

---

[74] Anonymous, *Quaestiones in De sensu et sensato Aristotelis*, q. 16: *Utrum organum tactus sit terrae per dominium*, corpus quaestionis, Oxford, Bodleian Library, Ms. Oriel 33, f. 195ra: "Praeterea, gustus et tactus immutantur maxime materialiter, organum enim tactus immutatur calefaciendo et frigefaciendo, organum vero gustus per humefactionem, unde non immutantur per medium extrinsecum. Ipsa enim organa gustus et tactus sunt maxime sensibilium retentiva. Oportet ergo eorum organa esse talia quae maxime possunt materialiter immutari et species receptas conservare et corruptivis resistere, talia sunt terrea, et ideo organum gustus et tactus sunt terrae per dominium".

[75] Aristotle, *De sensu et sensato*, 4, 440b30–441a2: "Huius autem causa quoniam peiorem habemus aliis animalibus odoratum et ipsis sensibus qui in nobis, tactum vero discretissimum aliis animalibus; gustus vero tactus quidam est". Cf. Averroes, *In De sensu et sensato* (ed. Shields and Blumberg, 1949a, p. 9): "Et ideo homo est inter omnia animalia melioris tactus; et maxime manus, scilicet caro palmae et maxime caro indicis".

[76] For example, Adam of Whitby, *Sententia libri De sensu et sensato Aristotelis*, Firenze, Biblioteca Nazionale Centrale, Conv. Soppr. G 3 464, f. 75va: "Nos autem habemus tactum certissimum, et causa est, quia corpus nostrum quod est quasi instrumentum tactus est compositum ex differentiis tangibilibus in maxima aequalitate, et ideo subtilissime discernit de illis differentiis. Gustus autem est quidam tactus, quare gustum habemus certiorem olfactu, et ideo manifestior est nobis natura saporis quam odoris: *Natura autem aquae* etc.".

[77] Adam of Bockenfield, *Cum intentio physici quod physicus*, Philadelphia, Free Library, Lewis European 53, f. 54rb; Madrid, Biblioteca Nacional de España 3314, f. 104rb–va; Milano, Biblioteca Ambrosiana H.105 inf., f. 108va; Oxford, Balliol College 313, f. 137vb: "Causa etiam quare nos habemus tactum certissimum est, quia instrumentum tactus est compositum ex differentiis tangibilibus, cum igitur inter omnia animalia corpus nostrum maxime accedat ad aequalitatem proportionis miscibilium propter aequalitatem proportionis, subtilissime discernit tactus in nobis de istis differentiis et forte discretius et certius et subtlius quam tactus araneae, non tamen subtiliora. Aranea enim rem subtiliorem discernit secundum tactum quam homo". The same remark concerning spiders' extraordinary accuracy of touch as compared to the human one can be found in the thirteenth century encyclopedia of Thomas of Cantimpré. Cf. Thomas of Cantimpré, *Liber de natura rerum* (ed. Boese, 1973, p. 106).

2.4 Conclusions

A short presentation of views on the nature of the sense of touch in relevant commentaries on *De sensu et sensato* written in Oxford in the thirteenth century shows that they differ in many respects: some of them were composed based on Aristotle's text alone, others included other sources too, their literary forms vary, so does the level of independence of thought displayed by their authors. All that notwithstanding, there is a remarkable level of coherence among them in the doctrinal sense. The most salient characteristic all commentaries have in common is that none of them focuses particularly on touch. Rather, remarks on the nature of touch are just glimpses in the long, elaborate passages dominated by sight. It is hard to avoid the impression that touch, along with hearing, is the least discussed sense in Aristotle's *De sensu*; consequently, it holds true also for the commentaries on this treatise. Of course, it is not surprising that early commentators barely give any attention to this topic, since they stick closely, sometimes even slavishly, to the commented text. As a result, if Aristotle does not discuss an issue at length, nor do they. What is strange, however, is that even such well-developed and detailed question-commentaries, such as the one by Geoffrey of Aspall, avoid this topic too. This disproportion is manifest when we see only one, sketchy and neglected question devoted to touch among almost a hundred questions devoted to other senses. The situation looks slightly better in Oriel 33, yet the focus is on the anatomical aspects of touch and physiological aspects of feeling rather than on explaining the sense's proper operations.

As a result, it is challenging to reconstruct Oxford scholars' doctrine of the role of touch in sense perception. Usually, we are left with a number of Aristotelian clichés, such as the idea that flesh is the medium for touch and that humans have an excellent sense of touch compared to other animals. One can see that these clichés are stated, but not analysed or problematised any further. Perhaps, Oxford scholars focused on touch more when commenting on other Aristotle's natural philosophical works, such as *De anima*, and found numerous urgent issues to discuss, for example the nature of sight, that marginalised this topic in the commentaries on *De sensu et sensato*. It appears that the Oxonian commentators on *De sensu* are uninterested in the perceptual process of touching and do not crave for further information about the object, medium, or organ of touch. From passing remarks in their commentaries, we get essentials of Aristotelian teaching. None of the examined commentaries presents a coherent theory of sense perception in which touch would play any significant role; actually, none even aims at one. The purpose of early Oxford commentaries on *De sensu*, regardless of their form, is only to reconstruct Aristotle's views. Therefore, it is not their authors' intention to formulate any theories of sensory perception of their own.[78]

[78] If one wants to find theories of this kind, one must look elsewhere. Future comparative studies of the commentaries on Aristotle's *De anima* and *De sensu et sensato* of the same milieu may shed some light on whether the theories of sense perception putting so little emphasis on touch are

Yet, the Oxford commentaries analyzed in this chapter still bring some original contributions with respect to their Aristotelian source, which indeed deserve attention: the opposition between the material and spiritual senses (and the concept of spiritual touch as a consequence), the general question whether the sense of touch alone distinguishes animals from other beings, and, to a lesser degree, the question of what it means to have a perfect touch. The concept of spiritual touch is perhaps the most significant among this group, as it allowed Oxford scholars to shed light on contact senses in particular and place tactile perception into a broader context of sense perception in general. In any case, these innovations should not be neglected when we consider that the commentaries were written by authors with so little access to source texts and even less access to secondary literature. In light of this, the credibility of their work and the foundation they provide for further interpretations is impressive.

Acknowledgements This work was supported by the National Science Centre, Poland (NCN) grant n. 2023/51/B/HS1/00308: Geoffrey of Aspall's Theory of Sensual Cognition. Critical Edition and Historical-Philosophical Analysis of His "Quaestiones super *De sensu et sensato* Aristotelis".

References

Primary Sources

Manuscripts

Adam of Bockenfield, *Cum intentio physici quod physicus*. In: Madrid, Biblioteca Nacional de España 3314, ff. 100r–110v; Milano, Biblioteca Ambrosiana H.105 inf., ff. 1r–18r; Oxford, Balliol College 313, ff. 132r–144v; Philadelphia, Free Library, Lewis European 53, ff. 52ra–57vb.

Adam of Whitby, *Commentum in De sensu et sensato Aristotelis*. In: Firenze, Biblioteca Nazionale Centrale, Conv. Soppr. G 3 464, ff. 73vb–78ra; München, Staatsbibliothek, Clm 14522, ff. 149rb–154vb; Paris, Bibliothèque Nationale de France, lat. 16149, ff. 62ra–67ra.

Anonymous, *Dubitationes in De sensu et sensato*, Metropolitan Chapter, M.80, ff. 131vb–132vb.

Anonymous, *Finito libro De anima*. In: Erfurt, Universitätsbibliothek, Dep. Ampl. F. 318, ff. 150r–161r; London, Wellcome Hist. Med. Libr. 3, ff. 53v–60r; Città del Vaticano, Biblioteca Apostolica Vaticana, Vat. lat. 5988, ff. 34ra–41va.

---

characteristic for *De sensu* commentaries alone or are they dominant also in other treatises on Aristotelian psychology and physiology. Commentaries on the *De anima* of the same milieu are, for example, Geoffrey of Aspall, *Quaestiones in De anima* (ed. Čizmić, 2010), and three anonymous question-commentaries on *De anima* (1971a, 1971b, 1971c, ed. Giele, van Steenberghen, and Bazán).

Anonymous, *Glossa Oxoniensis*. In: London, British Library, Ms. Royal 12.G.II, ff. 382v–397r; Cambridge, Gonville and Caius, Ms. 506 (384), ff. 282r–293v.
Anonymous, *Quaestiones super De sensu et sensato Aristotelis*, Oxford, Bodleian Library, Ms. Oriel 33, ff. 192ra–197vb.
Galfridus de Aspall, *Quaestiones in De sensu et sensato Aristotelis*. In: Cambridge, Gonville and Caius College 509/386, ff. 287ra–302rb; Oxford, Merton College Library 272, ff. 257ra–276ra; Oxford, New College 285, ff. 164–189; Todi, Biblioteca Comunale Lorenzo Leoni 15, ff. 99vb–123ra.

Modern Editions

Adam of Bockenfield. (2022). In De memoria et reminiscentia. In J. Brumberg-Chaumont & D. Poirel (Eds.), *Adam of Bockenfield and His Circle on Aristotle's* De memoria et reminiscentia. Oxford University Press.
Anonymous. (1971a). Quaestiones in Aristotelis libros I et II De anima. In M. Giele, F. van Steenberghen, & B. Bazán (Eds.), *Trois commentaires anonymes sur le Traité de l'âme d'Aristotle* (pp. 21–118). Peeters.
Anonymous. (1971b). Quaestiones in Aristotelis De anima. In M. Giele, F. van Steenberghen, & B. Bazán (Eds.), *Trois commentaires anonymes sur le Traité de* l'âme *d'Aristotle* (pp. 135–348). Peeters.
Anonymous. (1971c). Quaestiones super Aristotelis librum De anima. In M. Giele, F. van Steenberghen, & B. Bazán (Eds.), *Trois commentaires anonymes sur le* Traité de l'âme *d'Aristotle* (pp. 389–514). Peeters.
Aristotle. (1831). In A. I. Bekker (Ed.), *Opera omnia, I*. Apud Gregorium Reimerum.
Aristotle. (1984). In J. Barnes (Ed.), *Complete Works of Aristotle: The Revised Oxford Translation* (Vol. 1–2). Princeton University Press.
Aristotle. *De sensu et sensato*. In G. Galle & L. Peeters (Eds.), *Aristoteles Latinus XIII. 1. De sensu et sensato. Translatio anonyma*. Brepols, in preparation.
Averroes. (1949a). In De sensu et sensato. In A. L. Shields & H. Blumberg (Eds.), *Averrois Cordubensis Compendia librorum Aristotelis qui Parva Naturalia vocantur* (pp. 1–46). The Medieval Academy of America.
Averroes. (1949b). In De memoria et reminiscentia. In A. L. Shields & H. Blumberg (Eds.), *Averrois Cordubensis Compendia librorum Aristotelis qui Parva Naturalia vocantur* (pp. 47–72). The Medieval Academy of America.
Averroes. (1953). Commentarium magnum in De anima. In F. S. Crawford (Ed.), *Averrois Cordubensis Commentarium magnum in De anima libros* (pp. 3–546). The Medieval Academy of America.
Baldwin of Canterbury. (1855). De duplici resurrectione, quae per oboedientiae perseverantiam obtinetur. In J. P. Migne (Ed.), *Patrologiae Latinae cursus completus* (Vol. 204, pp. 430–455). D'Ambroise.
Geoffrey of Aspall. (2010). Quaestiones in De anima. In V. Čizmić (Ed.), *Galfridus de Aspall.* Questiones in De anima. *Kritische Edition* (Vol. 2). Universitäts-Bibliothek München.
Geoffrey of Aspall. (2017). In S. Donati & C. Trifogli (Eds.), *Questions on Aristotle's Physics*. Oxford University Press.
Hugh of Saint Victor. (1854). Soliloquium de arrha animae. In J. P. Migne (Ed.), *Patrologiae Latinae cursus completus* (Vol. 176, pp. 951–968). D'Ambroise.
Plato. (1969). Republic. In P. Sholrey (transl.), *Plato in Twelve Volumes*, vol. 5–6. Harvard University Press; William Heinemann Ltd.
Thomas of Cantimpré. (1973). Liber de natura rerum. In H. Boese (Ed.), *Thomas Cantimpratensis. Liber de natura rerum* (Vol. 1: Text). De Gruyter.

Secondary Sources

Adnès, P. (1991). Toucher. In A. Derville et al. (Eds.), *Dictionnaire de spiritualité ascétique et mystique* (Vol. 15: Taborin – Tyskiewicz, pp. 1073–1098). Beauchesne.

Beneduce, C. (2022). "Is Touch One Sense or Several?" A Late Medieval Scientific Question. In M. Gensler, M. Mansfeld, & M. Michałowska (Eds.), *The Embodied Soul: Aristotelian Psychology and Physiology in Medieval Europe between 1200 and 1420* (pp. 253–275). Springer.

Beneduce, C. (2023). John Buridan: The Human Body at the Intersection of Natural Philosophy and Medicine. In J. Zupko (Ed.), *The Philosophy of John Buridan* (pp. 161–182), special issue of the *American Catholic Philosophical Quarterly, 97*(2).

Black, D. (2011). Averroes on the Spirituality and Intentionality of Sensation. In P. Adamson (Ed.), *The Age of Averroes, Arabic Philosophy in the Sixth/Twelfth Century* (pp. 159–174). The Warburg Institute.

Bloch, D. (2008). Nicolaus Graecus and the *Translatio Vetus* of Aristotle's *De Sensu*. *Bulletin de Philosophie Médiévale, 50*, 83–103.

Brumberg-Chaumont, J., & Poirel, D. (2022). *Adam of Bockenfield and His Circle on Aristotle's* De memoria et reminiscentia. Oxford University Press.

Callus, D. A. (1943). Introduction of Aristotelian Learning to Oxford. In The British Academy (Ed.), *Proceedings of the British Academy* (Vol. 29, pp. 1–55). The British Academy.

De Leemans, P. (2020). *Parva Naturalia*, Commentaries on Aristotle's. In H. Lagerlund (Ed.), *Encyclopedia of Medieval Philosophy. Philosophy between 500 and 1500* (pp. 1385–1397). Springer.

Ebbesen, S., Thomsen Thörnqvist, C., & Decaix, V. (2016). Questions on *De sensu et sensato, De memoria* and *De somno et vigilia*: A Catalogue. *Bulletin de Philosophie Médiévale, 57*, 59–115.CFreeland1992Aristotle on the sense of touchMCraven NussbaumARorty Essays on Aristotle's De anima Oxford University Press227248

Freeland, C. (1992). Aristotle on the Sense of Touch. In M. Craven Nussbaum & A. Rorty (Eds.), *Essays on Aristotle's* De anima (pp. 227–248). Oxford University Press.

Galle, G. (2008a). Edition and Discussion of the Oxford Gloss on *De sensu* 1. *Archives d'histoire doctrinale et littéraire du Moyen Âge, 75*, 197–281.

Galle, G. (2008b). The Anonymous Translator of the *Translatio Vetus* of *De Sensu*. *Bulletin de Philosophie Médiévale, 50*, 105–150.

Galle, G. (2008c). The Dating and Earliest Reception of the *Translatio Vetus* of Aristotle's *De sensu*. *Medioevo, 33*, 1–90.

Galle, G. (2010). Interpretations of the *Translatio Vetus* of *De sensu 1* in Commentaries Attributed to Adam of Buckfield and in the "Oxford gloss". In P. Bernardini (Ed.), *I manoscritti e la filosofia* (pp. 47–66). Edizioni dell'Università.

Galle, G. (2023). The Order of the *Parva naturalia* in Three Commentaries on *De sensu* Associated with Adam of Bockenfield. Implications for the Authenticity Question. *Micrologus. Nature, Sciences and Medieval Societies, 31*, 15–44.

Gensler, M. (2020). The Concept of *Spiritus* in Walter Burley's *Parva naturalia* Commentaries. In R. Hofmeister Pich, A. Storck, & A. Culleton (Eds.), Homo, Natura, Mundus: *Human Beings and Their Relationships: Proceedings of the XIV International Congress of the S.I.E.P.M., July 24-28, 2017, Porto Alegre, Brazil* (pp. 805–815). Brepols.

Hasse, D. N. (2010). The Soul's Faculties. In R. Pasnau (Ed.), *The Cambridge History of Medieval Philosophy* (Vol. 1, pp. 305–319). Cambridge University Press.

Johansen, T. L. (1998). *Aristotle on the Sense-Organs*. Cambridge University Press.

Johnstone, M. A. (2021). Aristotle on the Unity of Touch. *Journal of the History of Philosophy, 59*(1), 23–43.

Kambaskovic-Sawers, D., & Wolfe, C. T. (2014). The Senses in Philosophy and Science: From the Nobility of Sight to the Materialism of Touch. In H. Roodenburg (Ed.), *A Cultural History of the Senses in the Renaissance*. Bloomsbury Publishing.

Köhler, T. W. (2017). Homo animal nobilissimum: *Konturen des Spezifisch Menschlichen in der Naturphilosophischen Aristoteleskommentierung des Dreizehnten Jahrhunderts* (Vol. 1). Brill.
Knuuttila, P. (2008). Aristotle's Theory of Perception and Medieval Aristotelianism. In: P. Knuuttila, S. & Kärkkäinen, (Eds.), *Theories of Perception in Medieval and Early Modern Philosophy*, Cham, (p. 1–22). Springer
Largier, N. (2004). *Tactus spiritualis*. Le sens du toucher et la volupté au Moyen Age. *Micrologus. Nature, Sciences and Medieval Societies, 13*, 233–249.
Lohr, C. H. (1967). Medieval Latin Aristotle Commentaries Authors A–F. *Traditio, 23*, 313–413.
Lohr, C. H. (1968). Medieval Latin Aristotle Commentaries Authors G-I. *Traditio, 24*, 149–245.
Mansfeld, M. (2021). The Middle Color: A History of a Problem in Thirteenth Century Oxford Commentaries on *De sensu et sensato. Analiza i Egzystencja, 54,* 127–154.
Mansfeld, M. (2022a). Anonymous Oxonian *Dubitationes* on Aristotle's *De sensu et sensato* (Prague, Metropolitan Chapter, Ms. M. 80, ff. 131vb–132vb). *Argument, 12*(2), 329–348.
Mansfeld, M. (2022b). The World of Senses. On the Process of Cognition in Walter Burley. In M. Gensler, M. Mansfeld, & M. Michałowska (Eds.), *The Embodied Soul. Aristotelian Psychology and Physiology in Medieval Europe between 1200 and 1420* (pp. 229–251). Springer.
Mantovani, M., Zambiasi, R., & Zuccolin, G. (2023). Aristotle's *De sensu* in the Latin Tradition 1250-1650, special issue *Micrologus. Nature, Sciences and Medieval Societies, 31*, VI–370.
Marenbon, J. (2020). Aristotelianism in the Greek, Latin, Syriac, Arabic, and Hebrew Traditions. In H. Lagerlund (Ed.), *Encyclopedia of Medieval Philosophy. Philosophy between 500 and 1500* (pp. 180–188). Springer.
Murphy, D. (2005). Aristotle on Why Plants Cannot Perceive. *Oxford Studies in Ancient Philosophy, 29*, 295–339.
Nelis, R. (2022). A Jumble of Writings: Commentaries on Aristotle's *De Longitudine et Brevitate Vitae* Attributed to Adam of Buckfield. *Early Science and Medicine, 27*(1), 1–56.
Peeters, L. (1996). *De sensu et sensato – Translatio vetus. Voorbereiding voor een kritische editie*. Leuven: unpublished MA dissertation, supervisor: F. Bossier.
Polansky, R. (2007). *Aristotle's* De Anima. *A Critical Commentary*. Cambridge University Press.
Redmond, D. (2019). Aristotle on the Organ and Medium of Touch. *Méthexis, 31*(1), 103–121.
Russell, J. C. (1936). *Dictionary of Writers of Thirteenth Century England*. Longmans.
Schriefl, A., & Seger, M. (2023). Aristotle on the Beginning of Animal Life and Soul Activities. *Apeiron. A Journal for Ancient Philosophy and Science, 56*(3), 587–619.
Sorabji, R. (1971). Aristotle on Demarcating the Five Senses. *Philosophical Review, 80*(1), 55–79.
Steiner Goldner, R. (2018). Touch and Flesh in Aristotle's *De Anima*. *Epoché: A Journal for the History of Philosophy, 15*(2), 435–446.
Thomsen Thörnqvist, C. (2022). Affected by the Matter: The Question of Plant Perception in the Medieval Latin Tradition on *De somno et vigilia*. In J. Toivanen (Ed.), *Forms of Representation in the Aristotelian Tradition, vol. 1: Sense Perception* (pp. 183–212). Brill.
Weijers, O. (1994). *Le travail intellectuel à la Faculté des arts de Paris: textes et maîtres (ca. 1200–1500), I. Répertoire des noms commençant par A–B*. Brepols.
Wood, R. (2019). Spirituality and Perception in Medieval Aristotelian Natural Philosophy. In E. Băltuţă (Ed.), *Medieval Perceptual Puzzles. Theories of Sense Perception in the 13th and 14th Centuries* (pp. 153–178). Brill.

Chapter 3
Labour Pain and Touch in Late Medieval Science

Chiara Beneduce

Abstract The nature of affective touch, and namely the pleasant and unpleasant character of tactile perceptual experience, is currently being investigated in both the biomedical sciences and the philosophy of science. Yet, the relationship of tactile perception to pleasure and pain was also an object of scientific study in the past. Within the framework of this volume on medieval and modern conceptualizations of touch, I will address late medieval scientific theories of the role played by touch in the perception of pleasure and pain. In the attempt to contribute to historical reconstructions of such medieval scientific theories, this chapter focuses especially on the phenomenon of labour pain. After providing an overview of medieval reflections on the connection between touch and pleasure and pain, it presents some examples of descriptions of labour pain and its physiology contained in—both edited and unedited—late medieval scientific works. It thereby explores to what extent touch was involved in the explanations of labour pain offered by late medieval science. The conclusion of this survey will be that, while perception of pleasure and pain were often connected to touch in late medieval philosophy and science, reference to touch was more sporadic in late medieval scientific discourses on labour pain and its physiology.

Keywords Touch · Affective touch · Gendered touch · Pleasure · Pain · Labour pain · Women and pain · Natural philosophy · Medicine · Middle Ages

---

C. Beneduce (✉)
KU Leuven, De Wulf-Mansion Centre for Ancient, Medieval and Renaissance Philosophy, Leuven, Belgium
e-mail: chiara.beneduce@kuleuven.be

C. Beneduce et al. (eds.), *The Sense of Touch*, Studies in the History of Philosophy of Mind 32, https://doi.org/10.1007/978-3-032-16308-0_3

3.1 Notes on Touch, Pleasure, and Pain in Late Medieval Philosophy and Science

We can gain access to pleasant and unpleasant experiences through touch. The pleasant or unpleasant component or character of a sensory experience is usually called "affective". The nature of *affective* touch is vastly studied in contemporary biomedicine, psychology, philosophy of mind, and in the philosophy of science more generally. It is moreover located at the intersection of all these disciplines.[1] Yet, reflections on the connection between the sense of touch and pleasurable and painful perceptual experiences are not new to science. The idea that perception of bodily pleasure and pain could occur through touch was already present, for example, in late medieval philosophy and science, namely in the span of the history of thought usually identified as in-between the rediscovery of Aristotle's works at medieval universities and the end of scholasticism (ca. 1100–1600, although the latter term is rather controversial).[2] Two notable examples of discussions of perception of pleasure and pain in relation to touch, for instance, appear in the work of Albert the Great (d. 1280) and Peter of Abano (d. 1316), two major masters of late medieval universities. The former is, *sans doute*, the greatest Latin commentator of Aristotle.[3] The latter is considered one of the most influential physicians of premodern thought.[4] In Albert the Great's *De anima* and *De bono*, one can find statements connecting both pleasure (*delectatio*) and grief (*tristitia*) to the sense of touch. In some straightforward sentences, we see Albert framing pleasure and pain as absolutely (*simpliciter, per se, proprie*) connected to the sense of touch.[5] In his most famous medical work, the *Conciliator*, Peter of Abano similarly refers to pain (*dolor*) and pleasure (*delitia*) as perceptual objects of touch.[6]

[1] The literature on affective touch is vast and growing. Here I will just mention the collective volume by Olausson et al. (2016), which offers an interdisciplinary perspective on affective touch, including philosophy, as well as the more recent Haggarty et al. (2023).

[2] See Kretzmann et al. (1984); Marenbon (2023).

[3] See Brumberg-Chaumont (2020).

[4] See Lagerlund (2020).

[5] Albert the Great, *De anima*, II, tr. 3, ch. 35 (ed. Stroick, 1968, p. 148): "Et ideo sensibilis delectatio et sensibilis tristitia non sunt nisi in tactu, in aliis autem sunt, inquantum vicinius vel distantius tactus indicant delectationes"; Albert the Great, *De bono*, tr. 1, q. 6, a. 2 (ed. Khüle et al., 1951, p. 80): "*Supra* enim probatum est in quaestionibus *De anima*, quod delectationes et tristitiae sunt simpliciter in tactu, secundum quid in aliis sensibus"; *Ibid*., tr. 3, q. 1, a. 2 (ed. Khüle et al., 1951, p. 122): "Item, in quaestione *De tactu* probatur, quod nullus sensuum per se delectatur nisi tactus"; *Ibid*., tr. 3, q. 5, a. 2 (ed. Khüle et al., 1951, p. 200): "Obicitur etiam contra illam divisionem quae dividit genera voluptatum secundum obiecta sensuum diversa. Probatum est enim *supra* in quaestione *De tactu*, quod nullus sensuum proprie facit delectationem nisi solus tactus. Ergo videtur, quod secundum numerum aliorum sensuum nulla sit delectatio vel voluptas". About Albert on pleasure, pain, and touch see also *infra*, 55–59.

[6] Peter of Abano, *Conciliator controversiarum, quae inter philosophos et medicos versantur*, diff. 77 (ed. Venezia, 1565, f. 118ra): "Propter quartum: vero dicendum ad primum quod dolor est una qualitatum et delitia, est namque in passione et passibili qualitate locandus. Si autem per qualitates

Undoubtedly, the late medieval idea of a connection between touch and at least the perception of bodily pleasure was influenced by a famous view expressed by Aristotle (d. 322 BC) in the *Ethica Nicomachea*. In this work, Aristotle had discussed temperance and intemperance, associating the latter with the bodily, brutish pleasures that human beings share with (other) animals: those perceived by touch and taste. In Aristotle's view, touch, like taste, was therefore tightly linked to the lowest forms of pleasures, first and foremost sexual pleasure, but also those derived from drinking and eating.[7] The discourse on temperance and intemperance, bodily pleasures, and touch was especially and explicitly taken up in late medieval commentaries on the *Ethica Nicomachea*, though not exclusively.[8] One eminent discussion of the link between touch and bodily pleasure coming in the wake of Aristotle's statements in the *Ethica Nicomachea* is Thomas Aquinas's (d. 1274)

Averroes intendat primas, aut secundas communiter solum expressas, ut ostensum, secum non consentio, cum dolor ab eis percipiatur alius, ac aliter. Neque hoc impediret, supposito est dolorem ad aliquid fore: quia primae qualitates directe in genere locatuntur qualitatis, grave, et leve in praedicamento substantiae, asperum et lene in genere positionis. 42 differentia ex Aristotele et Commentatore. Et tamen sunt unius sensus, puta tactus obiecta".

[7] Aristotle, *Ethica Nicomachea*, X.5, 1175b24–1176a3 (ed. Barnes, 1984b, vol. 2, p. 1858): "[…] Now sight is superior to touch in purity, and hearing and smell to taste; the pleasures, therefore, are similarly superior, and those of thought superior to these, and within each of the two kinds some are superior to others"; *Ibid.*, III.10, 1118a2–1118b7 (ed. Barnes, 1984b, vol. 2, 1765): "Temperance must be concerned with bodily pleasures, but not all even of these; […]. Temperance and self-indulgence, however, are concerned with the kind of pleasures that the other animals share in, which therefore appear slavish and brutish; these are touch and taste. […]". A similar discourse was reprised in the section on temperance and intemperance of the pseudo-Aristotelian *Problemata* and considerations on appetite, desire, pleasure, and pain as connected to touch are discussed also in Aristotle's *De anima*. See Pseudo-Aristotle, *Problemata*, book XXVIII (*Problems Connected with Temperance and Intemperance, Continence and Incontinence*), 2, 949b6–949b11 (ed. Barnes, 1984b, vol. 2, p. 1491): "Why is it that we speak of men as incontinent in connexion with two only of the senses, namely, touch and taste? Is it because of the pleasures that result from these in us and in the other animals? Being then shared by the animals, they are held in least honour and so are regarded as the only pleasures deserving of reproach, or at any rate more so than any others. So we blame a man who is a slave to them and call him incontinent and intemperate, because he is a slave to the worst pleasures"; Aristotle, *De anima*, III.11, 433b31–434a5 (ed. Barnes, 1984a, vol. 1, p. 689): "We must consider also in the case of imperfect animals, sc. those which have no sense but touch, what it is that in them originates movement. Can they have imagination or not? or desire? Clearly they have feelings of pleasure and pain, and if they have these they must have desire. But how can they have imagination? Must not we say that, as their movements are indefinite, they have imagination and desire, but indefinitely?".

[8] One example of commentary on the *Ethica Nicomachea* which discusses Aristotle's position on touch and brutish pleasures is the one by the fourteenth-century Parisian Master of Arts John Buridan (d. 1361 ca.). See John Buridan, *Quaestiones super decem libros Ethicorum*, III, q. 25 (ed. Paris, 1513, repr. 1968, f. 62va): "[…] delectationes tactus et gustus ut que sunt in comessationibus et potationibus et venereis actibus […]" and *Ibid.*, III, q. 25 (ed. Paris, 1513, repr. 1968, f. 62ra–b): "Item sequitur illud in quo conuenimus bestiis scilicet delectationes sensuales specialiter tactus, […]". On Buridan's discourse on temperance, see Grellard (2017). On late medieval commentaries on the *Ethica Nicomachea* and pleasure, see Robert (2021, pp. 235–248).

discourse in the *Summa Theologiae*, which directly addresses the relationship between temperance and the pleasures of touch.[9]

In late medieval science, reflections on the relationship between touch and experiences of pleasure and pain are also situated within discussions of the perceptual objects of touch— discussions that very often, albeit not exclusively, took as their point of departure the Aristotelian issue of whether touch constitutes one or several senses. In fact, in the second book of his *De anima*, Aristotle stated that "It is a problem whether touch is a single sense or a group of senses", but ultimately left answer to this problem undefined.[10] The first reason touch is a candidate for being considered a plural sense is that, unlike the other four senses, it is difficult to identify touch's proper or specific object of sensation, since touch is able to sense a plurality of perceptual objects. The issue of "whether touch is one or a plural sense" was addressed frequently in late medieval scholarly texts: mostly in commentaries on Aristotle's *De anima* but also in medical and even theological works.[11] In some of the texts addressing the issue of "whether touch is one or a plural sense", as well as in some others that discuss the objects of touch outside the issue of touch's unity or plurality, the possible status of pleasure and pain as objects of touch was considered. This is the case, for instance, in the *Liber de anima seu sextus de naturalibus* of Avicenna (Ibn Sīnā) (d. 1037), which includes pleasure and pain in its discussion

---

[9] Thomas Aquinas, *Secunda Secundae Summa Theologiae*, t. X, q. 141, a. 4 *Utrum temperantia solum sit circa concupiscentias et delectationes tactus* (ed. Roma, 1899, p. 126): "Maxime autem naturales animalibus sunt operationes quibus conservatur natura individui per cibum et potum, et natura speciei per coniunctionem maris et feminae. Et ideo circa delectationes ciborum et potuum, et circa delectationes venereorum, est proprie temperantia. Huiusmodi autem delectationes consequuntur sensum tactus. Unde relinquitur quod temperantia sit circa delectationes tactus". Aquinas links the perception of brutish pleasures to touch also in another fascinating passage of the *Summa Theologiae*, and namely Thomas Aquinas, *Prima Secundae Summa Theologiae*, t. VI, q. 31, a. 6 (ed. Roma, 1891, pp. 220–221): "Cum igitur delectatio tactus sit maxima ratione utilitatis, delectatio autem visus ratione cognitionis; si quis utramque comparare velit, inveniet simpliciter delectationem tactus esse maiorem delectatione visus, secundum quod sistit infra limites sensibilis delectationis. Quia manifestum est quod id quod est naturale in unoquoque, est potentissimum. Huiusmodi autem delectationes tactus sunt ad quas ordinantur concupiscentiae naturales, sicut cibi, et venerea, et huiusmodi. — Sed si consideremus delectationes visus, secundum quod visus deservit intellectui; sic delectationes visus erunt potiores, ea ratione qua et intelligibiles delectationes sunt potiores sensibilibus". On Aquinas, corporeal pleasure, touch, and temperance, see Vecchio (2009, p. 85) and Vecchio (2006, pp. 113–114).

[10] Aristotle, *De anima*, II.11, 422b20 (ed. Barnes, 1984a, vol. 1, p. 672): "Whatever can be said of what is tangible, can be said of touch, and vice versa; if touch is not a single sense but a group of senses, there must be several kinds of what is tangible. It is a problem whether touch is a single sense or a group of senses". Aristotle's position on this is sketched some lines later, at the beginning of the third book, where he states that there are only the five senses (sight, hearing, taste, smell, and touch) and adds somewhat tautologically, referring specifically to touch, that everything tangible can be perceived by touch. See *Ibid.*, III.1, 424b22–425a13 (ed. Barnes, 1984a, vol. 1, p. 675): "That there is no sense in addition to the five—sight, hearing, smell, taste, touch—may be established by the following considerations:—We in fact have sensation of everything of which touch can give us sensation (for all the qualities of the tangible qua tangible are perceived by us through touch); […]".

[11] On this, see Beneduce (2022).

of the possible perceptual objects of touch.[12] Touch and (sexual) pleasure are also together mentioned in the *Colliget* by Averroes (Ibn Rushd) (d. 1198).[13] Arguably, it cannot be ruled out that these two Arabic authors were the main sources for later Latin treatments of the perception of pleasure and pain in the context of discussions on the objects of touch and its unity or plurality. In the Latin tradition, examples of discussions of the topic of pleasure and pain as objects of touch are the following. First, in his *De homine*, Albert the Great firmly argues in favor of ascribing the perception of pleasure and pain to touch and explicitly invokes Avicenna as an authority for this position: the sense of pleasure, inasmuch as pleasure, and the sense of pain, inasmuch as pain, are nowhere else than in touch. Albert explains that touch's characteristic is to feel everything dissolving the continuity of the body and everything preserving it: so are, respectively, the sensation of pain felt when the body is whipped and the sensation of pleasure felt in intercourse. As far as the feeling of pleasure during intercourse is particularly concerned, Albert underlines that it is an increased sensation since generative seed *touches* nerves very subtly.[14] Another

[12] Avicenna, *Liber de anima seu sextus de naturalibus*, II, ch. 3 (ed. Van Riet, 1972, pp. 137–138): "Ergo dolor et remedium doloris etiam sunt de sensibilibus tactibilibus, sed differt tactus secundum hanc intentionem ab aliis sensibus. Aliorum etenim sensuum quidam sunt qui non habent voluptatem nec dolorem, et quidam sunt qui delectantur et dolent mediantibus aliis sensibus. Sed ille in quo non est voluptas est visus, qui non delectatur nec dolet, sed anima est quae delectatur intus propter illud aut dolet. Similiter etiam est dispositio auditus, quia auditus dolet ex forti sono, sicut visus ex forti colore sicut est lumen, nec dolet ex hoc quod audit aut videt, sed ex hoc quod tangit et accidit ei dolor tactibilis, et sic adventus huius afficit illum constrictione tactibili. Sed olfactus et gustus dolent et delectantur cum afficiuntur convenienti qualitate aut inconvenienti. Tactus vero, aliquando dolet propter qualitatem tactivam aut delectatur, aliquando dolet aut delectatur non mediante alia qualitate quae sit primum sensatum, sed propter solutionem continui et propter suam continuitatem".

[13] Averroes, *Averrois Cordubensis Colliget Libri VII*, II, ch. 13 (ed. Venezia, 1552, f. 25ra–b): "De proprio instrumento sensus tactus. Cap. 13. Et dicimus quod caro est proprium instrumentum sensus tactus. [...] Et istius virtutis quiddam est, quod communicat omnibus partibus carnis: et hic est sensus quatuor qualitatum contrariarum, sicut caliditas, frigiditas, siccitas, et humiditas et quiddam est, quod est proprium uni membro sicut est quando os stomachi sentit resolutionem factam in eo et iste sensus appellatur fames et sitis. Sed fames est sensus resolutionibus substantiae grossae, calidae et siccae et sitis est sensus resolutionis substantiae frigidae et humidae: ergo sensus proprius est duplex. Et sensus capitis membri de titillatione, quam habet in hora coitus, isti duo sunt connumerati in genere sensus tactus".

[14] Albert the Great, *De homine* (ed. Anzulewicz and Söder, 2008, pp. 254–256): "Quarto et ultimo quaeritur, quid est tangibile proprium? (1) Dicitur autem communiter quod primae qualitates, scilicet calidum, frigidum, humidum, siccum, sunt tangibilia. [...] quaeritur igitur, utrum omnes istae primo sensibiles sint secundum tactum vel non primo. [...] (2) Praeterea quaeritur de quibusdam aliis quae sentiuntur tactu, sicut est sensus divisionis continui in dolore percussionis, et sicut est sensus delectationis ex tactu aliquo, sicut in coitu, utrum illa sint tangibilia per se vel non. Et videtur quod per se, quia si detur quod per aliud, hoc esset per calidum et frigidum vel humidum et siccum; sed patet quod hoc est falsum, quia illae qualitates aequaliter sentiuntur in corpore, solutio autem continuitatis in sensu doloris sentitur inaequaliter. [...] Secundum autem quod tactus est perfectio corporis compositi in membris dissimilibus, sic sentit omne id quod est dissolvens continuationem talis corporis et quod est conservans ipsam; et sic est sensus doloris in verberibus et sensus delectationis in coitu. Sensus autem delectationis augmentatur, eo quod semen subtilissime

example can be found in the *Quaestiones in secundum librum Sententiarum* by Peter John Olivi (d. 1298), which instead denies that pleasure and pain are objects of touch and ascribes their perception to the common sense (*sensus communis*), namely to an ability of coordination and composition of the perceptual stimuli coming from the external world, which medieval authors counted among the psychic functions.[15] A third example, and the most extensive one I have found so far in the late medieval tradition, is that of John Buridan (d. 1361 ca). In q. 20 of the second book of his *Quaestiones de anima* (in the third redaction), Buridan explicitly questions whether the sensation felt in the genitals during ejaculation is caused by the sense of touch.[16] In a way, Buridan elaborates on the claim—previously found in Albert—that the sensation of pleasure in intercourse is due to the generative seed *touching* nerves.[17] Despite his own ultimate opinion that sexual pleasure is due to touch, Buridan opens discussion on the issue of whether sexual pleasure should be

---

tangit nervos. […] (2) Ex his autem quae habita sunt, patet quod sensus delectationis, inquantum est delectatio, et doloris, inquantum dolor, non est nisi in tactu, ut dicit Avicenna. […] Tactus autem proprium obiectum habet dolorem et delectationem in sensu, et propter hoc ipse solus per se delectatur et dolet, alii autem omnes per accidens. Hinc est etiam quod dicit Aristoteles in quarto *Ethicorum* quod castitas non est in gustu inquantum iudicium saporum est, sed inquantum est quidam tactus. Delectatio enim cibi non est obiectum gustus inquantum iudicium saporum est, sed inquantum est quidam tactus. Instrumentum autem tactus iudex est tangibilium dissimilium sibi, sicut medium iudex est extremorum excellentium; et illud instrumentum antequam tangat potentia est sicut tangibile, et cum tangit, fit actu huiusmodi".

[15] Peter John Olivi, *Quaestiones in secundum librum Sententiarum*, q. 61 (ed. Jansen, 1922–1926, vol. 2, pp. 574–587): "Quantum ad secundum, quod scilicet tactus in plures tactivas potentias dividatur". See the following passages in particular: *Ibid.*, q. 61 (ed. Jansen, 1922–1926, vol. 2, p. 574): "Primo, quia multa per tactum apprehenduntur quae non minus differunt genere quam obiecta diversorum sensuum, utpote, grave et leve, calidum et frigidum, humidum et siccum, durum et molle, densum et subtile, et item multiplex dispositio et indispositio proprii organi et totius corporis; nam gravedines indigestionum et inflationum et apostemationum et calores febriles et corporis inanitatem et indigentiam et satietatis plenitudinem et varios pruritus carnis membrorumque agilem mobilitatem vel contrariam tarditatem eorumque constans robur ac incostantem debilitatem eorumque scissuram vel integritatem ac dolores et delectationes ex his causatas videmur sensu tactus sentire, quae utique non minus ab invicem differunt quam differant a colore vel sono"; *Ibid.*, q. 61 (ed. Jansen, 1922–1926, vol. 2, p. 583): "Excipiunt tamen a praedictis dolorem et delectationem, quia sicut tactus non potest sentire suum proprium actum nisi forte valde semiplene, sic non potest sentire dolores vel delectationes consequentes et concomitantes suum actum, sed potius hoc spectat ad sensum communem, sicut infra tangetur". On this topic in Olivi, see also Yrjönsuuri (2008). On the medieval notion of "common sense" literature is vast. I just mention here Heller-Roazen (2007) as a general introduction to the concept of "common sense" through ancient, medieval, and modern times.

[16] John Buridan, *Quaestiones de anima (III red.)*, II, q. 20 (ed. Klima et al., 2023, p. 504): "Utrum sint tantum quinque sensus exteriores"; *Ibid.*, II, q. 20, [6] (ed. Klima et al., 2023, p. 506): "Item arguitur aliquantulum difficile quia membra genitalia per suas proprias naturas et complexiones sentiunt appropriatas qualitates quas alia membra non sentiunt nec alia organa. Igitur in eis est alius sensus et alterius rationis praeter illos qui in aliis membris inveniuntur. Consequentia patet quia aliter tu non posses habere rationem quod esset alius sensus tactus et gustus. Sed antecedens manifestum est quia membra genitalia sentiunt qualitates spermatum secundum quas delectantur, quas non sentirent alia membra, ut manus aut pes, quod patet quia non sic in illis delectantur".

[17] See *supra*, 55.

ascribed instead to a sense other than one of the five (external) senses identified by Aristotle.[18] These aforementioned cases—and that of Buridan in particular—might be considered the medieval background of a later, more widespread debate that exploded in the Renaissance and early modern era, which discussed the status of pleasure and pain as objects of touch and sexual pleasure as a potential "sixth sense". Reflections on pleasure and pain as the objects of touch are for example found in the work of Girolamo Fracastoro (d. 1553), Julius Caesar Scaliger (d. 1558), Gerolamo Cardano (d. 1576), and Giulio Cesare Vanini (d. 1619).[19] A long discussion of pleasure and pain as objects of touch is also found in the *De anima* by the so-called *Conimbricenses*, which, in its discussion of pleasure and pain as objects of touch, lists some of the medieval and later authors mentioned above.[20] It is relevant to mention here a passage of the Scottish philosopher William Hamilton (d. 1856), who, in the nineteenth century, wrote:

> That the recognition of the Locomotive Faculty, or rather, the recognition of the Muscular Sense as a medium of apprehension, is of a recent date and by psychologists of this country, is an opinion in both respects erroneous. – As far as I am aware, this distinction was originally taken by two Italian Aristotelians, some three centuries ago; and when the observation was again forgotten, both France and Germany are before Scotland in the merit of its modern revival. It was first promulgated by Julius Caesar Scaliger about the middle of the sixteenth century (1557). […] It should, however, be noticed[,] that Scaliger may have taken the hint for the discrimination of this and another sense, from Cardan. This philosopher makes Touch fourfold. One sense apprehending the four primary qualities, the Hot and Cold, the Dry and Humid; a second the Pleasurable and Painful; a third the Venereal sensations; a fourth the Heavy and Light. (*De subtilitate*, L. XIII).[21]

In later literature, and much more recent than Hamilton's nineteenth-century observation, we find the idea that the debate between Cardano and Scaliger on the objects of touch and sexual pleasure as a "sixth sense" constitutes a prehistory to recognition of the so-called "muscular sense" as distinct from tactile sensation.[22] However, as I noted, a prehistory of reflections on whether (or not) to count pleasure—including sexual pleasure—among the perceptual objects of touch is already to be identified in the late Middle Ages. Therefore, history of the recognition of the "muscular

---

[18] On this, see Beneduce (2022) and Beneduce (2024).

[19] On this, see Pogliano (2015, pp. 38–39 and pp. 48–49), who, however, does not mention Scaliger. On Scaliger, see Hamilton (quotation in this page) and Smith (*infra*, footnote 22).

[20] Conimbricenses, *In tres libros De anima*, book II, ch. XI, q. 2, art. I, II (ed. Coimbra, 1598, p. 180): "Quaestio II. Sit ne unus tactus, an plures"; "Dolor, et uoluptas, an ad tactum pertineant"; "Quae sit doloris, et uoluptatis causa"; "Quae ratione dolor, et uoluptas pertineant ad tactum, qua uero ad appetitum sensitiuum". The transcription is by Mário Santiago de Carvalho, as available at https://www.uc.pt/fluc/uidief/textos_publicacoes/de_anima (last accessed on November 9, 2024).

[21] Hamilton, *The Works of Thomas Reid, D.D. Now Fully Collected, with Selections from His Unpublished Letters. Preface, Notes and Supplementary Dissertations by Sir William Hamilton* (ed. Edinburgh, 1863, p. 867). Also see the discussion on touch in Hamilton, *Lectures on Metaphysics and Logic* (ed. Edinburgh and London, 1870, pp. 152–159).

[22] See, especially, Smith (2011).

sense" as distinct from tactile perception should look much further back than the sixteenth century to find its roots.[23]

In addition to (1) the reflections on the connection between touch and pleasure written in the wake of Aristotle's discourse on temperance and (2) the discussions of pleasure and pain as objects of touch found in discourses on the objects of touch and its unity or plurality, (3) we could search consideration of pleasure, pain, and touch in late medieval contemplations of the topics "what is pleasure", "what is pain", "the causes of pain", and "the causes of pleasure"—widespread issues in medieval writings, from ethical treatises to medical works.[24] By the late medieval period, scholarly texts sometimes even presented direct questions or topics such as "to which sense are pleasure and pain connected" or "through which among the five senses is pain felt". In Pietro Torrigiano's (d. 1320 ca.) *Plusquam commentum*, for example, we find the question "Whether there is pain in any of the senses" (*Utrum in quolibet sensu sit dolor*), where Torrigiano considers pain to be linked to the sense of touch and not to all the (other) senses.[25] The question about which sense is designated to feel pain is for example also widely found in late medieval commentaries on Hippocrates's *Aphorismi*.[26] A more comprehensive study of late medieval theories on the connection between touch and pleasure and pain should map out and analyze all these reflections. (4) Further research on the role of touch in the perception of pleasure and pain in medieval philosophy and science should also be carried out by selecting specific types of (corporeal) pleasures and forms of

[23] On muscular sense, proprioception and interoception, see further Alison Simmons and Antonia LoLordo in this volume.

[24] Examples of discussions of the nature of pleasure and pain and their causes in medieval philosophy and science are Thomas Aquinas discussing "quid sit delectatio" in Thomas Aquinas, *Sententia libri Ethicorum*, book 10, l. 6, n. 1 and 4 (ed. Roma, 1969, vol. II, pp. 568–569); or the treatise on virtues by Guido Vernani da Rimini (d. 1345 ca.), which extensively discusses "quid sit delectatio" and related questions about the nature of pleasure: "16.1 Quid sit delectatio; 16.2 De causis que faciunt delectationem; 16.3 De effectibus delectationis; 16.4 De bonitate et malitia delectationum". Guido Vernani da Rimini, *Liber de virtutibus*, pars V, dist. 16 (ed. Cova, 2011, pp. 294–300); or the topics on the nature of pain extensively discussed by Jacques Despars (d. 1458), while commenting on Avicenna's *Canon*: "De causis doloris absolute; de causis uniuscuisque doloris; de causis sedandi dolorem; de eis que proveniunt ex dolore; de causis delitiarum; quomodo ex motu provenit dolor; quomodo provenit dolor ex humoribus malis; quomodo provenit dolor ex ventis". Jacques Despars, *Canon medicinae* (Commentary on), *Tabula Avicennae in primum librum sui Canonis* (ed. Lyon, 1498, f. a3v). For another medical work, see the abovementioned question by Peter of Abano in the *Conciliator*: "Utrum dolor sentiatur, necne". Peter of Abano, *Conciliator controversiarum, quae inter philosophos et medicos versantur*, diff. 77 (ed. Venezia, 1565, ff. 117ra–118ra). See *supra*, 52. A recent article by Loviconi (2025) examines the theory of pain in the *Lilium medicinae* of Bernard of Gordon (d. 1330 ca.), by also mentioning Bernard of Gordon's approach to the issue of pain's perception.

[25] Pietro Torrigiano, *Plusquam commentum*, 2.88 (ed. Venezia, 1504, f. 72vb): "Adhuc autem etiam dubitabit aliquis utrum in quolibet sensu sit dolor. Et videtur primo quod sic. [...] Sed in contrarium videtur quod dolor non sit nisi ex duplici causa que superius dicta est quarum quelibet est comprehensibilis solo tactu sicut ostendemus. Propter quod dicendum quod dolor maxime proprie est in tactu".

[26] As indicated by Pesenti (2003, pp. 468–469).

pain to see if, to what extent, and in which ways touch emerges in their explanation. Strong first candidates to begin this investigation would be the "brutish" pleasures of the *Ethica Nicomachea*, especially the pleasures linked to food and drink and to coitus. As already reported, Albert the Great explains the peculiar intensity of pleasure felt in intercourse based on the very subtle way generative seed *touches* nerves (in the generative organs).[27] Touch and sexual pleasure are also connected by Peter of Abano in his commentary (in form of *expositio*) on Aristotle's *Problemata*.[28] That of Buridan on sexual pleasure provided above is an extended example of a late medieval reflection on sexual pleasure and whether it is perceived by touch.[29] As a further example, I also mention the *Quaestiones super librum De generatione animalium* by John Vath, an author active in Paris between the end of the thirteenth century and the beginning of the fourteenth century. Vath devotes to sexual pleasure several questions of his commentary on Aristotle's *De generatione animalium*.[30] In q. 19, Vath explicitly links sexual pleasure to touch[31] and in q. 21 he repeats that sexual pleasure is due to touch.[32] Then Vath explains that sexual pleasure occurs first in a specific part of the body, namely the genital organs, and secondly, and as a consequence, throughout the rest of the body. Pleasure during intercourse, he says, is caused by the *contact* of sperm with the nerve-filled organs. Since the organ of touch, involved by that contact, is located in the heart—as Vath claims in line with Aristotle's cardiocentric view—during sexual intercourse the heart is altered and

[27] See *supra*, 55.

[28] Peter of Abano, *Expositio problematum Aristotelis, IV, 15* (ed. Coucke, 2009, pp. 147–148): "Notandum quod Avicenna dicit in 9° *De animalibus* quod duplatur *delectatio feminarum*, quoniam ipse delectantur non solum *in motu sui spermatis et in motu orificii matricum in sugendo, sed etiam delectantur adhuc in diversitate motus matricis*. Quod etiam videtur tangere Galenus in 4° *De accidenti et morbo*. Sed ne ex hiis verbis errare contingat sine querela, breviter sciendum quod duplex est delectatio, scilicet intensiva et extensiva. Voco autem intensivam delectationem magis ultimatam et si brevioris sit temporis, extensivam autem oppositam. Nunc autem mulieres magis delectantur viris delectatione extensiva, pluri enim tempore in eis perdurat, cum causa delectationis earum longiori permaneat tempore. Delectatione autem intensiva magis viri delectantur quod ostenditur quia delectatio consistit in tactu. Ubi ergo tactus est perfectior et tactus maior, ibi et delectatio intensior, sed hoc est in viris, cum temperatioris sint complexionis muliere et cetera". On Peter of Abano, sexual pleasure, and touch, see also Cadden (1993, pp. 156–157).

[29] See *supra*, 56–57.

[30] Namely, q. 18 (*Queritur an in coitu sit delectatio*); q. 19 (*Queritur consequenter utrum delectatio coitus sit in nobis vehementior*); q. 20 (*Queritur consequenter utrum delectatio sit in nobis in coitu per hoc quod sperma spargitur super membra nervosa*); and q. 21 (*Queritur consequenter utrum delectatio que est in coeundo sit in toto corpore vel in aliqua parte corporis*). Cf. John Vath, *Quaestiones super librum De generatione animalium*, qq. 18–21 (ed. Cova, 1992, pp. 223–227).

[31] *Ibid.*, q. 19, p. 225: "Dicendum quod in coitu est maxima delectatio quam in aliqua alia, quia in illo actu vehementior est delectatio, in quo sensibilia magis imprimuntur in sensu ex quo delectatio est secundum cognitionem, ut tangit septimo *Ethicorum* et duodecimo *Metaphysice*. Sed in actu coeundi est huiusmodi, quia tactus a sensibili in coitu alteratur, ut dicit commentator secundo *De anima*".

[32] *Ibid.*, q. 21, p. 227: "Preterea, ibi est delectatio coeundi, ubi est tactus".

consequently so are all the other bodily parts, which depend on the heart.[33] Another commentary on the *De animalibus*, the anonymous contained in ms. Città del Vaticano, Biblioteca Apostolica Vaticana, Vat. Lat. 2164, presents extended reflections on sexual pleasure.[34] The text also links sexual pleasure with the sense of touch by stressing that the cause of sexual pleasure is to be individuated in the very subtle tactile feature of the nerves passing through the penis and the female cervix.[35] Yet, research on other, specific examples of bodily pleasure (than sexual pleasure) and pain might be promising. There are a number of very peculiar late medieval discussions of corporeal pleasure and pain, ranging from the sensations felt during defecation to the type of pain felt by Christ on the cross.[36] Among these discussions of specific types of pleasure and pain that are relevant for analysis with respect to touch perception, we find labour pain, the specific case study examined in this chapter.

3.2 Descriptions of Labour Pain in Some Late Medieval Scientific Texts

There is a passage by the fifteenth-century physician Michele Savonarola (d. 1468) in the *Regimine pregnantium* that encourages women to cry loudly during delivery, even if they do not actually feel much pain, in order to obtain delicacies like capons, sweets, and wine from their husbands and other members of the household.[37] This

[33] *Ibid.*: "Dicendum quod delectatio, que est in toto corpore, non est primo in toto corpore sed in parte, quia ibi est primo delectatio, in quo sperma digestum diffunditur super membra nervosa et reducit ea ad temperamentum de frigido. Et, si hoc est in pudendis et vasis seminariis, ergo etc. Tamen secundario et ex consequenti delectatio est in toto corpore, quia ex quo delectatio que est in coitu causatur ex contactu spermatis et membrorum nervosorum, et organum tactus est intra cor vel in cor ipsum, ut patet in *De sensu et sensato*, ideo in coitu oportet cor alterari et, ipso alterato, omnia membra per consequens alterantur, quia dependent ab eo".

[34] Anonymous, *Quaestiones de animalibus*, ms. Città del Vaticano, Biblioteca Apostolica Vaticana, Vat. Lat. 2164, f. 260vb: "Sed utrum in coytu debeat esse delectatio". *Ibid.*: "Consequenter utrum femina plus delectetur in coytu quam masculus". *Ibid.*, f. 261ra: "Utrum in coytu sit maior delectatio quam in cibo vel in potu. Certum quod in coytu maior". *Ibid.*, f. 262ra: "Utrum plus delectetur masculus quam femina, dicendum quod masculus". The transcription is mine. On this commentary, see also *infra*, 68.

[35] Anonymous, *Quaestiones de animalibus*, ms. Città del Vaticano, Biblioteca Apostolica Vaticana, Vat. Lat. 2164, f. 261va: "[...] una causa et principium delectationis in coytu est subtilissimus tactus nervorum qui sunt in virga et collo matricis [...]". The transcription is mine.

[36] I found an example of sensations felt in defecation in Anonymous, *Quaestiones de animalibus*, ms. Città del Vaticano, Biblioteca Apostolica Vaticana, Vat. Lat. 2164, f. 262ra: "Utrum in emissione fecum sit maior delectatio. Certum est quod non quia ibi non est ita subtilis tactus [...] Et videtur dolor et inordinatio propter duritiem stercorum [...]". The transcription is mine. On Christ's pain, see *infra*, 64–65, for an example.

[37] Michele Savonarola, *Ad mulieres ferrarienses de regimine pregnantium et noviter natorum usque ad septennium* (ed. Belloni, 1952, p. 121): "E sappi, frontosa, chel cridar forte in tal caxo ti è molto

suggestion is accompanied by reference to Avicenna's claim that screaming during delivery helps women in labour. This Avicennian "Et clamet" is found in the third book, fen. XXI, second treatise, chapter XXIII (*De regimine eius quam ledunt partus accidentia*) of the *Canon*.[38] In fact, Avicenna writes extensively in the *Canon* about the causes of difficulties in delivery and labour pain, as well as about how to confront difficult deliveries.[39] In chapter XXI (*De difficultate partus*), Avicenna lists a set of reasons that might cause labour to become difficult. These causes are attributed to the pregnant women's constitution and attitudes; to the characteristics of the fetus (e.g., female or male, too big or too small, dead or alive, etc.); to the conditions of the uterus; to the conditions of the placenta; to the conditions of nearby organs, such as the bladder and intestines; to the circumstances of delivery (e.g., if the fetus descends too quickly); and to other causes, such as a delivery occurring in cold climates or if the mother has the habit of smelling aromatics.[40] Before Avicenna's work, a section on labour pain and on detailed practical tips for facilitating delivery

---

giovativo, imperò dice Avicena *et clamet*: e se anco non ti dolesse tanto, pur te conseglio crida forte, a ziò chel te sia creduto il tuo male, havendote compassione il marito e li altri di caxa, asmorzendo tal fuoco grande cum capuoni, confecti et vini avantezati". A recent English translation of this work, based on a revised edition of the original text, is Michele Savonarola, *A Mother's Manual for the Women of Ferrara* (ed. Zuccolin, transl. Marafioti, 2022, p. 156 for the relevant passage). I thank Gabriella Zuccolin for having directed me to this quotation from Savonarola's text, of which she talks about in Zuccolin (2011).

[38] Avicenna, *Liber canonis Auicenne reuisus et ab omni errore mendaque purgatus summaque cum diligentia impressus*, book III, fen. XXI, tr. 2, ch. XXI (ed. Venezia, 1507, f. 369vb): "Cumque presens est partus et incipiunt accidentia eius comedat rem pauce quantitatis et multi nutrimenti et bibat desuper vinum odoriferum: deinde oportet vt sedeat mulier hora vna: et tendat pedes suos: deinde resupinetur super dorsum suum hora vna: postea surgat subito et ascendat gradus et descendat et clamet". The transcription here and below, from the 1507 edition, is by Lenka Jiroušková, as available in the "Online Arabic and Latin Corpus" at https://www.arabic-latin-corpus.philosophie.uni-wuerzburg.de/text/Avic_Canon_la.index.xhtml (last accessed on November 9, 2024).

[39] See Avicenna, *Liber canonis Auicenne reuisus et ab omni errore mendaque purgatus summaque cum diligentia impressus*, book III, fen. XXI, tr. 2, ch. XXI–XXXI (ed. Venezia, 1507, ff. 369rb–370va).

[40] *Ibid.*, book III, fen. XXI, tr. 2, ch. XXI (ed. Venezia, 1507, f. 369rb–va): "Difficultas partus aut sit causa pregnantis: aut causa fetus: aut causa matricis: aut secundine: aut causa vicinantium et communicantium: aut causa hore partus: aut causa obstetricis aut propter causas primitiuas. Facta vero causa pregnantis est vt sit stricta: aut passa egritudines: aut famem: aut sit timida aut non consueta impregnationi […] Facta autem causa fetus: aut est propter genus eius. Femina enim penitus est difficilioris partus quam masculus aut est propter magnitudinem eius: […] et quandoque est difficultas propterea quod est mortus: Facta vero causa matricis est […] Facta autem propter secundinam est […] Facta vero causa vicinantium est vt sit in vesica apostema: aut nocumentum aliud ex fixione vrine et alijs: aut sit in intestino fex sicca plurima: aut apostema: aut colica alterius generis: aut emorroydes: aut ragadie ani: […] Facta autem causa hore partus est vt sit fetus velox in administratione partus […] Facta autem propter causas primitiuas est sicut si forte fiat frigus: quare fortis fiat constrictio membrorum partus: et propter illud multiplicatur in regionibus septemtrionalibus et in ventis septemtrionalibus. […] et sicut si mulier sit plurimum vtens aromaticis et odoramento aromatum: et fiat matrix eius assidue attractionis ad superiora […]".

and alleviating labour pain appeared in another Arabic text, namely Rhazes's (Abū Bakr al-Rāzī) (d. 925) *Liber ad Almansorem*.[41]

Reflections on labour pain were already present in Greek Antiquity: the *Corpus Hippocraticum,* as a first example, includes reflections on childbirth, and on difficulties and sufferings related to it.[42] Labour pain is further addressed in Aristotle's zoological thought, particularly in book VII of the *Historia animalium.* Aristotle explained differences in labour pain according to the sex of the child; he attributed the pain that may directly precede labour to the movement of fetus's head and also discussed the different parts and organs of the body affected by pain during delivery. His text compares the labour pain of women to the labour pain experienced by other animals, concluding that the pain is much stronger in women. Aristotle also described the factors which may lead to stronger pain during delivery for women, including a sedentary lifestyle or having shortness of breath (he identified labour as especially difficult when women cannot hold their breath while exerting force during labour).[43] Childbirth and the pain related to it was addressed and described at

[41] See Rhazes, *Liber ad Almansorem*, IV.28 (ed. Lyon, 1510/11, ff. 1r–216v): "De alleuiatione partus et regimine parturientis". The transcription is by Lenka Jiroušková, as available in the "Online Arabic and Latin Corpus" at https://www.arabic-latin-corpus.philosophie.uni-wuerzburg.de/text/Rhaz_Alman_B_la.index.xhtml (last accessed on November 9, 2024). The ninth book of Rhazes's *Liber ad Almansorem* was popular in the Latin Middle Ages and frequently commented upon. I have never heard about extant medieval Latin commentaries on the fourth book, instead.

[42] An example is the work *Nature of the Child*, which in Latin is generally known as *De natura pueri* or *De natura fetus*. See especially the following passages in Hippocrates, *Nature of the Child* (ed. Potter, 2012, vol. 10, p. 47 and pp. 89–90): "Women suffer the greatest difficulties during childbirth and the period of the lochia in their first pregnancy, and those who have borne less children experience greater difficulty than those who have borne more"; "If when a fetus's membranes are torn off, it succeeds in turning in the direction of its head, its birth will be easy; but if a fetus lies crosswise or moves toward its feet—this happens if it happens by chance to turn in this direction, or due to narrowness of its mother's uterus, or if the mother does not remain at rest at the beginning of her birth pangs—as it moves like this the woman will have difficulty giving birth. In fact, many mothers have died in this case, or their babies have, or both have. In giving birth, women suffer most when it is for the first time, due to their inexperience with the pangs: they suffer through their whole body, but most especially in their loins and hips, for their hips separate. Those who have gained more experience in giving birth suffer less than those who are doing so for the first time, and those who have given birth many times suffer much less".

[43] Aristotle, *Historia animalium*, VII.4, 584a25–587a5 (ed. Barnes, 1984a, vol. 1, pp. 915–920): "And labour in the case of female children is apt to be protracted and sluggish, while in the case of male children it is acute and by a long way more difficult. Women who have connexion with their husbands shortly before childbirth are delivered all the more quickly. Occasionally women seem to be in the pains of labour though labour has not in fact commenced, what seemed like the commencement of labour being really the result of the foetus turning its head. […] When women are in labour, their pains occur in many different parts of the body, and in most cases to one or other of the thighs. Those are the quickest to be delivered who experience severe pains in the region of the belly; and parturition is difficult in those who begin by suffering pain in the loins, and speedy when the pain is abdominal. If the child about to be born be a male, the preliminary flood is watery and pale in colour, but if a girl it is tinged with blood, though still watery. In some cases of labour these latter phenomena do not occur, either one way or the other. In other animals parturition is unaccompanied by pain, and the dam is plainly seen to suffer but moderate inconvenience. In women,

other times in Antiquity, with Soranus of Ephesus's (late 1st/early 2nd cent. AD) *Gynaecia* being perhaps the most famous example.[44]

Given this lively background of discussions in Ancient and Arabic science, it is not surprising that delivery and labour pain were treated as a scientific topic also in medieval Latin scholarly texts, becoming an object of theoretical reflection in the medieval science produced at the universities.[45] This science was especially—albeit indeed not exclusively—elaborated by *commenting* on previous works (Ancient and Arabic) which Latin authors considered authoritative.[46] In this chapter, I discuss the descriptions of delivery and labour pain found in some late medieval Latin scholarly commentaries to see if and to what extent touch was involved in late medieval scientific explanations of labour pain. Labour pain was discussed in Latin commentaries on Hippocratic works, on Aristotle's *De animalibus*, on pseudo-Aristotle's *Problemata*, and on Avicenna's *Canon*. The preliminary study undertaken in this chapter presents examples taken from all these commentary traditions. Since Aristotle was indisputably immensely dominant at medieval university, my overview starts from and especially looks upon the few nearly completed commentaries on the *De animalibus* that remain from the late Latin Middle Ages.[47]

however, the pains are more severe, and this is especially the case in persons of sedentary habits, and in those who are weak-chested and short of breath. Labour is apt to be especially difficult if during the process the woman while exerting force with her breath fails to hold it in".

[44] See Soranus of Ephesus, *Gynecology*, book II (ed. Temkin, 1991, pp. 73–75, pp. 176–178, pp. 183–183, p. 186). According to Monica Green, the name of Soranus in not mentioned as a gynaecological authority in medieval West. What circulated in medieval West was a shortened translation in Latin of Soranus's *Gynaecia* made by Muscio in the sixth century, which does not seem to have been very much fortunate after the first half of the thirteenth century in the face of the circulation of the *Trotula* treatises. See Green (2021, unpublished). In the *Trotula* treatises there is a focus on difficulties in delivery and on how to face them (*De difficultate partus*), more than on labour pain. Post-partum pain is also addressed (*De dolore matricis post partum*). See Trotula (so-called, various authors), *Trotula* (ed. Green, 2009, p. 166, p. 174, and p. 202).

[45] In line with a general trend in studies of medieval theories of generation as discussed by van der Lugt (2021), my impression is that the topic of childbirth and labour pain during the Middle Ages has mostly been tackled by the perspective of cultural and social history. An example of this is the book by Foscati (2023), in which medical sources are involved to account for medieval views of labour pain but these sources are used to assess the physician's attitude towards a female condition and not to explore scientific doctrines of labour pain *tout court*, namely how labour pain was discussed as a biological phenomenon. More recently, Foscati (2025) has published an article entirely devoted to medieval attitudes towards labour pain. In this article, Foscati explores some medieval sources (mostly medical, but not exclusively) to point out that alleviation of labour pain was a medieval desideratum despite the biblical claim that women are destined to suffer during childbirth. What is still lacking in literature is a comprehensive study of how labour pain was addressed as a biological topic in late medieval natural philosophical and medical texts produced at the university during the core centuries of the scholastics. This chapter is also an attempt to begin filling this gap.

[46] On the centrality of commentaries to development of medieval philosophy and science, see, for example, Del Punta (1997) and Fioravanti et al. (2002).

[47] On the medieval *De animalibus* tradition, see, among others, Perfetti (2012).

My survey starts, more particularly, with two texts by Peter of Spain (thirteenth century), and namely from a passage of his *Questiones super libro De animalibus* which has a parallel in his *Problemata*.[48] At the start of the eleventh book of his commentary on the *De animalibus*, Peter of Spain takes up Aristotle's discourse to ask why more pain is felt in delivery by women than by other animals. Peter of Spain identifies two causes. First, he states that, when something is nobler, then it entails more difficulty, and since human beings are nobler than animals, a woman feels more pain than a female animal. As the second cause, he highlights the passibility or susceptibility (*passibilitas*) of a woman's substance, namely her capacity to feel and suffer (pain, in this case), *plus* the good quality (*bonitas*) of her sense or faculty of sensing (*sensus*).[49] Thus, according to Peter of Spain, the peculiarly good quality of women's capacity for sensing is one of the reasons they experience more pain in labour than other animals. Which sense or capacity of sensing that Peter of Spain has precisely in mind becomes clear in a passage from his *Problemata*, where we read that the implicated sense is touch.[50] So, in describing labour pain, Peter of Spain explains the experience in terms of tactile perception. More specifically, he links the experience of labour pain to touch to the extent that the *good quality of touch* makes women *more affected by labour pain* than other animals.

The idea that a better sense of touch implied feeling more pleasure or pain is not only found in Peter of Spain. Still in the context of the commentary tradition on the *Problemata*, after Peter of Spain, another Peter, the abovementioned Peter of Abano, stated that the better a complexion, or bodily constitution, and sense of touch, the more significant the capacity for pleasure—a suggestion that men feel more sexual pleasure than women since they are said to have a better complexion and sense of touch.[51] Another medieval example of this sort can be found in a theological work, namely in Thomas Aquinas's *Scriptum super libros Sententiarum*. When discussing Christ's pain while in the cross, Aquinas links (the force of) Christ's pain to touch.

---

[48] The medieval author called "Peter of Spain" and the corpus attributed to him are full of mysteries. Even to whom it is we are referring when we speak about "Peter of Spain" has been much debated and is still doubtful. On Peter of Spain, see Lagerlund (2020b).

[49] Peter of Spain, *Quaestiones super libro De animalibus*, book XI, q. 1 (ed. Navarro Sánchez, 2015, pp. 275–276): "Questio prima. [...] Circa undecimum restant quedam querenda. 1. Primum est in quo mense plus doleant pregnantes [...] Et quare maior est dolor in partu mulieris quam in partu aliorum animalium. [...] Ad secundum dicendum quod quanto res est nobilior tanto cum maiori difficultate habetur, et propter hoc homo sicut nobilius est quolibet animali, plus laborat mulier in partu quam femine aliorum animalium. Secunda causa est passibilitas substantie sue et bonitas sui sensus".

[50] Peter of Spain, *Problemata*, n. 102 (ed. de Asúa, 1991, p. 392): "Quare maior est dolor in partu mulieris quam in ceterorum. Respondeo. Res quanta est nobilior tanta maior difficultate habetur. Unde cum in generante animalium homo sit nobilior idcirco difficilius ad lucem producit quam alia animalia. Secunda causa est passibilitas substantie mulieris et bonitas tactus".

[51] See the text from Peter of Abano's *Expositio problematum Aristotelis* reported *supra*, footnote 28. For the notion of "complexion" in medieval philosophy and science, see, among others, Beneduce (2023).

The extreme intensity of Christ's pain is explained as due to the perfectness of Christ's touch, based on Christ's perfect bodily constitution or complexion.[52]

We can draw some conclusions from these passages by Peter of Spain, Peter of Abano, and Thomas Aquinas. First, their descriptions of pain or pleasure are all connected to touch. Second, they all argue that where there is a better sense of touch, more pain or pleasure follow. Third, a better sense of touch is linked to and presumed to be caused by a more refined or nobler nature. More precisely, Peter of Abano and Thomas Aquinas explicitly associate a more tempered, and therefore better, complexion to a superior sense of touch; in the case of Peter of Spain, the link between the superiority of touch and the superiority of bodily constitution is much more implicit. It cannot be excluded, however, because of two of the reasons Peter of Spain gives and pairs regarding why women experience more labour pain than animals: a good sense of touch and the nobler nature of human beings compared to animals.

As Thomas Aquinas suggested in his description of Christ's pain, the idea that a better sense of touch is connected to a better constitution and also to a *better intellect* roots in Aristotle's *De anima*, where a good sense of touch and soft flesh are linked to an efficient intellect. In medieval Latin it became the adage "Molles carne, apti mente".[53] A question might actually follow here, as a short digression, and namely whether women, which are mostly considered moister and softer than men in ancient philosophy and science, are to be ascribed with a better touch, a better complexion, and a better intellect.[54] Actually, as studied by Sandra Plastina[55] and by Gabriella Zuccolin,[56] this claim was made in the Renaissance debate on women's

[52] Thomas Aquinas, *Scriptum super Sententiis magistri Petri Lombardi*, III, dist. 15, q. 2, q. 3, sol. III (ed. Moos, 1933, p. 500): "[…] magnitudo doloris sensibilis Christi potest ex tribus considerari. Primo ex ipsa natura passionis; et sic habuit magnam acerbitatem: tum ex complexione patientis quae erat temperatissima; unde habebat optimum tactum et per consequens erat in eo vehemens sensus laesionis—bonitas enim tactus attestatur etiam bonitati complexionis et bonitati mentis, ut dicitur in II *De anima* […]".

[53] The *Auctoritates Aristotelis* have: "Molles carne mente aptos dicimus, duros vero ineptos". Anonymous, *Auctoritates Aristotelis*, AA 6:99 (ed. Hamesse, 1974, p. 182). The corresponding passage in Aristotle is Aristotle, *De anima*, II.9, 421a7–421a26 (ed. Barnes, 1984a, vol. 1, p. 670): "That is why man is the most intelligent of all animals. This is confirmed by the fact that it is to differences in the organ of touch and to nothing else that the differences between man and man in respect of natural endowment are due; men whose flesh is hard are ill-endowed with intellect, men whose flesh is soft, well-endowed". On this adage in Thomas Aquinas (and further in Pietro Pomponazzi), see Graciotti (2023).

[54] On the softness of women's body and flesh, see for example the Hippocratic text *Diseases of Women*: "This is so for the reason I first indicated when I contended that a woman is more porous and softer than a man; […]". Hippocrates, *Diseases of Women 1*, 1 (ed. Potter, 2018, vol. 11, p. 11). On Aristotle on females having softer flesh, see Connell (2021). For the medieval tradition, cf. also Albert the Great, *De animalibus libri XXVI*, book IX, tr. 1, ch. 4 (ed. Stadler, 1916, p. 683): "Mulier etiam plus omnibus aliis feminis animalium habet menstruum propter humiditatem et mollitiem maiorem sui corporis […]".

[55] See Plastina (2017).

[56] See Zuccolin (2019).

dignity: women's superiority was sometimes justified by saying that women's better touch and softer nature implied women's intellectual superiority. In Peter of Abano's passage, there is no reflection on the softer nature of women, which could possibly imply that they have a better sense of touch and a greater perception of sexual pleasure. On the contrary, Peter of Abano clearly states that men have a better bodily constitution and, therefore, a better sense of touch than women, which entails a greater perception of sexual pleasure in men than in women.[57] As far as Peter of Spain is concerned, the text does not definitively clarify whether the author is referring merely to the good sense of touch that women possess *in so far as they are human beings* (*versus* animals) or whether he also means to imply that a good sense of touch is a *female* characteristic. On the one hand, since Peter of Spain is clearly comparing human beings to animals, we might opt for the first hypothesis. However, in the *Problemata*, in the same short sentence he pairs the good sense of touch (*bonitas tactus*) with the susceptibility of a woman's substance (*passibilitas substantie mulieris*), so we might interpret "bonitas tactus" as well as the "passibilitas substantie" as specifically "mulieris" (of a woman, typical of women, or female), and thus as a characteristic distinctive to the woman. Notably, a passage found in another medieval work, namely in Tommaso del Garbo's (d. 1370) *Summa medicinalis*, unlike the other medieval texts mentioned so far, more explicitly makes the relevant inference, stating that if Aristotle's adage holds true, then women should have better intellect than men—something which, in Tommaso's view however, is absurd (*quod est absurdum*).[58]

This noted, one key element so far has thus been that Peter of Spain links labour pain to touch.

My survey then proceeds with another text in the late medieval *De animalibus* tradition, namely the *Quaestiones super de animalibus* commonly ascribed to Albert the Great. Albert discusses labour pain in the question about delivery contained in that work, asking why women feel more pain than other animals during delivery. His position is the following: those who feel more pleasure during intercourse, feel more pain during delivery; among all animals, the human woman is the one that feels the most pleasure during conception, *because she has a subtler sense*; she also therefore suffers more pain than other animals during labour. Like Peter of Spain (albeit less directly, since the "subtle sense" is explicitly referred to perception of pleasure in Albert's discourse), Albert makes reference to women's sense or faculty of sensing to explain their heightened suffering during delivery with respect to other animals. However, there is no mention of the specific sense Albert supposes to be subtler in women. The other cause Albert identifies for women's greater experience of pain is an abundance of spermatic blood that makes human fetuses bigger, thereby causing more pain.[59]

---

[57] See *supra*, footnote 28.

[58] Tommaso del Garbo, *Summa medicinalis*, q. 82 (ed. Venezia, 1531, f. 79va): "Amplius, si molles carne sint apti mente, ergo mulieres erunt plus apte quam viri, quod est absurdum".

[59] Albert the Great, *Quaestiones super de animalibus*, qq. 19–23 *De partu mulieris eisque circumstantiis* (ed. Filthaut, 1955, pp. 211–212): "[...] Tertio quaeritur, utrum mulier magis laboret in

The other zoological treatise by Albert, *De animalibus libri*, which was written between the end of the 1250s and the beginning of the 1260s, also contains an extended account of labour pain. *De animalibus*, book IX, treatise 1, chapter 8 closely follows Aristotle's writings in *Historia animalium*, VII. As in the case of Aristotle's text, Albert's discusses the parts of the body where pain is most felt during delivery. Here, Albert also states that women suffer more labour pain than other animals, but he does not enter into the reasons for this. He goes on to reprise Aristotle's observations on good and bad respiration and its impact on a smoother or more difficult delivery.[60] Like Aristotle, Albert does not mention touch.

In *De animalibus*, book X, treatise 2, chapter 4 Albert again takes up the discourse on labour pain, this time with a very detailed lists of causes of difficulties and pain during delivery that overlaps almost entirely with Avicenna's in the *Canon*. Difficulties and pain during delivery are said to be dependent on the size of a child's head, the bodily constitution of the mother, the fetus itself, the characteristics of the uterus or of the placenta, the season, and the behaviour of the mother.[61] As in

partu quam alia animalia. Et videtur, quod non. Quia finis debet esse proportionalis principio; sed mulier maxime delectatur in coitu et in conceptu; ergo proportionaliter minus dolebit in partu. Oppositum dicit Philosophus. [...] Ad tertium quaesitum dicendum, quod dolor est sensus contrarii per Aristotelem IV *Ethicorum* et secundum Galenum, et delectatio causatur ex sensu convenientis. Sed nunc est ita, quod inter omnia animalia mulier magis delectatur in conceptione, et hoc non est, nisi quia subtiliorem sensum habet. Ex motu enim spermatis et transitu super membra sensibilia causatur delectatio in coitu. Sed ubi est maior delectatio, ibi est maior dolor, cum inest. Quod patet in singulis membris, quia maximus est dolor in oculo, quia est sensus subtilior; et ideo mulier magis dolet in partu. Et universaliter quae magis delectantur in coitu, magis dolent in partu. Et alia causa est, quia mulier abundat in sanguine spermatico respectu aliarum feminarum, et ideo fetus eius maior est respective, et magnitudo fetus est causa doloris in partu".

[60] Albert the Great, *De animalibus libri XXVI*, book IX, tr. 1, ch. 8 *De partu et exitu embrionis et de accidentibus communiter in partu animalium et de emundatione post partum et de lactatione pueri* (ed. Stadler, 1916, pp. 702–703): "Aput partum autem augmentantur dolores in pluribus locis: plures enim habent dolorem tunc in una coxarum eo quod fetus in illa matricis parte iacet et comprimit os coxae aut musculos aut nervos. Quaecumque autem dolores sentiunt inferius circa ventrem acutos, citius pariunt, habentes autem dolorem in pariete qui est supra matricem et antaron sive dyafracma vocatur, non cito pariunt. Adhuc autem si fetus fuerit mas, prius ante exitum pueri descendit humor vinolentus citrinus. Si autem fuerit femina, descendit humiditas aquea sanguinea. In quibusdam autem aliquando forte non contingunt ista quae diximus, neque eveniunt dolores magni in tempore partus aliis feminis animalium praeter mulierem. Manifestum enim est, quod partus alia animalia non valde gravat. Sed mulieribus proprie accidit dolor fortis, et praecipue mulieribus quae non habent costas amplas et magnas, eo quod illae propter plenitudinem arti uteri non bene possunt tenere anhelitum, ne respirent, quod maxime confert ad enitendum et exprimendum partum ex utero: has enim partus multum gravat, eo quod inter pariendum anhelant propter gravitatem et angustiam spiritus quam patiuntur et tunc in respiratione illa quando afferunt et emittunt spiritum, matrix sursum attrahitur et impedit exitum partus. [...] Haec autem scientia obstetricandi indiget magno exercitio, [...]".

[61] *Ibid.*, book X, tr. 2, ch. 4 *Et est digressio declarans horam partus, et qualitatem natorum in horis illis et causam difficultatis in partu* (ed. Stadler, 1916, pp. 758–759): "[...] Causa autem communis difficultatis partus et dolorum in partu in muliere est magnitudo capitis nati, quod in utero fere totum suum accipit incrementum: et cum sit rotundum ut spera, in omnem partem inguen matris extendit et lacerat: et hoc est doloris causa, quia inguen sensibilis locus est valde propter concur-

Avicenna's text, touch does not play an explanatory role. Yet, at the beginning of his discourse, Albert explains labour pain in terms of the physiology of sensation, stating that pain is caused by the descent of the fetus's head in the groin, because the groin is a particularly sensitive part of the body due to the inflow of nerves found there. We can easily infer that it is the head of the child *touching* these nerves that causes the pain, but this is not explicitly stated by Albert.

Peter of Spain's and Albert the Great's commentaries are the only complete late medieval commentaries on the *De animalibus*, aside from the already mentioned later anonymous commentary contained in ms. Città del Vaticano, Biblioteca Apostolica Vaticana, Vat. Lat. 2164, which is believed to have possibly been composed in an early fourteenth-century Faculty of Arts.[62] At first glance, this manuscript does not seem to contain extended discussions of labour pain. Yet some partial late medieval commentaries on the *De animalibus* do. A discussion of labour pain is present, for example, in Pedro Gallego's (d. 1267 ca) *Liber de animalibus*, a work written between the 1250–60s (namely approximatively at the same time as Albert the Great's *De animalibus*), which covers only the *Historia animalium* and part of the *De partibus animalium*. The treatment of labour pain in Pedro Gallego is almost a literal paraphrase of the corresponding passages in Aristotle's *Historia animalium*, book VII,[63] and therefore in part overlaps with Albert the Great's *De animalibus*, book IX. As in those texts, in Gallego's touch is not mentioned in relation to labour pain.[64] Another example is the set of *Determinationes* written at the end of thirteenth century in Paris and ascribed to a master called "Vericus" or "Ulricus", which are contained in ms. Paris, Bibliothèque nationale de France, Cod. Lat. 16089. This is not a direct commentary on the *De animalibus* but belongs to a collection of questions on naturalistic topics, which, in part, have been recognized as patently linked

sum nervorum in loco illo. Una tamen mulier magis difficulter parit quam alia et magis laborat in partu uno quam in alio: et huiusmodi difficultatis causa est multiplex. Aliquando enim est dispositio ipsius mulieris in causa, […] Aliquando autem causa difficultatis est ipse fetus: […] Aliquando autem causa difficultatis est propter matricem, […] Aliquando autem est difficultas propter secundinam […] Membra etiam vicina aliquando faciunt difficultatem, […] Aliquando etiam est difficultas propter horam in qua descendit, eo quod in illa hora mater forte est debilis et non potest confortari. Ex tempore etiam est difficultas, […]. Id autem, quod etiam valde nocet, est si mulier consueta sit odorare aromatica et uti eis: […]".

[62] Anonymous, *Quaestiones de animalibus*, ms. Città del Vaticano, Biblioteca Apostolica Vaticana, Vat. Lat. 2164.

[63] The work by Gallego is in fact considered a translation *sui generis* of Aristotle's text. See Pedro Gallego, *Liber de animalibus* (ed. Martínez Gázquez, 2000, p. 69).

[64] Pedro Gallego, *Liber de animalibus*, book VII (ed. Martínez Gázquez, 2000, pp. 117–118): "Et dolent mulieres in pluribus locis. Quedam in una coxarum, quedam in pariete, qui est supra matricem. Et que dolent fortiter in uentre, cito pariunt. Et si fuerit mas descendit prius uirus citrinum, si femina descendit humiditas sanguinea. Animalia alia non dolent multum in partu quod apparet, quia non multum grauantur. Et grauantur multum mulieres, que non habent costas magnas, quia non possunt bene retinere anelitum et grauantur nimis si anelant in medio partus propter spiritum. Et descendit primo humiditas et funditur tela et exit embrio. Et si transponitur ostetrices exercitate debent iuuare et sapienter scindere umbilicum et ligare cum filo lane fortiter, quia si solueretur, exiret sanguis et moreretur fetus".

to and dependent on the tradition of the *De animalibus*.[65] This work discusses the question of whether women living in northern regions suffer more labour pain than women in southern regions, which becomes an occasion for Vericus to more extensively discuss the causes of labour pain.[66] The text looks like as a condensation of Avicenna's discourse in the *Canon* and therefore also itself bears similarity to Albert the Great's discourse in the *De animalibus*, book X. Touch is not mentioned in Vericus's discourse on labour pain nor is sensation more generally.

Leaving the tradition of the *De animalibus* and its satellite texts and turning to late medieval Latin commentaries on Avicenna's *Canon*, my survey goes on with the two most famous Latin commentaries on this Avicennian work, namely the texts by Gentile da Foligno (d. 1348) and Jacques Despars (d. 1458). They address the topic of delivery and its difficulties by commenting on the third book, second treatise, fen. XXI of Avicenna's work.[67] Both texts follow Avicenna's reasoning in the *Canon* very closely, commenting on each of the chapters one by one.[68] The relevant sections of Gentile's and Despars's texts, like the one they comment upon, do not mention touch in relation to labour pain.

Another account of difficult delivery and labour pain, the last of my survey, can be found in Dino del Garbo's (d. 1327) *De natura fetus*, a commentary on an

---

[65] On the *Determinationes* by Master Vericus, see Cova (1997).

[66] Master Vericus (or Wericus or Ulricus), *Determinationes*, Paris, Bibliothèque nationale de France, Cod. Lat. 16089, f. 76va: "Consequenter quaeritur utrum mulieres septentrionales plus laborant in partu quam meridionales. Dico quod dolor in partu causatur ex quatuor: ex parte matris, ex parte fetus, ex parte loci et ex parte temporis. Ex parte matris, si ipsa sit stricta in partibus inferioribus; etiam si sit puer et iuvenis, si sit infirma, si sit passa famem, si sit timida, si sit pinguis, quia talis est carnosa et ratione carnis illa pellis vel inguen non potest ita bene extendi. Etiam ex parte fetus plura impediunt et aggravant, ut si fetus sit femella, quia non se iuvat; etiam si sit aborsus, tum quia se non iuvat tum quia talis partus est innaturalis; etiam si ordine converso exeat, ut pedes ante et primo; etiam si fetus habeat rotundum caput, quia figura rotunda inepta est ad exitum vel ad motum rectum. Ex parte loci regionis vel temporis eodem modo si regio sit frigida et tempus hyemale vel frigidum, quia frigiditas loci vel temporis constringit poros et non possunt meatus ita elargari. Dicendum ergo quod mulieres septentrionales plus laborant in partu quam meridionales, quia ratione frigidi continentis habent poros magis strictos, et frigiditas illa impedit extensionem inguinis et aliorum membrorum ibidem, quia frigidi est constringere. In meridionalibus <autem est e converso>, quia illa regio calida est multum, et calidi est aperire; ideo facilius pariunt. Dico etiam quod septentrionales facilius concipiunt, quia calidiores sunt ratione frigidi continentis calidum in eis unitum (?); et ideo citius concipiunt. In meridionalibus autem est e converso, quia tardius recipiunt, eo quod sunt frigidiores". The transcription is mine. I thank Paul Bakker for some suggestions.

[67] On medieval commentaries on Avicenna's *Canon*, see especially Chandelier (2017).

[68] I report just the titles of the relevant sections, due to the extensiveness of the texts. Gentile da Foligno, *Canon medicinae* (Commentary on), book III, fen XXI, tr. 2, ch. 21–24, *De difficultate partus*; *De signis*; *De regimine eius quam ledunt partus accidentia*; *De regime eius in qua fit difficilis partus absque semita medicinarum* (ed. Venezia, 1520, f. 509va–510ra; f. 510ra; f. 510ra; f. 510rb–510va) and Jacques Despars, *Canon medicinae* (Commentary on), book III, fen XXI, tr. 2, ch. 21–25, *De figuris naturalibus et non naturalibus partus*; *De difficultate partus*; *De signis difficultatis et facilitatis partus*; *De regimine eius quam ledit partus*; *De regimine hiuius in qua difficilis fit partus absque semita medicinarum* (ed. Lyon, 1498, ff. Qqq7v–Rrr2r).

embryological work ascribed to Hippocrates. Dino's text is based on some brief reflections on labour pain contained in Hippocrates's text, on which Dino elaborates extensively.[69] Dino discusses Hippocrates's claim that women suffer more pain at their first delivery and when the delivery is long. He first says that pain should not in principle be involved in delivery, because delivery is a natural operation and in natural operations pain should not be involved. Yet, he adds, experience shows that there is a significant amount of pain in delivery, and its causes are various. He especially remarks upon two causes of labour pain. One is that delivery entails a strong motion of expulsive force that implies a confluence of blood and, as a consequence, a strong tension and filling up—which are at the basis of pain. The second cause he identifies is a strong disjunction of the ligaments at the joint of the hips; this disjunction, due to the exit of the fetus, causes pain.[70] Dino's explanation of labour pain is therefore mainly based on two anatomo-physiological considerations that however do not explicitly involve touch as explanatory cause.

3.3 Concluding Remarks. Touch and Labour Pain: a "Hapax"?

Though the connection between touch and the perception of pleasure and pain seems to be often affirmed in late medieval philosophy and science, the role medieval scientists ascribed to touch in experiences of pleasure and pain should be further evaluated by looking at a wider range of textual traditions, different types of issues, and at single phenomena of bodily pleasure and pain. To help develop this research line, this chapter has taken the phenomenon of labour pain as its case study.

The late medieval Latin scholarly discourses on difficulties in delivery and on labour pain analyzed in this chapter closely follow their sources, at times Hippocrates, Aristotle, or Avicenna. Dino del Garbo expands on the text he is

[69] On Hippocrates's text, see *supra*, footnote 42.

[70] The relevant question is "Utrum in partu debeat esse dolor" in Dino del Garbo, *Expositio Jacobi supra capitulum de generatione embrionis* [...] *Dinus supra librum Ypocratis de natura fetus* (ed. Venezia, 1518, f. 89va). See especially the following passage, *Ibid.*, f. 58va: "Maxime autem dolent mulieres in partu et in purgatione partus primo parientes et in mora parientes magis dolent. [...] Et causa huius doloris est multiplex. Una est quia in partu est vehemens motus virtutis expulsive modo in vehementi motu virtutis expulsive accidit concursus sanguinis ad locum et per consequens vehemens tensio et repletio que est causa doloris vehementis. Secunda causa est quia in partu oportet esse fortem disiunctionem ligamentorum iuncturarum ex initium circa anchas si debet fetus exire ex matrice hac autem disiunctio fortis non potest esse sine dolore vehementi quare etc.". On labour pain in Dino's text, see also *Ibid.*, ff. 86v–87r.

commenting upon, namely the Hippocratic *De natura fetus*. Albert the Great's *De animalibus* (book X) and the text by Vericus are patently modelled on Avicenna's *Canon*, as are, unsurprisingly, Gentile da Foligno's and Jacques Despars's commentaries on the *Canon*. The text by Pedro Gallego is instead a paraphrase of Aristotle's discourse on labour pain and Aristotle is also closely followed by Albert the Great in book IX of his *De animalibus*. Peter of Spain's *De animalibus* and *Problemata* and the *Quaestiones de animalibus* ascribed to Albert the Great address the discourse on labour pain by taking Aristotle's comparison between human and animal labour pain as the relevant question to address. Unlike their source, however, Peter of Spain and the *Quaestiones de animalibus* by Albert the Great incorporate a discourse on pain (and pleasure) sense perception into their reflections on difficult deliveries and labour pain. The "good quality" of sensing or "more subtle" ability to sense were used to explain women's more intense experience of labour pain. The *De animalibus* by Albert the Great also contains a reference to the physiology of pain perception, identifying the cause of labour pain in the descent of the fetus's head in the groin—a very sensitive part of the body due to the local inflow of nerves. The only case in which sense perception in relation to labour pain is explicitly linked to the sense of touch in the abovementioned texts, however, is Peter of Spain's *Problemata*. More research on labour pain as a biological subject in late medieval scholarly texts might inform us about a) if labour pain is also discussed in terms of its *perception* in other natural philosophical and medical texts of the late Middle Ages, and b) if the reference to *the sense of touch* as related to labour pain, the one found in Peter of Spain, is also found in some other texts or whether it is a "hapax" in medieval science. In the latter case, we would have an example, in late medieval Latin scholarly production, of a type of pain perception not prevalently linked to touch.

Acknowledgements A first draft of this chapter was prepared during my "Veni" project "The Sense of Touch. From Anomaly to Paradigm" (project number: VI.Veni.191F.005), financed by the *Nederlandse Organisatie voor Wetenschappelijk Onderzoek* (NWO), and has been presented at the occasion of the conference "The Sense of Touch. Medieval and Modern Debates in Philosophy and Science" (Nijmegen, June 19–21, 2023). A previous version of this chapter's second section ("Descriptions of Labour Pain in Some Late Medieval Scientific Texts") was presented at the occasion of the *Journée d'étude* "Corps féminins et douleurs de l'enfantement", an event supported by the *Société d'Histoire de la Naissance* (Paris, April 5, 2024). The contents of this chapter's first section ("Notes on Touch, Pleasure, and Pain in Late Medieval Philosophy and Science") are also displayed, in Italian, in my book Beneduce (2024). The final writing of this chapter has occurred during my postdoctoral position at the CNRS Laboratoire "SPHERE" (UMR7219) in the framework of the individual project "FEMALE BODIES, FEMALE SENSATIONS. A History of Female Corporeity and Sense-Perception in Medieval Science", supported by the "ANR Access ERC Stg – édition 2023" grant (project number: ANR-23-AERC-0017), awarded by the *Agence nationale de la recherche*.

References

Primary Sources

Manuscripts

Città del Vaticano, Biblioteca Apostolica Vaticana, Vat. Lat. 2164.
Paris, Bibliothèque nationale de France, Cod. Lat. 16089.

Early Modern Printed Editions and Modern Editions

Albert the Great. (1916). H. Stadler (Ed.), *De animalibus libri XXVI* (Vol. 1, books I–XII). Aschendorff.

Albert the Great. (1951). De bono. In H. Khüle et al. (Eds.), *Alberti Magni Opera Omnia, XXVIII*. Aschendorff.

Albert the Great. (1955). Quaestiones super de animalibus. In E. Filthaut (Ed.), *Alberti Magni Opera Omnia, XII*. Aschendorff.

Albert the Great. (1968). De anima. In C. Stroick (Ed.), *Alberti Magni Opera Omnia, VII.1*. Aschendorff.

Albert the Great. (2008). De homine. In H. Anzulewicz & J. R. Söder (Eds.), *Alberti Magni Opera Omnia, XXVII*. Aschendorff.

Anonymous. (1974). Auctoritates Aristotelis. In J. Hamesse (Ed.), *Les* Auctoritates Aristotelis. *Un florilège médiéval. Étude historique et édition critique*. Publications Universitaires.

Aristotle. (1984a). J. Barnes (Ed.), *The Complete Works of Aristotle. The Revised Oxford Translation* (Vol. 1). Princeton University Press.

Aristotle. (1984b). J. Barnes (Ed.), *The Complete Works of Aristotle. The Revised Oxford Translation* (Vol. 2). Princeton University Press.

Averroes. (1552). *Averrois Cordubensis Colliget Libri VII*. Venezia.

Avicenna. (1507). *Liber canonis Auicenne reuisus et ab omni errore mendaque purgatus summaque cum diligentia impressus*. Venezia.

Avicenna. (1972). Liber de anima seu sextus de naturalibus II. In S. Van Riet (Ed.), *Avicenna latinus. Liber de anima seu sextus de naturalibus, I–II–III*. Peeters; Brill.

Conimbricenses. (1598). *In tres libros De anima*. Coimbra.

Dino del Garbo. (1518). *Expositio Jacobi supra capitulum de generatione embrionis* […] *Dinus supra librum Ypocratis de natura fetus*. Venezia.

Gentile da Foligno. (1520). *Canon medicinae* (Commentary on). Venezia.

Guido Vernani da Rimini. (2011). Liber de virtutibus. In L. Cova (Ed.), *Il* liber de virtutibus *di Guido Vernani da Rimini: una rivisitazione trecentesca dell'etica tomista (con l'edizione del testo)*. Brepols.

Hippocrates. (2012). Nature of the Child. In P. Potter (Ed.), *Generation. Nature of the Child. Diseases 4. Nature of Women and Barrenness*. Harvard University Press.

Hippocrates. (2018). Diseases of Women 1. In P. Potter (Ed.), *Diseases of Women 1–2*. Harvard University Press.

Jacques Despars. (1498). *Canon medicinae* (Commentary on). Lyon.

John Buridan. (1513). *Quaestiones super decem libros Ethicorum*. Paris. Repr. 1982. Minerva.

John Buridan. (2023). Quaestiones de anima (III red.). In G. Klima, P. G. Sobol, P. Hartman, & J. Zupko (Eds.), *John Buridan's Questions on Aristotle's* De Anima – *Iohannis Buridani Quaestiones in Aristotelis De Anima*. Springer.

John Vath. (1992). Quaestiones super librum De generatione animalium. In L. Cova (Ed.), Le questioni di Giovanni Vath sul *De generatione animalium. Archives d'histoire doctrinale et littéraire du Moyen Âge, 59*, 175–287.

Michele Savonarola. (1952). Ad mulieres ferrarienses de regimine pregnantium et noviter natorum usque ad septennium. In L. Belloni (Ed.), *Il trattato ginecologico-pediatrico in volgare* Ad mulieres ferrarienses de regimine pregnantium et noviter natorum usque ad septennium *di Michele Savonarola*. Industrie grafiche italiane Stucchi.

Michele Savonarola. (2022). A Mother's Manual for the Women of Ferrara. In G. Zuccolin (Ed.), M. Marafioti (transl.), *Michele Savonarola.* A Mother's Manual for the Women of Ferrara. *A Fifteenth-Century Guide to Pregnancy and Pediatrics*. Iter Press.

Pedro Gallego. (2000). Liber de animalibus. In J. M. Gázquez (Ed.), *Petri Galleci Opera Omnia quae extant. Summa de astronomia, Liber de animalibus, Regitiva domus*. SISMEL–Edizioni del Galluzzo.

Peter John Olivi. (1922–1926). Quaestiones in secundum librum Sententiarum. In B. Jansen (Ed.), *Quaestiones in secundum librum Sententiarum* (Vol. 2). Collegium S. Bonaventurae.

Peter of Abano. (1565). *Conciliator controversiarum, quae inter philosophos et medicos versantur*. Venezia.

Peter of Abano. (2009). Expositio problematum Aristotelis, IV, 15. In G. Coucke (Ed.), *Non adeo est honesta ut delectabilis*. Sexual Pleasure in Medieval Medicine: the Case of Petrus de Abano. In C. Casagrande & S. Vecchio (Eds.), *Piacere e dolore. Materiali per una storia delle passioni nel Medioevo* (pp. 117–148). SISMEL–Edizioni del Galluzzo.

Peter of Spain. (1991). Problemata. In M. de Asúa (Ed.), *The Organization of Discourse on Animals in the Thirteenth Century. Peter of Spain, Albert the Great, and the Commentaries on* De animalibus. Dissertation. University of Notre Dame.

Peter of Spain. (2015). Questiones super libro De animalibus Aristotelis. In F. Navarro Sánchez (Ed.), *Peter of Spain,* Questiones super libro De animalibus Aristotelis*: Critical Edition with Introduction*. Ashgate Publishing.

Pietro Torrigiano. (1504). *Plusquam commentum*. Venezia.

Rhazes. (1510/11). Liber ad Almansorem. In *Opera parua Abubetri filii Zacharie filii arasi* (Vol. 1). Lyon.

Soranus of Ephesus. (1991). Gynecology. In O. Temkin (Ed.), *Soranus'* Gynecology. Johns Hopkins University Press.

Thomas Aquinas. (1891). Prima Secundae Summa Theologiae. In *Sancti Thomae Aquinatis doctoris angelici Opera omnia iussu impensaque Leonis XIII P. M. edita, Prima Secundae Summa Theologiae, a quaestione I ad quaestionem LXX*, t. VI. Typographia Polyglotta.

Thomas Aquinas. (1899). Secunda Secundae Summa Theologiae. In *Sancti Thomae Aquinatis doctoris angelici Opera omnia iussu impensaque Leonis XIII P. M. edita, a quaestione CXXIII ad quaestionem C*, t. X. Typographia Polyglotta.

Thomas Aquinas. (1933). In R. P. M. F. Moos (Ed.), *Thomae Aquinatis Scriptum super Sententiis magistri Petri Lombardi*, t. III. P. Lethielleux.

Thomas Aquinas. (1969). Sententia libri Ethicorum. In *Sancti Thomae Aquinatis doctoris angelici Opera omnia iussu impensaque Leonis XIII P. M. edita, Sententia libri Ethicorum, volumen II, libri IV–X – indices*, t. XLVII. ad Sanctae Sabinae.

Tommaso del Garbo. (1531). *Summa medicinalis*. Venezia.

Trotula (so-called, various authors). (2009). M. H. Green (Ed.), V. Brancone (transl.), Trotula. *Un compendio medievale di medicina delle donne*. SISMEL–Edizioni del Galluzzo.

William Hamilton. (1863). *The Works of Thomas Reid, D.D. Now Fully Collected, with Selections from His Unpublished Letters. Preface, Notes and Supplementary Dissertations by Sir William Hamilton*. Maclachlan and Stewart.

William Hamilton. (1870). *Lectures on Metaphysics and Logic* (Vol. 2). William Blackwood and Sons.

Secondary Sources

Beneduce, C. (2022). "Is Touch One Sense or Several?" A Late Medieval Scientific Question. In M. Gensler, M. Mansfeld, & M. Michałowska (Eds.), *The Embodied Soul: Aristotelian Psychology and Physiology in Medieval Europe between 1200 and 1420* (pp. 253–275). Springer.

Beneduce, C. (Ed.). (2023). *Complexio.* Across Disciplines. *Early Science and Medicine* (special issue) *28*(3–5).

Beneduce, C. (2024). *La scienza del tatto. Un percorso tra filosofia naturale e medicina.* Editrice Bibliografica.

Brumberg-Chaumont, J. (2020). Albert the Great. In H. Lagerlund (Ed.), *Encyclopedia of Medieval Philosophy* (pp. 86–91). Springer.

Cadden, J. (1993). *Meanings of Sex Difference in the Middle Ages: Medicine, Science, and Culture.* Cambridge University Press.

Chandelier, J. (2017). *Avicenne et la médecine en Italie: le* Canon *dans les universités (1200–1350).* Honoré Champion.

Connell, S. M. (2021). *Aristotle on Women: Physiology, Psychology, and Politics.* Cambridge University Press.

Cova, L. (1997). Il corpus zoologico di Aristotele nei dibattiti tra gli "artisti" parigini alle soglie del XIV secolo. In O. Weijers & L. Holtz (Eds.), *L'enseignement des disciplines à la Faculté des arts (Paris et Oxford, XIIIe–XVe siècles)* (pp. 281–302). Brepols.

Del Punta, F. (1997). The Genre of Commentaries in the Middle Ages and its Relation to the Nature and Originality of Medieval Thought. In J. Aertsen & A. Speer (Eds.), *Was ist Philosophie im Mittelalter? Qu'est-ce que la philosophie au Moyen Âge? What is Philosophy in the Middle Ages?: Akten des X. Internationalen Kongresses für Mittelalterliche Philosophie der Société Internationale pour l'Etude de la Philosophie Médiévale* (pp. 138–151). De Gruyter.

Fioravanti, G., Leonardi, C., & Perfetti, S. (Eds.). (2002). *Il commento filosofico nell'Occidente latino, secoli XIII–XV.* Brepols.

Foscati, A. (2023). *Le meraviglie del parto. Donare la vita tra Medioevo ed Età Moderna.* Einaudi.

Foscati, A. (2025). "In Pain You Shall Bring Forth Children". Humble Acceptance of the Pain of Childbirth? A Brief Overview from the Middle Ages to the Sixteenth Century. *Medicina nei Secoli, Journal of History of Medicine and Medical Humanities, 37*(1), 81–100.

Graciotti, L. (2023). *Utrum molles carne sint mente apti*: Tommaso d'Aquino e Pietro Pomponazzi su *De anima*, II, 9. *Bruniana & Campanelliana, 29*(2), 383–398.

Green, M. H. (2021) (unpublished). Cliff Notes' on the Circulation of the Gynecological Texts of Soranus and Muscio in the Middle Ages. https://www.academia.edu/49293690/Green_Cliff_Notes_on_the_Circulation_of_the_Gynecological_Texts_of_Soranus_and_Muscio_in_the_Middle_Ages_2021_ Last accessed on November 8, 2024.

Grellard, C. (2017). Le philosophe et le glouton. Le plaisir de boire et de manger dans les commentaires de l'*Éthique À Nicomaque* de Jean Buridan et Nicole Oresme. In C. Grellard (Ed.), *Miroir de l'amitié. Mélanges offerts à Joël Biard* (pp. 371–385). Vrin.

Haggarty, C. J., Makdani, A., & McGlone, F. (2023). Affective Touch: Psychophysics, Physiology and Vicarious Touch Perception. In N. P. Holmes (Ed.), *Somatosensory Research Methods* (pp. 109–128). Humana.

Heller-Roazen, D. (2007). *The Inner Touch: Archaeology of a Sensation.* Zoone Books.

Kretzmann, N., Kenny, A., Pinborg, J., & Stump, E. (Eds.). (1984). *The Cambridge History of Later Medieval Philosophy: From the Rediscovery of Aristotle to the Disintegration of Scholasticism, 1100–1600.* Cambridge University Press.

Lagerlund, H. (2020). Peter of Abano. In H. Lagerlund (Ed.), *Encyclopedia of Medieval Philosophy* (pp. 1438–1439). Springer.

Lagerlund, H. (2020b). Peter of Spain. In H. Lagerlund (Ed.), *Encyclopedia of Medieval Philosophy* (pp. 1453–1458). Springer.

Loviconi, L. (2025). Conceptions et significations de la douleur dans le *Lilium medicinae* (1303–1305) de Bernard de Gordon. *Medicina nei Secoli. Journal of History of Medicine and Medical Humanities, 37*(1), 59–80.

Marenbon, J. (2023). Medieval Philosophy. In *The Stanford Encyclopedia of Philosophy* (Spring 2023 Edition), Edward N. Zalta & Uri Nodelman (Eds.), https://plato.stanford.edu/archives/spr2023/entries/medieval-philosophy. Last accessed on November 8, 2024.

Olausson, H., Wessberg, J., Morrison, I., & McGlone, F. (Eds.). (2016). *Affective Touch and the Neurophysiology of CT Afferents*. Springer.

Perfetti, S. (2012). *Animali pensati nella filosofia tra Medioevo e prima Età Moderna*. ETS.

Pesenti, T. (2003). *Marsilio Santasofia tra corti e università: la carriera di un* monarcha medicinae *del Trecento*. Antilia.

Plastina, S. (2017). *Mollezza della carne e sottigliezza dell'ingegno. La natura della donna nel Rinascimento europeo*. Carocci.

Pogliano, C. (2015). *Senso lato. Il tatto e la cultura occidentale*. Carocci.

Robert, A. (2021). *Épicure aux Enfers: hérésie, athéisme et hédonisme au Moyen Âge*. Fayard.

Smith, R. (2011). "The Sixth Sense". Towards a History of Muscular Sensation. *Gesnerus, 68*(1), 218–271.

Van der Lugt, M. (2021). Generation in the Middle Ages. Past, Present, Future. In A. Paravicini Bagliani & D. Jacquart (Eds.), *Le Moyen Âge et les sciences* (pp. 365–387). SISMEL–Edizioni del Galluzzo.

Vecchio, S. (2006). Il discorso sulle passioni nei commenti all'*Etica Nicomachea*: da Alberto Magno a Tommaso d'Aquino. *Documenti e studi sulla tradizione filosofica medievale, 17*, 93–119.

Vecchio, S. (2009). Il piacere da Abelardo a Tommaso. In C. Casagrande & S. Vecchio (Eds.), *Piacere e dolore. Materiali per una storia delle passioni nel Medioevo* (pp. 67–86). SISMEL–Edizioni del Galluzzo.

Yrjönsuuri, M. (2008). Perceiving One's Own Body. In S. Knuuttila & P. Kärkkäinen (Eds.), *Theories of Perception in Medieval and Early Modern Philosophy* (pp. 101–116). Springer.

Zuccolin, G. (2011). Nascere in latino ed in volgare. Tra la *Practica* ed il *De regimine*. In C. Crisciani & G. Zuccolin (Eds.), *Michele Savonarola. Medicina e cultura di corte* (pp. 137–210). SISMEL–Edizioni del Galluzzo.

Zuccolin, G. (2019). Tommaso d'Aquino sulla complessione corporea. *Revue des Sciences philosophiques et théologiques, 103*(4), 625–648.

Chapter 4
Nicholas of Autrecourt on the Superiority of Touch and the Ontology of Sensible Qualities

Aurélien Robert

Abstract For Aristotle and most of his medieval commentators in the Latin West, sight is regarded as superior to the other senses, insofar as it affords us the most knowledge about the things that constitute the sensible world. Nicholas of Autrecourt, however, suggests that the sense of touch should be given greater value, demonstrating that it is through touch that we can attain certainty of something's real existence—namely, something that exists in a subject (in *esse subiectivum*) and that does not continually vary through time (something that has an *esse fixum*). The aim of this chapter is to demonstrate that Autrecourt's theory of touch is grounded in a comprehensive analysis of sensible appearances, which suggests that such appearances are largely dependent upon the local motions of atoms (either changes in the atomic structure of the object known or the local motion of atoms between the object and the knowing subject). If this reconstruction of Autrecourt's position is correct, it would follow for those sensibles that are not tangible objects that they are mere appearances with mutable being—that is, mere effects upon our cognitive faculties caused by the position and motion of atoms, which lie beyond the reach of our sensory perception. Our hypothesis is that Autrecourt is not merely advocating Democritus' atomism, but also his theory of sensible qualities. Without this framework, it would be difficult to understand the primacy given to touch over the other senses in Autrecourt's responses to philosophical scepticism.

Keywords Nicholas of Autrecourt · Illusions · Certainty · Scepticism · Atomism

---

A. Robert (✉)
SPHERE UMR7219 (CNRS - Université Paris Cité - Université Paris 1 Panthéon Sorbonne), Paris, France
e-mail: aurelien.robert1@u-paris.fr

C. Beneduce et al. (eds.), *The Sense of Touch*, Studies in the History of Philosophy of Mind 32, https://doi.org/10.1007/978-3-032-16308-0_4

4.1 Introduction

Nicholas of Autrecourt (1295/98–1369) has long been regarded as the "medieval David Hume",[1] owing to his critical stance towards the very possibility of attaining certain knowledge of reality and his challenge to the validity of some existential inferences from sensory data to entities not directly cognized, or from effects to their causes. However, contemporary scholarship is unanimous in rejecting the view that Autrecourt was a sceptic. To the contrary, it is now clear that he aimed to delineate the conditions under which true and stable knowledge of the external world might be attained.[2] In the few letters of his that have come down to us, he even refers to his opponent, Bernard of Arezzo (*fl.* 1324–1342), as the true sceptic:

> And, as it seems to me, from your position there follow things that are more absurd than follow from the position of the Academics. And, therefore, in order to avoid such absurdities, I have upheld in disputations in the Aula of the Sorbonne, that I am evidently certain of the objects of the five senses and of my own acts.[3]

He thus opposes scepticism with a form of empiricism, asserting, on the one hand, that sensory cognition provides the subject with some evidence, even a certain truth, and, on the other hand, that the subject's reflexive evidence about his own cognitive acts—we know that we are sensing, thinking, knowing, and so forth—enables us to obtain some solid cognition of, at least, some objects of thought. Yet, in a surprising turn, Autrecourt also asserts that certain and evident knowledge is, in fact, exceedingly rare. In most cases, we must content ourselves with knowledge that is merely probable. This raises the question: what, then, can we be certain of when it comes to the objects apprehended by the five senses?

Though the issue of the reliability of the senses is not extensively addressed in his *Correspondence with Master Giles and Bernard of Arezzo*, Autrecourt engages with this topic in considerable depth in his sole surviving treatise (written before 1340), which is generally referred to by its incipit (*Exigit ordo*), but whose real title is: *A useful treatise to see whether the arguments of the Aristotelians were demonstrative*.[4] In this work, Autrecourt places particular emphasis on the role of sensible appearances (*apparentiae*)—those phenomena presented to us through the five senses—as distinct from other types of *apparentiae*, such as those found in the imagination or in dreams. This is the central concern of the chapter titled "Does everything that appears exist?".[5] In this chapter Autrecourt also aims to demonstrate

[1] Rashdall (1907).

[2] Thijssen (2000) and Grellard (2005).

[3] Nicholas of Autrecourt, *Correspondence with Master Giles and Bernard of Arezzo* (ed. De Rijk, 1994, pp. 55–57).

[4] Nicholas of Autrecourt, *Exigit ordo* (ed. O'Donnell, 1939). For the Latin text, I use the edition by Musu, namely Nicholas of Autrecourt, *Exigit ordo* (ed. and transl. Musu, 2009), which contains some corrections introduced by Zénon Kaluza. An English translation of the *Exigit ordo* can be found in Nicholas of Autrecourt, *Universal Treatise* (transl. Kennedy *et al.*, 1971).

[5] Nicholas of Autrecourt, *Exigit ordo* (ed. O'Donnell, 1939, corrected by Musu, 2009, pp. 202–241).

that everything that appears is true. His argument, however, does not seek to establish that we are capable of knowing reality as it truly is, but rather that reality cannot appear to us otherwise, even though the same thing may present itself differently to different individuals.[6] For there is no real error in the simple act of sensory perception; error arises only in the judgments we make about our perceptions, such as when one attributes properties to a subject which it does not possess, or when one draws faulty inferences from sensory cognition.

The question is now whether the things that appear to us exist independently of our minds, and whether these appearances correspond to some extra-mental reality. Here again, Autrecourt appears to display considerable confidence in the foundational role of sensory cognition.

> Such conformity to the thing outside [the perceiver] is not naturally found in the appearance properly speaking, if not by its term by which things are reached, for example when things exist in act in themselves and in so far as they are cognized by the external senses.[7]

Sensible appearance, in and of itself, is insufficient to determine whether a thing exists or to apprehend the nature of what appears to us. It is true that every appearance is the appearance of something. For it to be the appearance of something external to the perceiver, however, it must be apprehended by the external senses. From a phenomenological standpoint, what appears in the imagination or in dreams is distinct from that which is perceived through the senses, even though Autrecourt asserts that both, in a way, are true.[8] What is true in such cases is that something appears objectively within the mind, not that it exists subjectively in extra-mental reality. Direct empirical knowledge would therefore enjoy a certain privilege, giving access not only to appearances, but in a certain way also to the thing itself.

Autrecourt acknowledges that sensory cognition can fluctuate and that two individuals may perceive the same object differently, depending on their respective situation and point of view. To preclude the sceptical consequences of such a claim, he introduces the criterion of clarity, specifically what he terms cognition "in full light" (*in pleno lumine*).[9] The reliability of sensory cognition, therefore, rests upon a degree of clarity beyond which further sensory knowledge is unattainable. In some cases, it becomes necessary to compare different perspectives in order to correct the apparent variability of the *apparentiae* and to stabilize our understanding of the sensible world.[10] However, despite this variability in appearances, sensory knowledge seems to provide reliable access to the existence of the thing that appears, even

---

[6] Robert (2023).

[7] Nicholas of Autrecourt, *Exigit ordo* (ed. O'Donnell, 1939, corrected by Musu, 2009, p. 206): "Nam talis conformitas ad rem extra non est nata esse in apparentia proprie nisi per extremum secundum quod magis attenditur, puta secundum quod res actualiter existent in se et ut cognitae per sensus exteriores".

[8] Nicholas of Autrecourt, *Exigit ordo* (ed. O'Donnell, 1939, corrected by Musu, 2009, p. 238 and p. 250, for instance).

[9] Kaluza (1997); Grellard (2005).

[10] Salvestrini (2022).

if it does not inform us reliably about its nature. As for inferences drawn from sensory appearances, their validity must be measured against the principle of non-contradiction.[11] Beyond sensory cognition "in full light" and a few inferences that are often quasi-tautological, our certainty concerning the external world is ultimately quite limited.

The thesis according to which "everything that appears is true" was condemned at Autrecourt's trial in Avignon. In the documents resulting from his conviction, we first find a negative formulation:

> I once asserted in Rue du Fouarre that it cannot be evidently shown that not all that appears if true. I recant this as false.[12]

This same proposition can also be found in the articles confessed by Autrecourt in front of the censors:

> That by natural appearances we can acquire almost no certitude about things; yet that a modest amount of certitude can be obtained in a short span of time, if people direct their intellect towards things, rather than towards the Philosopher's intellect and that of his Commentator. I recant this as false and erroneous.[13]

While everything that appears to us is true, there is little certainty offered by those things that appear to us. How can we reconcile this apparent tension, or even conflict, between the truth of sensible appearances and the rare certainty of knowledge? To address this, it is essential to distinguish between truth and evidence, on the one hand, and certainty, on the other. While everything revealed to us by the senses may be true, in the sense of a pre-propositional truth approaching a form of irrefragable evidence, certainty requires a form of certification, most often achieved through reasoning.

Our aim in this chapter is to demonstrate that, in contrast to Aristotle and most of his followers, Autrecourt attributes a crucial role to the sense of touch in his phenomenological approach to empirical knowledge. Of the five senses, it is touch, in his opinion, that provides the most reliable access to reality, and that induces, moreover, the least doubt concerning the existence and nature of sensible objects. If this interpretation is correct, it suggests that Autrecourt accords an epistemic superiority to the sense of touch over the other senses—a perspective that was rather unconventional at a time when the visual paradigm was becoming increasingly dominant.[14]

---

[11] Weinberg (1948, pp. 9–30); Grellard (2005, pp. 67–92); Krause (2009).

[12] Nicholas of Autrecourt, *Correspondence with Master Giles and Bernard of Arezzo* (ed. De Rijk, 1994, pp. 182–183, translation modified): "Dixi semel in vico Straminum quod non potest evidenter ostendi quin omnia que apparent sint vera. Revoco tanquam falsum".

[13] Nicholas of Autrecourt, *Correspondence with Master Giles and Bernard of Arezzo* (ed. De Rijk, 1994, pp. 168–169, translation modified): "Quod de rebus per apparentia naturalia quasi nulla certitudo potest haberi, illa tamen modica potest in brevi haberi tempore, si homines convertant intellectum suum ad res, non ad intellectum Philosophi et sui Commentatoris. Revoco tanquam falsum et erroneum".

[14] See, for instance, Dennery (2005). Even in texts that feature in-depth and original discussions of the sense of touch, it is extremely rare to find any such reversal of the hierarchy of the senses; see Beneduce (2024).

More broadly, we would like to suggest that this unusual valorization of the sense of touch may be seen as part of a larger effort to rehabilitate the philosophy of Democritus, and more particularly his ontology of sensible qualities.

4.2 The Epistemic Value of Touch from an Aristotelian Perspective

In his treatise *On the Soul*, Aristotle stated that while all animals possess the sense of touch, human beings surpass all others in the acuity of their tactile perceptions, while being weaker than the others in terms of the other four senses.[15] But this superiority was expressed in terms of acuity, sometimes of utility, not in purely epistemic terms. Indeed, as stated in the opening words of his *Metaphysics*:

> By nature, all men long to know. An indication is their delight in the senses. For these, quite apart from their utility, are intrinsically delightful, and that through the eyes more than the others. [...] And the reason for this is that sight is the sense that especially produces cognition in us and reveals many distinguishing features of things.[16]

This assertion of the primacy of sight, based on the fact that it would enable us to distinguish more properties in the thing, seems to be contradicted by other texts in which Aristotle says that touch cognizes several properties,[17] such as hot, cold, dry and wet (*On the Soul*, II.11, 422b18–34), but also heavy and light (*On Sense and Sensibles*, 6, 445b3–7) or viscous and brittle, rough and smooth, coarse and fine (*On Generation and Corruption*, II.1 329b7–16). A large majority of medieval readers of Aristotle in the Latin West, however, assume the validity of the claim that sight contributes more than the other senses to our knowledge of the external world.[18]

This is the case, for example, with John Buridan (*ca.* 1300–1361). Commenting on the first sentences of Aristotle's *Metaphysics*, probably at a time when Autrecourt was also at the University of Paris, Buridan wonders "Whether sight should be preferred to the other senses".[19] Among the mentioned arguments *contra* the primacy of sight is the idea that touch is more useful than the other senses, particularly for life and for preserving life.[20] We can live blind, but we can hardly survive without touch. According to Aristotle, all animals, moreover, have the sense of touch, while some cannot see or smell.[21] John Buridan also considers the possibility that touch

---

[15] Aristotle, *On the Soul*, II.9, 417a20–26 (transl. Schields, 2016, p. 41).

[16] Aristotle, *Metaphysics*, A.1, 980a21–23 (transl. Lawson-Tancred, 1998, p. 4).

[17] Steiner Goldner (2018).

[18] See, for instance, Mantovani (2021).

[19] John Buridan, *In Metaphysicen Aristotelis*, I, q. 6 (ed. Paris, 1588 [1518], ff. VIrb–VIvb): "Queritur sexto utrum visus inter ceteros sensum debeamus magis diligere." For a presentation of Buridan's views on touch, see Beneduce (2019, 2022, 2024).

[20] Aristotle, *On the Soul*, II.3, 414a1–4 (transl. Schields, 2016, p. 26).

[21] Aristotle, *On the Soul*, II.3, 415a3–5 (transl. Schields, 2016, p. 26).

might be preferable from a cognitive point of view, because its findings would be more certain (*certius*) and because it would allow us to experience more differences in things (not only hot, cold, dry, wet, but also heavy and light for instance). But while he acknowledges, much like Aristotle, that the sense of touch is indeed most useful for our survival, he resolutely denies to the sense of touch any cognitive superiority. In his estimation, it is the sense of sight that allows us to judge with the utmost certainty (*certissime iudicat*), due to the organ's subtlety and its ability to provide finer access to common sensibles such as motion, magnitudes, and figures, as well as its greater capacity to nourish the higher faculties, ranging from imagination to intellect.

Autrecourt, as we shall see, adopts a position contrary to that of John Buridan, defending the superiority of touch from the standpoint of knowledge rather than utility. In his discussion of sensory illusions, for instance, he assigns a prominent role to the sense of touch. Indeed, regarding visual knowledge, it is often challenging to discern whether we are perceiving the object itself or merely its image. The visual qualities of an object, moreover, can fluctuate rapidly due to movement, lighting, perspective, distance, and so forth. Such variability can lead us to doubt whether we are truly apprehending the object or merely its transient appearance or representation. There is no such variability with touch, in Autrecourt's opinion:

> In the case of touch, indeed, there does not seem to be any reason for doubting, for it does not seem to be concerned sometimes with an image, but always with a fixed thing (*res fixa*).[22]

It is precisely this assertion that we seek to elucidate in the pages that follow.

4.3 Touch and Illusions

For ancient and medieval philosophers, instances of sensory illusions primarily pertain to sight, as it is these illusions that pose the most significant challenges within a framework that privileges this sense. The other senses are equally prone to deception, however. Regarding touch, Aristotle consistently employs the same example of crossed fingers, which create the illusion of two distinct objects being touched where there is only one.[23] In his treatise *On Dreams*, wherein he addresses several cases of illusions—for instance, when sensations persist even after the object has vanished, or when we are misled by slight resemblances, or by various ailments and afflictions—, he also uses the example of tactile illusions. Generally speaking, illusions may arise when there exists a dissonance among different sensory faculties, particularly concerning common sensibles, as opposed to the proper sensibles of

---

[22] Nicholas of Autrecourt, *Exigit ordo* (ed. O'Donnell, 1939, corrected by Musu, 2009, p. 216): "In tactu vero non sic videtur esse causa dubitandi, nam non videtur esse aliquando ad imaginem, sed semper ad rem fixam".

[23] Freeland (1995).

each sense. Nevertheless, it is through such disagreements among the senses that we are also able to rectify these perceptual errors.

> The reason why these things happen is that the ruling part and that by which appearances (*phantasmata*) occur do not judge on the basis of the same faculty. Proof of this is the fact that the sun appears only one foot across, and yet frequently something else contradicts the appearance (*phantasia*). Again, by crossing of the fingers a single object appears as two, but even so we still deny that there are two things. For sight has more authority than touch. If touch stood alone, we should actually judge the single object to be two.[24]

For Aristotle, the senses never make mistakes (go wrong) about their proper objects (*On the Soul*, III.3, 428b18–19), whereas mistakes can occur with common sensibles such as number, shape, size, movement and rest, etc. Although there is no single proper object of touch, it always seems to involve properties of the body *qua* body and properties related to the qualities of the four elements that make up its matter (*On the Soul*, II.11, 423b27–29). While touch can sometimes be a good guide to the size or weight of an object, this is not always the case for numbers. In the case of tactile illusion, it is the sight that must correct the error of touch about the number of objects, because of its intrinsic epistemic superiority relative to this common sensible.

We also find the example of crossed fingers in book IV of the *Metaphysics* when Aristotle discusses the thesis of the "Ancients," who affirmed that everything that appears is true (a claim which, as has been mentioned before, is also supported by Autrecourt). The target of this passage is a form of relativism attributed to Protagoras and Democritus which holds that, if everything that appears is true, two contradictory appearances could be true at the same time. The consequence would therefore be that all knowledge is relative to the knowing subject, since each subject can see things appear in different and sometimes even contradictory ways.

> For neither do all subjects share the same appearances nor does the same subject at all times. In fact, the same subject often experiences opposite appearances at the same time (as touch says that there are two things in a crossing of fingers and sight one). But there is nothing that appears differently to the same sense and in the same respect and in the same way and at the same time. So at least this claim would be true.[25]

For Aristotle, when two individuals perceive the same object differently, or when a single individual experiences two seemingly contradictory perceptions, there exists a method for correction through the accumulation of experiences to refine the judgments of the intellect. In this context, the application of the principle of non-contradiction may be confined to a single sensory faculty (as one perception cannot contradict itself), taking into account the identity of the subject and the object, as well as the modality of knowledge and time (at a given instant, the same sense cannot perceive the same object in two different and contradictory ways).

Autrecourt largely concurs with this characterization of sensible knowledge. At any given moment, he asserts, we indeed see, smell, taste, or touch a particular

---

[24] Aristotle, *On Dreams*, 2, 460b16–22 (transl. Gallop, 2015, pp. 95–97).

[25] Aristotle, *Metaphysics*, Γ.6, 1011a29–36 (transl. Lawson-Tancred, 1998, p. 105).

object, and it presents itself to us in a specific manner.[26] In fact, Autrecourt employs this reliability of the senses in relation to their proper objects as a foundational argument for his thesis that all appearances are true: a sensible object cannot manifest itself in any way other than how it appears to us at a specific moment.[27] The error, he contends, invariably resides in the subsequent judgment of the intellect, which may mistakenly attribute a perceived property to the object when the property may actually reside elsewhere (such as in the sensory organ or within an image retained in the mind).

Concerning the well-known case of the sun's apparent size, a magnitude that seems much smaller to us than it truly is, Autrecourt, like Epicurus, asserts that this appearance of the sun is indeed true. This assertion does not imply that the sun is, in reality, only two feet in height, nor does it preclude the possibility of inferring—through the insights of mathematicians and astronomers—that it is, in fact, much larger than it appears. The essential point remains that the sun cannot manifest itself to us in any manner other than it does. Consequently, this "true" appearance does not entail a true judgment about the sun itself. According to Autrecourt, we apprehend the sun through an image that preserves its proportions but does not accurately reflect the actual size of the object.[28] Thus, the error lies in attributing these dimensions to the object itself rather than to its image within our cognitive faculties. If we were to attend carefully to the manner in which the sun presents itself to us, we would recognize that our knowledge is not complete knowledge ("in full light") and that it possesses the nature of knowledge derived from an image.

Autrecourt endeavours to explain most cases of sensory illusions along the same lines. When we see a broken-looking stick in water, we are not seeing the stick itself, directly, but its distorted image.

> The same answer would be given concerning a stick appearing broken in the water. I say that the appearance has to do with an image, which is exactly like what appears to us. I would say the same concerning a river-bank which, to a man in a ship, appears to move.[29]

It is worth noticing that, in this specific case, some medieval thinkers were ready to acknowledge that the sense of touch can help to correct our vision and to assure ourselves that the stick in the water is not actually broken.[30]

Autrecourt hastens to add the example of taste: in this case too, when one individual perceives a bitter flavour while another detects sweetness, it may be attributed to an inherent quality—either genuinely bitter or sweet—present in one of their

---

[26] Grellard (2005–2006).

[27] Robert (2023).

[28] Nicholas of Autrecourt, *Exigit ordo* (ed. O'Donnell, 1939, corrected by Musu, 2009, p. 214).

[29] Nicholas of Autrecourt, *Exigit ordo* (ed. O'Donnell, 1939, corrected by Musu, 2009, p. 214; transl. Kennedy *et al.*, 1971, p. 110): "Idem responderetur de baculo qui appareret fractus in aqua, et dico quod apparitio est ad quamdam imaginem quae est talis dispositionis, sicut apparet. Et idem dicerem de motu ripae qui apparet existenti in navi".

[30] See Tachau (1998); Grellard (2004).

sensory organs, rather than in the thing itself.[31] In other words, we are not fundamentally mistaken, and there exists a degree of truth in such sensible appearances, even though the external object that is the focus of our perception may not possess the characteristic we ascribe to it according to the testimony of our senses. Autrecourt's overall strategy involves generalizing the well-known case of the stick appearing broken in water: only touch can provide the assurance that we truly know something in reality rather than merely its image.

Evidently, this does not address the potential for tactile illusions. While illusions in the other senses may extend to the very existence of sensible objects, touch at least guarantees the existence of a real subject external to ourselves. If all sensations are indeed true, then the sensations of touch carry with them a greater weight of evidence for the existence of the object and certain of its properties. It nevertheless remains a fact that errors can still arise in the judgments formulated by the intellect by way of a response to what we touch.

A sceptic would readily counter these explanations with a classic argument: since we are unable to touch every sensible object and since our cognition of the world occasionally amounts to the cognition of mere images or representations, how can we be certain that there exists a corresponding entity in external reality (*in esse subiectivum*) for that which is apprehended as an object (*in esse obiectivum*)?

> It would follow that no one could speak of the true subjective existence of whiteness or anything else. For one cannot say [anything about it] except according to its appearance. Now, one will say that the appearance terminates at the image of the thing and not at something subjectively existing outside.[32]

Autrecourt begins his answer to the sceptic with a first statement: when we see a whiteness, we are at least certain that we are seeing something and that there is something that is seen.

> It must be said that it is certain that when sight sees whiteness it sees something. Its appearance vouches for this; and also that it is outside the eye and in such and such a place. I grant that all these things are true. Now, when vision does not change, no matter where it turns or what its state, and whiteness appears to it, one imposes that name "whiteness" and says that there is a true whiteness there, possessing a fixed or subjective being (*esse fixum vel subiectivum*). When this is not the case, one imposes the name and calls that thing an image, as when a man is seen in a mirror, or when a man sees the river-bank move, whereas he would not see it move if he were in another spot, provided that he would see it most fully.[33]

---

[31] Nicholas of Autrecourt, *Exigit ordo* (ed. O'Donnell, 1939, corrected by Musu, 2009, p. 208).

[32] Nicholas of Autrecourt, *Exigit ordo* (ed. O'Donnell, 1939, corrected by Musu, 2009, p. 216): "[...] sequeretur quod nullus posset dicere de existentia vera subjectiva albedinis vel alicujus rei. Non enim potest dicere nisi per suam apparentiam; nunc dicetur quod illa apparentia terminatur ad imaginem rei et non ad aliquid existens subjective in re extra".

[33] Nicholas of Autrecourt, *Exigit ordo* (ed. O'Donnell, 1939, corrected by Musu, 2009, p. 216): "Dicendum quod immo visus videns albedinem, quod aliquid videat certum est. Hoc dicit sua apparentia et quod est extra oculum et in tali loco; et concedo quod haec omnia sunt vera. Nunc, quando non mutatur visio ad quodcumque se divertat et quoquo modo existens, et realiter illud sibi apparet, imponit illud nomen et dicit quod ibi est vera albedo habens esse fixum vel subjectivum; quando non sic, imponit nomen et appellat illud imaginem, sicut quando homo videtur in speculo

Contrary to imaginations or dreams, visual appearances would indicate that something exists outside the eye—and therefore outside the perceiver—and that this something exists in a specific place. But how do we know that we are not under an illusion? In addition to this first phenomenological certification of the existence of an act of perception and of its object, Autrecourt adds the following criterion: if sight is not changed or modified, wherever it turns its gaze, it is possible to name this object—a whiteness, for instance—in order to fix the reference to the external object. We can therefore say—even though we are not yet absolutely certain—that this sensible quality most probably has an *esse fixum vel subiectivum*. When our visual knowledge of an object is changeable, however—as when we see colours on the neck of a dove, for example, or when we see the same thing differently from our neighbour—, then what appears has no *esse fixum vel subiectivum*, for it is a mere image or representation of that fixed thing. Up to this point, Autrecourt's position is not far removed from the responses by William of Ockham, Walter Chatton and Adam Wodeham to Peter Auriol's theory of the "apparent being" (*esse apparens*).[34] But Autrecourt seems to limit his claim to the affirmation that we can impose a name because the phenomenon seems stable, not that sensible qualities really have such an fixed and subjective being (*esse fixum et subiectivum*).

What are we to make of the classic sceptical argument about dreams? How can we be sure that we are not in a dream? Can we not still consider it possible to move around an object while dreaming and for it to appear in a stable way in that dream? Generally speaking, Autrecourt maintains that there is some phenomenological difference between the way something appears to us in ordinary sensible experience and the way something appears in a dream, as if something inside us is telling us that we are dreaming. According to him, we know we are dealing with reality when we can see the thing "in full light" (*in pleno lumine*), which basically means that there are no shadows left, i.e., when we can see every aspect of the thing, so that we can be sure that we are perceiving the *res* and not a mere image.

For Autrecourt, such clarity of knowledge is attainable only through a cognition of the external senses. In some cases at least, the external senses enable us to assess directly the conformity or discrepancy between appearance, understood as representation, and the thing as it exists. In the case of perceptible illusions, one sense can correct the other and thus enable us to correct our judgement. According to Autrecourt, one could never measure such a discrepancy in the case of the dream; although he does not really give an argument for this, it is probably because one is too immersed in the images constructed by the imagination. However, this enclosing of dreams upon themselves would make dreamt objects distinguishable from the sensible experience of really existing objects. Moreover, dreamed appearances would never be truly complete, so that it is always possible, simply by examining what appears, to realise that we are dreaming. To say that a dreamed image is in

---

vel quando homo videt ripam moveri, alibi existens non videret ipsam moveri, esto quod plenissime videret eam".

[34] Wood (1982).

some sense *index sui*, that it shows itself as an image forged by the imagination, may seem philosophically weak, or at least puzzling. But we should probably see this as a kind of transcendental argument: we have little choice but to resort to these weak criteria. Just as we are forced, according to Autrecourt, to accept the truth of appearances precisely because external things cannot appear to us otherwise in a particular situation, so we must accept the truth of dreams, since we cannot fight against these nocturnal apparitions. Their truth corresponds to the reality of their effects. But at the level of judgements produced by the intellect, the appearances of the external senses and those of the internal senses alone, as in dreams, do not give rise to the same types of judgements. For the certainty of the existence of an object outside ourselves can only be derived from the empirical knowledge of something outside ourselves.

In Autrecourt's own words, this "full light" is like the measure or standard of certainty; we have no other criteria for that. As a last resort, either we accept that the appearance is true, or there is no longer any standard of truth or certainty. As for dreams, Autrecourt adds the following remark:

> We call it "full light" because a light which is not full does not have the nature of a measure. Hence they argue frivolously who say: "In sleep it appears that I am running through a camp, or that I am in heaven, and yet this is not so". I answer: "It did not appear to you with a full appearance, for a full appearance of the truth of this statement is an appearance through the external senses when you see the motion through sight".[35]

In other words, although dreams operate on the basis of prior sensations, of imaginations and memories, there exists a qualitative distinction between ordinary sensation and dreams. In dreams, appearances are never complete; things do not present themselves "in full light," thereby preventing any complete apprehension. There remain areas of shadow, which signal that we are encountering an image rather than the actual object. How, then, can we ascertain whether we truly know the object "in full light"? For Autrecourt, we must inevitably employ this maximum of clarity as a standard, positing that appearance serves as an *index sui* regarding its epistemic value.

This does not imply that we do not frequently question the reality of sensible properties. Indeed, sensible appearances can change rapidly, allowing the same object to present itself differently from various perspectives, at different times, and so forth. According to Autrecourt, however, this variability never occurs with touch, as this singular sense directly apprehends the object in its *esse subiectivum et fixum*. Thus, touch emerges as the one sense among the five that enables us to differentiate reality from dreams, and, more generally, to assuage our doubts regarding the existence of external material objects.

---

[35] Nicholas of Autrecourt, *Exigit ordo* (ed. O'Donnell, 1939, corrected by Musu, 2009, p. 214): "Dictum est autem lumen plenum, quia lumen non plenum <non> habet rationem mensurae, propter quod frivole arguunt dicentes: apparet dormiendo me currere per castra vel esse in caelo et tamen ita non est. Respondeo: non apparebat tibi apparentia plena; nam apparentia plena veritatis hujus propositionis est apparentia sensuum exteriorum, quando per visum vides motum".

4.4 The Objects of Touch

Other authors of Autrecourt's time had already suggested that touch could rectify optical illusions. What Autrecourt did differently was to establish touch as the universal guarantee of the existence of objects beyond their mere appearances. According to him, the authentic criterion for determining the fixed and subjective existence of a sensible object is not solely the coherence of our empirical experiences, but rather the fixation of that object—specifically, its presence within a physical body and, even more significantly, within corporeal matter. But what, precisely, do we apprehend through touch?

At first glance, Autrecourt appears to advocate for an Aristotelian theory regarding the proper objects of the five senses. For instance, he somewhat naively posits that if someone inquires whether the water is hot, he would rely on touch to ascertain this.[36] Similarly, if someone questions whether a certain wall is white, he would rely on sight. In other passages, however, he introduces additional elements.

As is well known, Autrecourt's physics is opposed to that of Aristotle. Indeed, he espouses a form of atomism akin to the physics of Democritus, according to which—and this was a rarity during Autrecourt's time—all things are composed of atoms and void. Furthermore, he reduces all causality to efficient causality, positing that all physical phenomena can be reduced and explained by movements of atoms and their arrangements in material entities.[37] As a consequence, when he defends his atomistic interpretation of rarity and density, he asserts that for a body to be more or less dense depends on the existence of void spaces between the atomic bodies composing the whole body.[38] If the density of an object can be perceived through sight, its "solidity"[39] is only determined through touch. In this regard, there is no superiority of touch over the other senses; rather, there exist different capacities or functions alongside various objects. Each sense appears to possess its proper object, although touch encompasses several: elemental qualities, textures, and weight.

Most of the time, indeed, Autrecourt treats sight and touch as if they are on an equal footing. For instance, when explaining how we form the concept of "continuity", he explains that we first form the concept of "contiguity", and if there is no intermediary between the contiguous parts, we conceive that they are continuous.

> Therefore let us form concepts about our problem, and see how they come to the intellect. On the part of the subject, there comes a concept of the continuum, common to it and the

[36] Nicholas of Autrecourt, *Exigit ordo* (ed. O'Donnell, 1939, corrected by Musu, 2009, p. 48).

[37] Grellard (2009).

[38] Grant (1971).

[39] The text says "solid" (*solidum*) as opposed to "light" (*levis*), but it is not clear whether it means only some form of heaviness or also some form of three-dimensional corporeality: Nicholas of Autrecoutrt, *Exigit ordo* (ed. O'Donnell, 1939, corrected by Musu, 2009, p. 162).

> contiguous, that is: having parts with nothing between them. This [concept] comes to the intellect by means of sight and touch.[40]

If some objects can be cognized by different senses, if their abilities to distinguish some properties are different, though equally important in the formation of a picture of sensible reality, there are also significant differences. In the case of colours, the eye does not have direct access to something fixed and sometimes not directly to the coloured object. It may only see a reflection or image of the object in a mirror. It takes something like a phenomenological reduction—to borrow a more recent expression—to truly determine the object seen, a process of multiplying perceptions and points of view. Touch, on the other hand, gives direct access to the thing understood as a material subject. Although tactile illusions do exist, they only concern certain properties (such as the number of objects, for instance). But for Autrecourt, there can be no doubt that what we touch exists, or even that it is a material body made up of parts.

Does Autrecourt intend to go further and assert that only corporeal and material properties exist, i.e., properties that depend on the atomic structure of material bodies, and that sensible qualities such as colour are pure phenomena, entirely dependent on our relationship with material entities? While Joseph Lappe[41] interpreted Autrecourt's theory as a radical form of phenomenalism, Christophe Grellard[42] tried to play down this interpretation. If the reconstruction presented above is correct, then we can indeed speak of phenomenalism. This means that phenomena can be seen as good guides to our knowledge of the external world, even if the actual atomic structure of material entities eludes sensible apprehension. In fact, sensible appearances do not contradict Autrecourt's hypothesis—albeit one that remains merely probable—of atomism. By drawing analogies with observable phenomena, such as the evaporation of water or the flight of dust in sunlight, we can conceive of atomistic physics operating at an infra-sensible level.[43]

It is therefore quite plausible that our sensory cognition is grounded in impressions caused by a flow of atoms emanating from the material entities we encounter. Consequently, it would be tempting to suggest that Autrecourt is not merely adopting Democritus' atomism to reinterpret the various forms of change and causality in Aristotle's *Physics*, but that he is also embracing the Democritean notion that all senses can ultimately be reduced to touch.[44] If all sensations arise from the same type of (efficient) causality and are directly contingent upon the positions and movements of atoms, one might be inclined to conclude that even sight is fundamentally

---

[40] Nicolas of Autrécourt, *Exigit ordo* (ed. O'Donnell, 1939, corrected by Musu, 2009, p. 128): "Formemus igitur conceptus circa quaesitum et videamus qualiter veniunt ad intellectum; venit quidem ex parte subjecti conceptus de continuo qui est communis pro ipso et pro contiguo, videlicet habere partes inter quas non est medium, et iste venit ad intellectum mediante visu vel tactu".

[41] Lappe (1908).

[42] Grellard (2005).

[43] Robert (2023).

[44] See Aristotle, *On Sense and Sensibles*, 4, 442a29–442b3 (ed. Beare, 1908, no page numbers).

nothing more than contact with atoms from material objects. Autrecourt never advances his reasoning to this conclusion, however, and appears to maintain sharper differences among the five senses, specifically to preserve a certain epistemic privilege for touch over the others.

The one aspect of Autrecourt's phenomenological approach to empirical knowledge that remains to be elucidated is the relationship between the atomic structure of bodies and their sensible properties. Might his observation that sensible appearances can vary so significantly indicate that he is defending a distinction akin to that between primary and secondary qualities, or perhaps between atomic organization and emergent sensible properties? Evidence of such a distinction can be found in his employment of the vocabulary of fixity in contrast to motion, and of objects in contrast to their images or representations. This lexicon is entirely foreign to Aristotle, as a cursory examination of the *Aristoteles Latinus* reveals. I propose to suggest a few potential sources for this terminology in order to illuminate these concepts of *res fixa* and *esse fixum*.

4.5 The *Res Fixa* and the Changing Sensible Properties of Extramental Things

The concept of "fixity" can be traced to Averroes's commentary on Aristotle's *De anima*, specifically to the chapter concerning the sense of touch. The passage in question addresses whether a sensible quality can act upon a subject that is not disposed to receive its action. For instance, can a smell impart its effect upon a being incapable of smelling? For Aristotle, this is not only impossible but is also nonsensical. Beyond the requirement that the subject must be properly disposed to receive the effect of a sensible quality, Aristotle further clarifies that qualities do not act independently; rather, they act by virtue of a substance, and more precisely, by virtue of the material body that underlies these qualities. Additionally, certain bodies exert their action through a medium, while others act directly. For example, light, sound, and smell necessarily operate through a medium, whereas the tangible qualities of a body, as well as flavours, act directly upon another body.

> Neither light and darkness nor sound nor odour affects bodies in any way: rather, that [body] in which they are does. For example, it is the air accompanying thunder which splits timber. Still, tangible things and flavours do affect bodies. For if they did not, what would affect or alter things without souls?[45]

Aristotle appears to suggest that, in the case of touch—and by extension, tastes, which are related to tactile properties—, the interaction occurs not merely through direct contact, but operates through a form of body-to-body causality. By contrast, other sensible qualities act at a distance, via a medium, wherein the medium itself transmits the action. Averroes, in his commentary on this passage, concludes that

[45] Aristotle, *On the Soul*, II.9, 424b9–16 (transl. Shields, 2016, p. 48).

bodies affected by tangible qualities are "determined and fixed" (*determinata et fixa*), while something perceived in the air, for instance, can be influenced by wind, which is a body neither determined nor fixed (*non determinatum neque fixum*).[46] For Averroes, the fixity of tangible qualities thus stands in opposition to all other sensible properties, which involve some form of motion or change within a medium.

This vocabulary of *res fixa* also appears in Averroes's commentary on Aristotle's *On the Soul*, book III, where he distinguishes between the speed and non-fixity of images and ideas in contrast to their objects, i.e., the *res fixa* outside the soul.[47] The fixity of the body as the subject or object of sensory perception is therefore contrasted, on one hand, with all that entails distance and motion, and on the other, with images, whose nature is characterized by change and motions as well. This interpretation aligns with Autrecourt's treatment of sensory illusions.

Most of the Latin commentators on Aristotle's treatise *On the Soul* tended to ascribe fixity to sensible qualities, with the notable exception of sounds.[48] Indeed, they struggled to account for the diffusion and operation of sound without invoking local motion. Other sensible qualities, such as colours, odours, and tactile properties, were generally regarded as real entities possessing an *esse fixum* within their subject. Thomas Aquinas (ca. 1225–1274), for example, affirms: "Colour, odour, taste, and the tangible qualities possess a fixed and permanent existence in their subjects (*esse permanens et fixum in suo subiecto*)."[49] By contrast, "sound", he observes, "is produced by change and lacks a fixed and stable existence in a subject (*non habet esse fixum et quiescens in subiecto*), but actually consists in a form of motion."[50]

For Thomas Aquinas, as for Roger Bacon (*ca.*1214–1292),[51] only sounds are transmitted through the medium via local motion, whereas the other sensibles affect our faculties by means of a distinct process of alteration, one that occurs instantaneously and without the need for local motion. In this framework, the sense of touch enjoys no unique privilege, as tactile properties are regarded as no more fixed than any other sensible qualities, apart from sound.

Yet, even if we accept that sensible qualities other than sound possess a fixed being in the real object, might we not still contend that what we perceive are merely emanations, or even representations, of those qualities? Put differently, while the quality itself may possess an *esse fixum*, its appearance might only have a transient and mutable being. As Peter Auriol succinctly encapsulated the debate:

---

[46] Averroes, *Averrois Cordubensis Commentarium Magnum in Aristotelis De Anima Libros*, II, 12, com. 127 (ed. Crawford, 1953, pp. 322–323).

[47] Averroes, *Averrois Cordubensis Commentarium Magnum in Aristotelis De Anima Libros*, III, 8, com. 39 (ed. Crawford, 1953, p. 505).

[48] Pasnau (2000).

[49] Thomas Aquinas, *Commentary on Aristotle's* De anima (transl. Foster *et alii*, 1951, p. 439).

[50] *Ibidem*.

[51] Roger Bacon, *Liber de sensu et sensato* (ed. Steele, 1937, p. 133): "Videtur quod posset dici quod licet sonus non sit bene et multum et dui permanensn habens esse fixum sicut color et odor in suo subiecto, habet tamen esse adeo permanens ad tempus [...]".

> Those who ask concerning the colours which are in a dove's neck, an image which appears in a mirror, or a candle appearing somewhere other than its location, whether these have real being or only intentional being, mean to ask whether these have only objective and fictitious or apparent being, or whether they have *real and fixed being* externally in the nature of things, independent of any apprehension. From which it is clear that [in the senses] intentional being is nothing other than objective vision or appearance.[52]

With these considerations in view, it becomes easier to understand what Autrecourt intended when he claimed that only touch apprehends a *res* with an *esse fixum* in a subject. For Autrecourt, much like Epicurus and Lucretius before him, it appears that the sensibles affect us in a manner analogous to sound—through the local motion of atoms—except in the case of tangible qualities, which affect us through direct contact. Thus, if only touch attains the *res fixa*, this implies that the other senses invariably perceive moving and changing entities, and in some cases, merely their representation—*apparentiae* in motion and flux, possessing only an objective, non-fixed being, as Peter Auriol suggests. The crucial question, then, concerns what is to be understood by the non-fixed and objective being of sensible properties other than tactile ones.

As previously indicated, Autrecourt's atomistic physics sought to reduce all physical changes to the local motion of atoms.[53]

> Thus, in natural things there is nothing but local motion; and when from such motion there follows the coming together of natural bodies which assemble with each other and receive the nature of a subject, it is called generation; when they separate, it is called corruption; and when atomic bodies are joined to a subject by local motion and are such that their joining produces no change in the subject or in what is called its natural operation, then this is called alteration.[54]

Alteration is defined here as the addition or removal of atoms that do not affect the identity of the subject. The case of light is somehow different, because when we see something at a distance thanks to light, a local motion of atoms is always involved, from the source of light to the thing illuminated and our senses.

---

[52] Peter Auriol, *Scriptum in libros Sententiarum*, I, d. 23 (ed. Pinborg, 1980, pp. 133–134; quoted and translated by Tachau, 1998, p. 96): "Sed quaerentes de coloribus iridis, aut de coloribus qui sunt in collo columbae, aut de imagine quae apparet in speculo, aut de candela apparente extra situm, utrum habeant esse reale aut intentionale tantum, intendunt quaerere utrum habeant esse obiectivum tantum et ficticium seu apparens, aut habeant esse reale et fixum extra in rerum nature absque omni apprehensione. Per quod patet quod esse intentionale non est aliud quam visio aut apparitio obiectiva".

[53] Grellard (2009).

[54] Nicholas of Autrecourt, *Exigit ordo* (ed. O'Donnell, 1939, corrected by Musu, 2009, p. 200): "Sic ergo in rebus naturae non est nisi motus localis : sed quando ad talem motum sequitur congregatio corporum naturalium quae colliguntur ad invicem et sortiuntur naturam unius suppositi dicitur generatio: quando segregantur dicitur corruptio, et quando per motum localem corpora atomalia <conjunguntur> cum aliquo supposito quae sunt talia quod nec adventus ipsorum fieri videtur ad motum suppositi nec ad illud quod dicitur operatio naturalis ejus, tunc dicitur alteratio". (My translation)

> It can be said that the ray of light itself is nothing but bodies that are naturally inclined to follow the movement of the Sun or another luminous body. If it is argued against this that a ray of light is generated instantaneously, we must reply that although it appears to be generated instantaneously, because it is generated suddenly, its production nevertheless takes time to occur. In the same way, sound, according to the common doctrine, multiplies in the medium successively, because it does so by local motion; it nevertheless appears to be generated suddenly. It is therefore not difficult to imagine that there are subtler and more penetrating bodies that seem to spread as if suddenly throughout the medium.[55]

Like sound, light is diffused through local motion of particles, even though it seems instantaneous. Thus, as he explained in relation to the error of the senses, distance can be the cause of erroneous judgements, even if the perception itself can still be considered true.[56] Autrecourt also explains, moreover, visual appearances and their variations by the local motion of particles.

> They say, indeed, that if we see a mountain twenty leagues away, this is because it causes certain realities which they call *species*. These *species* multiply themselves throughout the intervening medium, so that they are infinite in number and [can be produced] an infinite number of times; and thus they go on multiplying until they reach the sight. And the whole process takes place in an instant. In sound the production is successive and yet it seems to occur suddenly. [...] For our part, however, we do not posit that such entities are caused by objects from scratch (*de novo*). We simply say that, when an object is present to the vision, and the eyes are open, etc., and all the requisite conditions are met, a certain reality becomes present to the soul which previously was not present [to it], though it did exist beforehand. Nor do we claim that [this reality] is reduced to such atoms, but simply that, by a certain particular motion that we imagine there, it is sometimes present to this thing, sometimes not.[57]

---

[55] Nicolas of Autrecourt, *Universal Treatise* (transl. Kennedy *et al.*, 1971, p. 63, slightly modified). Cf. Nicholas of Autrecourt, *Exigit ordo* (ed. O'Donnell, 1939, corrected by Musu, 2009, p. 56): "Posset dici quod ipsum lumen nihil aliud est quam quaedam corpora quae nata sunt assequi motum solis seu etiam alicujus alterius corporis luminosi. Si autem dicatur hic instando quod lumen generatur in instanti, dicendum quod licet videatur generari in instanti, quia quasi subito, in tempore tamen est. Unde sonus secundum doctrinam communem multiplicat se in medio successive quia cum quodam motu locali; videtur tamen fieri quasi subito. Sic quoque non est difficile imaginari quod aliqua sint corpora subtiliora et magis penetrantia quae quasi subito videntur se diffundere per totum medium [...]".

[56] Nicholas of Autrecourt, *Exigit ordo* (ed. O'Donnell, 1939, corrected by Musu, 2009, p. 248): "Septima conclusio est quod si unus respiciat aliquod visibile de prope et alius a remotis magis, poterit dicere alteri: tu non vides illud quod ego video, et erit hoc verum, licet reputare possint se idem videre, eo quod visiones sunt propinquae. Similiter de auditu et similibus dicendum est". This, of course, is not true for touch.

[57] Nicolas of Autrecourt, *Universal Treatise* (transl. Kennedy *et al.*, 1971, pp. 44–45, slightly modified). Cf. Nicholas of Autrecourt, *Exigit ordo* (ed. O'Donnell, 1939, corrected by Musu, 2009, p. 90): "Dicunt siquidem si sit unus mons qui videatur a viginti leucis, hoc est quia causat quasdam realitates quas vocant species et illae multiplicant se per totum medium, ita quod sunt infinitae et infinities, et semper sic multiplicant se, donec veniant ad visum et totum fit in instanti; in sono fit successive et tamen quasi subito videtur fieri. [...] Nunc ex nostra parte non ponimus aliquas tales res de novo causari ab objectis (abiectis *ms.*), sed solum dicimus quod objecto praesente visui, oculo aperto, etc., accipe in hoc concurrenda, aliqua realitas est nunc praesens animae quae prius non erat praesens; erat tamen. Nec dicimus quod resolvatur (resolvantur *ed.*) particularia atomalia, sed solum quasi per quemdam motum specialem ibi imaginatum aliquando est huic rei praesens,

While it is clear that, for Autrecourt, vision and hearing function in the same way and are based on a local motion, two interpretations of this text can nevertheless be offered that are not mutually exclusive.

First, by "particular motion", Autrecourt may well be referring to the local motion of atoms rather than to the multiplication of *species*. The multiplication of a *species*, indeed, is not a local motion but more closely resembles the replication of cells in present-day biological thought. In this view, each *species* is a distinct entity, newly generated (*de novo*), endlessly replicating its predecessor. Autrecourt denies the production of such new entities, however, asserting instead that only local motion occurs among the eternally existing entities, namely the atoms. The second plausible interpretation of this passage would be that Autrecourt refers specifically to the local motion of atoms within the object being perceived. In this reading, the internal motion of the object's atoms would determine how the body appears, independent of any possible motion of atoms between the object itself and the observer's cognitive faculties.

These two interpretations are not mutually exclusive because both rest upon the reduction of all physical changes to the local motions of atoms. The first interpretation emphasizes that sensible objects somehow emit atoms that eventually reach the sense organs, much as we find in the accounts of Epicurus and Lucretius. The second, by contrast, attributes sensible phenomena (*apparentiae*) and their variations to changes in the arrangement and positions of the atoms within the object itself, akin to Democritus' theory of sensible qualities. Local motion is central to each case, and we remain uncertain as to whether the properties we perceive can be firmly anchored in the subject we are aiming to apprehend. The second interpretation, moreover, seems to challenge any robust form of realism about qualities, as Robert Pasnau has already observed.[58] Without attempting to conclusively adjudicate between these two readings, I would propose that it is plausible that Autrecourt aligns himself—at least in part—with Democritus' ontology of sensible qualities.

4.6 A Democritean View of Sensible Qualities?

Some commentators have tried to make Autrecourt a relativist in the tradition of Protagoras.[59] The appearances of sensible objects, it is argued, are relative to the observer's perspective, such that the same entity may present itself as bitter to one person and sweet to another. If, then, all that appears is to be deemed true, it would follow that two contradictory perceptions could simultaneously be true. It is precisely for this reason that Aristotle rejects such a position in Book IV of the

aliquando non." We modified the Latin with the ms. Oxford, Bodleian Library, Canon. Misc. 43, f. 3va.

[58] Pasnau (2011), pp. 412–414.

[59] Dennery (2009).

Metaphysics, asserting that it stands in direct opposition to the principle of non-contradiction, a principle upon which all coherent philosophical discourse must necessarily depend. This passage from the *Metaphysics* obviously serves as a major source for Autrecourt's treatment of *apparentiae*, as already noticed by O'Donnell[60] and Weinberg.[61] But Aristotle, in this text, was attacking not only Protagoras but also Democritus. When Autrecourt criticizes Aristotle's position, might we not entertain the possibility that he is endeavouring to revive the stance of Democritus, rather than that of Protagoras, as he does elsewhere, particularly in his defence of atomistic physics?[62]

To substantiate this hypothesis, one must read in conjunction the other Aristotelian texts that address the thesis ascribed to Democritus concerning the truth of appearances, particularly the first book of *On Generation and Corruption*. This text played a significant role in the fourteenth-century debates on atomism, as has been pointed out by John E. Murdoch,[63] and particularly in debates on Autrecourt's atomism, as shown by Stefano Caroti.[64] Its importance for the discussion on the ontology of sensible qualities, however, has been less emphasized.

In *On Generation and Corruption*, Aristotle explains that there is always—for Democritus—a physical explanation for the diversity of appearances of a thing that is considered to be unique and stable.

> [For Democritus] changes in the compound were thus thought to give the same thing contrary appearances to different observers. The admixtures of a small particle would effect a transposition, and if one component were transposed the compound would appear utterly different, just as "tragedy" and "trugedy" come into existence from the same letters.[65]

The Democritean theory of sensible qualities is notoriously difficult to interpret, as the extant testimonies often seem contradictory.[66] Some sources ascribe to Democritus a form of phenomenalism, suggesting that sensible qualities are appearances grounded in material reality, while others attribute to him a theory of conventionalism, holding that what we call qualities are merely linguistic conventions for designating phenomena whose underlying reality remains hidden from us. One point is clear: his theory is founded on a radical disjunction between what truly exists—matter composed of atoms and void—and what appears to us of this reality. In the Middle Ages, the conventionalist reading of this theory was not discussed, as it was primarily transmitted by figures such as Galen, Plutarch, and Sextus Empiricus, who were not widely read. However, from the late thirteenth century onwards, an interpretive approach emerged that sought to distinguish between the *esse fixum* of material bodies and the transient being of appearances, without neces-

---

60 O'Donnell (1942).

61 Weinberg (1948).

62 Grellard (2005, pp. 211–218); Robert (2023).

63 Murdoch (1999).

64 Caroti (2002, 2006).

65 Aristotle, *On Generation and Corruption*, I.2, 315b10–15 (transl. Williams, 1982, p. 4).

66 Pasnau (2007).

sarily reducing those appearances to mere subjective illusions, as seen in Peter Auriol's theory of *esse apparens*. According to this view, *apparentiae* would rather result from imperceptible changes in the internal arrangement of the object and the specific conditions of the perceiving subject at any given moment.

In commenting on this passage from *On Generation and Corruption*, some medieval philosophers therefore ask whether the qualities that appear to us in sensible experience have an *esse fixum*. This is the case, in particular, with Giles of Rome (*ca.* 1243–1316) and with an anonymous commentator who, for the most part, draws upon Giles of Rome's text.[67]

According to Giles of Rome, the Democritean theory of alteration must be understood as a form of radical reductionism. Any qualitative change that appears to us as such is merely the effect of a change in the position and order of the atoms in the thing:

> [Aristotle] first shows how [Democritus and Leucippus] saved alteration, then how they saved generation. As far as alteration is concerned, as we have said, they saved it on the basis of order and position. The fact that the atoms will be ordered in such and such a way and will have such and such a position, for example, when a higher atom becomes lower, one will say that an alteration occurs. This was reasonable enough once the principles were assumed. For if the true resides in what appears, by a displacement in the compound the same thing can appear contrary to another, like the neck of a dove which, at the slightest displacement and movement appears of this or that colour [...] He thus affirmed that as soon as a small atom arrives and is displaced the compound is changed and in a general way is altered and appears differently.[68]

All it takes is a minimal change in the thing, even at the level of a single atom, for it to appear differently to the same person. Similarly, two individuals may perceive the same object differently, as they do not perceive precisely the same atoms from their distinct perspectives. The implication of this theory of the sensible is that appearances may fluctuate constantly, even when the object itself remains virtually intact.

---

[67] Giles of Rome wrote a lecture (*sententia*) on the *De generatione et corruptione* as well as some questions (*quaestiones*) on the *De generatione et corruptione*, both edited in Venice in 1518. Cf. Giles of Rome, *Sententia et Quaestiones super De generatione et corruptione* (ed. Venezia, 1518). The anonymous commentary is found in Anonymous, *De generatione et corruptione*, Ms. Montpellier, Bibliothèque Interuniversitaire, Section de Médecine, H. 293, ff. 132ra–146rb (Q.14: *Utrum subiectum existat in apparendo, ita quod omne quod apparet sit subiectum* [ff. 137vb–138rb]; Q.15: *Utrum color fuerit aliquod ens fixum in rebus* [f. 138rb–vb]). Here we will take into consideration only Giles of Rome, because it is likely that Nicholas of Autrecourt had access to Giles of Rome's commentary at the College de Sorbonne (see Ms. Paris, BnF, Lat. 16158, which was given to the library of the College in 1310 by Nicolas de Bar-le-Duc).

[68] Giles of Rome, *Quaestiones super De generatione et corruptione* (ed. Venezia, 1518, f. 7vb): "[...] primo ostendit quomodo salvabant alterationem, secondo quomodo salvabant generationem. [...] Alterationem autem ut dictum est salvabant ex ordine et positione, quia eo quod ordinabantur aliter et aliter athomi et habebant aliam et aliam positionem, ut quia athomus superior fiebat inferior, dicebat alterationem fieri. Hoc autem erit satis rationabile supposito suo principio, nam si verum est in apparendo, tamen per translationem compositi videtur contrarium alii, ut collum columbe translatum et motum alterius et alterius coloris apparet [...]. Ideo dicebat quod uno athomo parvo adveniente et transmutato transmutatur compositum et universaliter(?) alteratur et apparet aliter".

It is entirely conceivable that two atoms within a compound might exchange positions (one previously at the top shifting beneath), without any alteration in the compound's overall figure. Yet, this would still result in a change in how the object appears to us. Thus, the material unity of an object can maintain a degree of fixity in its configuration (it remains a solid body, retaining the same quantity, figure, and weight), while its appearance changes. This raises the issue of the non-fixity of appearances, and especially of what Aristotle refers to as "sensible qualities."

According to Giles of Rome, indeed:

> When these atoms vary in their position and order, this leads to this or that quality; and from the fact that by position or order alone the quality and colour vary, Democritus and Leucippus do not mean that there is colour here, i.e., that it is something fixed in nature (*aliquid fixum in natura*), but only that by this mutation of place the thing is coloured and altered.[69]

In other words, when something is altered and undergoes a qualitative change, there is not something like a real quality inherent in a substance that would change in nature or intensity, but only a change in the atomic structure of that piece of matter.

In his questions on Aristotle's *On Generation and Corruption*, having first addressed the question of "whether everything that appears to the senses is true", Giles of Rome asks "whether colour is something fixed in nature or whether it has some entity according to truth".[70] In this context, he attempts to reformulate the Democritean thesis in Aristotelian terms. Thus, in the *quod non* section, he recalls that, like the other accidents, sensible qualities only have their being in relation to a substance, so that their unity is only a unity of attribution, not of predication; when speaking of sensible qualities (of red, for instance), we refer not to the same common nature, but only to something relative to that which truly possesses being, i.e., to a subject possessing a stable essence. If this is the case, then "neither colour nor any accident possesses any real being or fixed being, since in themselves they do not have a nature or entity".[71]

Added to this first argument, Giles of Rome introduces a second and a third argument of the same type. The second starts from the definition of colour as the limit of the diaphanous in Aristotle's *On Sense and Sensibles* (3, 439b11–13). If this definition of colour is correct, then it becomes difficult to assess the presence of colour beyond the coloured surface—in its depth, for example. It could be, therefore, that colour is merely an appearance that says little about the thing itself and that it has

---

[69] Giles of Rome, *Quaestiones super De generatione et corruptione* (ed. Venezia, 1518, f. 8rb): "Prout vero illi athomi diversificabantur in positione et ordine inducebatur alia et alia qualitas, et quia ex solu situ et ex solo ordine diversificabantur qualitas et color, Democritus et Leucippus non inquiunt colorem esse hic, id est esse aliquid fixum in natura, sed sola conversione inquiunt rem colorari et alterari".

[70] Giles of Rome, *Quaestiones super De generatione et corruptione* (ed. Venezia, 1518, respectively at ff. 52ra–rb and 52rb–vb).

[71] Giles of Rome, *Quaestiones super De generatione et corruptione* (ed. Venezia, 1518, f. 52va): "[…] si sic est, nec color nec aliquod accidens habet aliquid esse reale vel esse fixum, cum in se non habeant naturam vel entitatem aliquam".

no fixed being in nature.[72] Similarly—and this is the third argument—, colours seem to be generated by the cutting up of the bodies, as when we discover layers of superimposed colours by scraping away the surface of a painting. In this case, how can we be sure that it is not the very act of cutting that generates the colours, since they did not appear beforehand?

Giles of Rome stands in opposition to these arguments and affirms the reality of colours. According to him, colours are, first and foremost, properties associated with light, present within the diaphanous medium, but also manifest in a more substantial manner when found in an opaque and determinate body. He further suggests that they may rightly be termed *apparentiae*, if by this term we refer to the diaphanous medium itself, through which we perceive colour, which is itself quite real. Nevertheless, the phenomenon of colour remains a reality, dependent both upon the properties of the object and, to some extent, on the medium and the organ of perception.

At the time when Autrecourt was composing his *Exigit ordo* and was corresponding with Bernard of Arezzo and Giles Fieno in Paris, the Democritean theory of sensible qualities was not only known through the texts of Aristotle, but was also discussed in certain commentaries.[73] Although the textual evidence is insufficient to establish any direct link between these commentaries and Autrecourt's works, it nevertheless suggests that he may have been aware of these debates and may have drawn inspiration from them in his critique of Aristotle, thereby enacting a return to the position of the *antiqui*, among whom Democritus undoubtedly holds an important place. In any case, the vocabulary used by Autrecourt is very similar to that found in Giles of Rome. As for his own position, it closely resembles that of Democritus as reconstructed by Giles of Rome.

4.7 Conclusion

In his examination of phenomenal appearances within the philosophical debates of the fourteenth century, Oleg V. Bychkov suggests distinguishing between three main positions concerning the nature of sensible phenomena or *apparentiae*.[74] Realism upholds the genuine existence of sensible qualities, while anti-realism regards them as merely subjective illusions. Relationalism, on the other hand, defines sensible qualities—and, by extension, appearances—as relational properties arising from the interaction between external reality and our cognitive apparatus. Autrecourt likely advocates a form of relationalism, asserting that appearances are contingent upon the properties of bodies and the situational context of the perceiver.

[72] Giles of Rome, *Quaestiones super De generatione et corruptione* (ed. Venezia, 1518, f. 52va): "[...] color secundum naturam suam nihil esse aliud nisi apparentia quadam, sed quod est tale non videtur habere esse fixum in natura".

[73] For the wider context of this reception, see Robert (2024).

[74] Bychkov (2018).

The nature of this cognitive relation differs, however, depending on whichever one of the senses is brought into play. In the case of touch, the absence of distance and the immediacy of contact with the material body impart a distinct character to tactile perceptions, setting them apart from other sensible manifestations. As Autrecourt asserts, unlike other types of sensible perceptions, touch pertains not to an image but to something that exists externally in a subject (*in esse subiectivum*). For other senses, distance invariably entails local motion, resulting in what appears to us as often being merely an image or representation of reality.

In revisiting the insights of ancient atomists, Autrecourt ascribes new significance to the sense of touch, not solely in terms of its universality and utility, as Aristotle suggested, but with regard also to its epistemic value for our understanding of the sensible world. Generally, he contends that tactile knowledge provides assurance of the existence of entities outside ourselves. Yet, it is touch that furnishes us with the certainty that what we perceive is not merely an image but a material body with a fixed reality.

Admittedly, the qualities experienced through touch (such as heat or hardness) may fluctuate, just like other qualities. We can at least be assured nonetheless of the existence of the material body that gives rise to these qualitative phenomena. This re-evaluation of touch and tangible properties may imply a depreciation of the other senses and their corresponding qualities, but this is not entirely the case, as Autrecourt appears to maintain confidence in all five senses. When he employs the terms *esse fixum* and *esse subiectivum* to delineate the objects of touch, however, he contrasts this fixity not only with the local motion of sounds or the atoms of light, but also with the representations or images that result from such motions in our minds, thereby defining other sensible appearances as impermanent and evanescent.

If this interpretation is correct, it is plausible to think of Autrecourt as subscribing to a theory akin to the Democritean understanding of colours, as interpreted by medieval philosophers like Giles of Rome: that sensible qualities lack *esse fixum*. In this framework, material bodies that influence our senses are real and possess a physical and objective structure, enabling them to affect other bodies either through direct contact or via a medium through local motions of atoms. Consequently, the only certainty we can affirm is the existence of solid bodies in the external world, each possessing certain properties tied to the organization of their components and their material essence. Sensible appearances, by contrast, are nothing more than properties that emerge, as it were, from material bodies and that affect us in different ways depending on the context, i.e., the distance, the point of view, and the particular dispositions of our own bodies and our environment.

References

Primary Sources

Manuscripts

Anonymous, *De generatione et corruptione*. In Ms. Montpellier, Bibliothèque Interuniversitaire, Section de Médecine, H. 293, ff. 132ra–146rb.
Ms. Paris, Bibliothèque nationale de France, Lat. 16158.
Nicholas of Aurecourt, *Exigit ordo*. In Ms. Oxford, Bodleian Library, Canon. Misc. 43.

Early Modern Printed Editions and Modern Editions

Aristotle. (1908). J. I. Beare (ed. and transl.), *On Sense and Sensibles*. Clarendon Press.
Aristotle. (1982). C. J. F. Williams (transl.), *On Generation and Corruption*. Clarendon Press.
Aristotle. (1998). H. Lawson-Tancred (transl.), *Metaphysics*. Penguin Books.
Aristotle. (2015, first edition 1996). D. Gallop (transl.), *On Dreams*. Oxbow Books.
Aristotle. (2016). Ch. Shields (ed.), *On the Soul*. Oxford University Press.
Averroes. (1953). F. S. Crawford (ed.), *Averrois Cordubensis Commentarium Magnum in Aristotelis De Anima Libros*. The Medieval Academy of America.
Giles of Rome. (1518). *Sententia et Quaestiones super De generatione et corruptione*. Venezia.
John Buridan. (1588 [1518]). *In Metaphysicen Aristotelis*. Paris.
Nicholas of Autrecourt. (1939). *Exigit ordo*. In J. R. O'Donnell (Ed.), Nicholas of Autrecourt. *Mediaeval Studies*, *1*, 179–280.
Nicholas of Autrecourt. (1971). L. A. Kennedy, R. E. Arnold, and A. E. Millward (transl.), L. A. Kennedy (intr.). *The Universal Treatise*. Marquette University Press.
Nicholas of Autrecourt. (1994). Correspondence with Master Giles and Bernard of Arezzo. In L. M. De Rijk (ed. and transl.), *Nicholas of Autrecourt. His Correspondence with Master Giles and Bernard of Arezzo. A Critical Edition from the Two Parisian Manuscripts with an Introduction, English Translation, Explanatory Notes and Indexes*. Brill.
Nicholas of Autrecourt. (2009). *Exigit ordo*. In A. Musu (ed. and transl.), *Nicola di Autrecourt, Il* Trattato. Edizioni ETS.
Roger Bacon. (1937). R. Steele (ed.), *Liber de sensu et sensato*. Clarendon Press.
Thomas Aquinas. (1951). K. Foster *et alii* (transl.), *Commentary on Aristotle's* De anima. Yale University Press.

Secondary Literature

Beneduce, C. (2019). La fisiologia del tatto nel XIV secolo. Il caso di Giovanni Buridano. In O. Grassi & G. Catapano (Eds.), *Rappresentazioni della natura nel Medioevo* (pp. 207–220). SISMEL-Edizioni del Galluzzo.
Beneduce, C. (2022). "Is Touch One or Several?" A Late Medieval Scientific Question. In M. Gensler, M. Mansfeld, & M. Michalowska (Eds.), *The Embodied Soul. Aristotelian Psychology and Physiology in Medieval Europe between 1200 and 1430* (pp. 253–275). Springer.

Beneduce, C. (2024). *La scienza del tatto. Un percorso tra filosofia naturale e medicina*. Editrice Bibliografica.

Bychkov, O. V. (2018). The Status of the Phenomenal Appearance of the Sensory in Fourteenth-Century Franciscan Thought After Duns Scotus (Peter Aureol to Adam Wodeham). *Franciscan Studies, 76*, 267–286.

Caroti, S. (2002). *Or i est, or n' i est une*. Nicolas d'Autrécourt et le *ludus baterellorum* ou de la génération et de la corruption. In P. J. J. M. Bakker, E. Faye, & C. Grellard (Eds.), *Chemins de la pensée médiévale. Études offertes à Zénon Kaluza* (pp. 135–168). Brepols.

Caroti, S. (2006). Nicolas d'Autrécourt, la génération, la corruption et l'altération. In S. Caroti & C. Grellard (Eds.), *Nicolas d'Autrécourt et la faculté des arts de Paris (1317–1340). Actes du colloque de Paris. 19–21 mai 2005* (pp. 219–252). Stilgraff Editrice.

Dennery, D. (2005). *Seeing and Being Seen in the Late Medieval World: Optics, Theology and Religious Life*. Cambridge University Press.

Dennery, D. (2009). Protagoras and the Fourteenth-Century Invention of Epistemological Relativism. In L. Bitel (Ed.), *Visualizing the Invisible: Visionary Techniques in Religious and Cultural Contexts* (pp. 29–51), special issue of *Visual Resources, 25*(1–2).

Freeland, C. (1995). Aristotle on the Sense of Touch. In M. Nussbaum & A. Rorty (Eds.), *Essays on Aristotle's* De anima (pp. 227–248). Oxford University Press.

Grant, E. (1971). The Arguments of Nicholas of Autrecourt for the Existence of *Interparticulate Vacua. In XIIe Congrès international d'histoire des sciences* (Vol. III.A, pp. 65–68). Blanchard.

Grellard, C. (2004). Comment peut-on se fier à l'expérience? Esquisse d'une typologie des réponses médiévales au problème sceptique. *Quaestio, 4*, 113–135.

Grellard, C. (2005). *Croire et savoir. Les principes de la connaissance selon Nicolas d'Autrécourt*. Vrin.

Grellard, Ch. (2005–2006). *Sine specula sine macula*. La perception et son objet chez Nicolas d'Autrécourt. *Chôra, 3–4*, 229–250.

Grellard, C. (2009). Nicholas of Autrecourt's Atomistic Physics. In C. Grellard & A. Robert (Eds.), *Atomism in Late Medieval Philosophy and Theology* (pp. 107–126). Brill.

Kaluza, Z. (1997). Voir: la clarté de la connaissance chez Nicolas d'Autrécourt. *Micrologus, 5*, 89–105. Repr. in Kaluza, Z. (2023). *L'ontologie de Nicolas d'Autrécourt* (pp. 109–123). Brill.

Krause, A. (2009). Nikolaus von Autrecourt über das erste Prinzip und Gewißheit von Sätzen. *Vivarium, 47*(4), 407–420.

Lappe, J. (1908). *Nicolaus von Autrecourt. Sein Leben, seine Philosophie, seine Schriften*. Druck und Verlag der Ascendorffschen Buchhandlung.

Mantovani, M. (2021). The First of all Natural Sciences: Roger Bacon on *Perspectiva* and Human Knowledge. *Vivarium, 59*(3), 186–214.

Murdoch, J. E. (1999). Aristotle on Democritus's Argument against Infinite Divisibility in *De generatione et corruptione*, book I, chapter 2. In J. M. M. H. Thijssen & H. A. G. Braakhuis (Eds.), *The Commentary Tradition on Aristotle's De generatione et corruptione: Ancient, Medieval, and Early Modern* (pp. 87–102). Brepols.

O'Donnell, J. R. (1942). The Philosophy of Nicholas of Autrecourt and His Appraisal of Aristotle. *Mediaeval Studies, 4*, 97–125.

Pasnau, R. (2000). Sensible Qualities: The Case of Sounds. *Journal for the History of Philosophy, 38*(1), 27–40.

Pasnau, R. (2007). Democritus and Secondary Qualities. *Archiv für Geschichte der Philosophie, 89*(2), 99–121.

Pasnau, R. (2011). *Metaphysical Themes. 1274–1671*. Clarendon Press.

Pinborg, J. (1980). Radulphus Brito on Universals. *Cahiers de l'Institut du Moyen Âge Grec et Latin, 35*, 56–142.

Rashdall, H. (1907). Nicholas de Ultricuria, a Medieval Hume. *Proceedings of the Aristotelian Society (New Series), 8*, 1–27.

Robert, A. (2023). Tout ce qui apparaît est vrai: Nicolas d'Autrécourt, l'épicurien malgré lui. *Cahiers philosophiques, 2*, 81–101.

Robert, A. (2024). Atoms, Minima, and Points: Democritean Corpuscularism in the Thirteenth Century. In N. Polloni & S. Roudaut (Eds.), *Hylomorphism into Pieces. Elements, Atoms, and Corpuscles in Natural Philosophy and Medicine, 1400–1600* (pp. 27–65). Palgrave Macmillan.

Salvestrini, A. (2022). *Apparentia* in the Thought of Nicholas of Autrecourt: Intentionality, Intersubjectivity, and Probabilism in the Status of Mental Being. *British Journal for the History of Philosophy, 30*, 953–972.

Steiner Goldner, R. (2018). Aristotle and the Priority of Touch. In A. Purves (Ed.), *Touch and the Ancient Senses* (pp. 50–63). Routledge.

Tachau, K. (1998). *Vision and Certitude in the Age of Ockham. Optics, Epistemology, and the Foundations of Semantics (1250–1345).* Brill.

Thijssen, H. (2000). The Quest for Certain Knowledge in the Fourteenth Century: Nicholas of Autrecourt against the Academics. In J. Sihvola (Ed.), *Ancient Scepticism and the Sceptical Tradition*, special issue of *Acta Philosophica Fennica, 66*, 199–223.

Weinberg, J. (1948). *Nicolaus of Autrecourt. A Study in the 14th Century Thought*. Greenwood Press.

Wood, R. (1982). Adam Wodeham on Sensory Illusions with an Edition of *Lectura secunda Prologus, quaestio 3. Traditio, 38*, 213–252.

Chapter 5
With Fingers and Tongue: The Primacy of Touch and Tactile Qualities in Renaissance Galenic Medicine

Elisabeth Moreau

Abstract In pre-modern medicine, touch played a central role in assessing the nature of bodies. Physicians used the palm of the hand to diagnose a patient's state of health or "temperament", a concept drawn from Galenic medicine. Temperament referred to the balance of primary qualities—hot, cold, dry, and moist—which could be sensed through touch. Interestingly, Renaissance physicians applied this framework to the analysis of drug temperament to unveil a broader understanding of touch and tactility. While their views emphasized the role of additional tactile qualities such as texture and taste, they also put forward the role of the tongue, alongside the hand, in discerning temperament through touch. In this essay, I explore related conceptions of touch in the physiological and pharmacological treatises of two French physicians, Jean Fernel (1497–1558) and his commentator Jean Riolan the Elder (1539–1605). I argue that their conceptions of touch, temperament, and tactile qualities relied as much on Galen's medical texts as on Aristotle's natural philosophy about the notion of "contact senses". In doing so, they articulated a novel understanding of touch as a gateway to the knowledge of bodies at the level of their elements and material particles.

Keywords Medicine · Body · Complexion · Remedies · Drugs · Qualities · Properties · Galen · Aristotle · Taste · Touch · Senses · Sensation

5.1 Introduction

In 1567, the final version of Jean Fernel's masterpiece, the *Universa medicina*, was posthumously published as an up-to-date medical summa, which tacitly aimed to supplant previous textbooks such as Avicenna's *Canon*. Written in a didactic style and sophisticated Latin prose, the book celebrated, among other themes, the

E. Moreau (✉)
Department of History, FNRS – Université libre de Bruxelles, Brussels, Belgium
e-mail: elisabeth.moreau@ulb.be

C. Beneduce et al. (eds.), *The Sense of Touch*, Studies in the History of Philosophy of Mind 32, https://doi.org/10.1007/978-3-032-16308-0_5

fundamental role of the senses in scientific knowledge by referring to several verses of Lucretius' *On Natural Things* (*De rerum natura*).[1] While this elegant Latin poem was a fashionable reference in the Renaissance, it also suggested a materialistic and atomistic view on nature through sensory experience.[2] Following this view, touch played a prominent role in grasping the properties of bodies, which gave insight, ultimately, into their atomic building blocks. According to Pablo Maurette, such Lucretian-inspired ideas on touch and the senses were emblematic of the Renaissance and "the epistemological shift from authority-based evidence to firsthand experience, the new conception of the human body and its place in the universe, and the aesthetic sensibilities that make this period exceptional".[3] Strikingly, Maurette's claim reflects the historiographical perspective on the "scientific revolution" as the emergence of innovative conceptions of the body. Such an epistemological shift supposedly drew on alternative figures of authority—in this case, Lucretius—as opposed to the "tradition" based on Aristotelian physics and Galenic medicine. And yet, the corollary of this claim about a new emphasis on touch and the senses requires contextualization and clarification in light of Renaissance medical learning.

As will be argued in this chapter, Renaissance physicians did emphasize touch as an epistemic tool to get insight into the material composition of bodies. However, most of them followed Galen's texts and the so-called "medical tradition" in adopting this perspective. In this regard, Fernel, as a university-trained physician, shared Lucretius' emphasis on the epistemic role of touch in the investigation of nature, especially the materiality and primary components of bodies. And yet, his quotations from *On the Nature of Things* remained philological and aesthetic devices in his argument. If the French physician was attentive to the elements and properties of nature, from the human body to the ingredients of *materia medica*, he did so by referring to the ancient authorities in Renaissance medicine and the sciences: Galen and Aristotle. Before going further into this theme, it is necessary to briefly recall the role of sensory experience in the medical tradition.

In early medicine, physicians relied on their senses by following the traditional method of treatment based on reason and experience. According to Galen, the notion of experience was essential to diagnose the patient, identify illness through a series of "signs" and symptoms, and establish the active powers of remedies.[4] To do so, Galenic physicians needed to train their senses. Among these senses, touch had a prominent role in determining the constitution of patients, for instance, by examining their fluids and pulse. Such a view has been advanced in the recent scholarship on pre-modern sciences and the five senses in the Renaissance.[5] This current has counterbalanced most studies on early modern knowledge and the sciences, which have stated the prevalence of sight following the Platonic and post-Cartesian episte-

[1] Fernel, *Physiologia* (ed. Paris, 1567, reprinted in Forrester, 2003, p. 230 and p. 232).

[2] Moshenska (2014, pp. 201–205).

[3] Maurette (2014, p. 311).

[4] Van der Eijk (2008).

[5] Bynum and Porter (2004); Kambaskovic-Sawers and Wolfe (2014).

mology of the senses. In aiming to consider additional senses to sight, historians of early modern medicine have surveyed touch by focusing on medical practice at the bedside of the patient, as well as the relationship between skin, anatomy, and dissection.[6] In this chapter, I propose to consider some related theories on the nature and role of touch in Galenic medicine and Aristotelian natural philosophy.

A usual assumption in Galenic medicine was that physicians diagnosed the patient's state of health with the palm of their hand.[7] To investigate this claim, I will examine to what extent Renaissance Galenic physicians regarded touch as revealing the "nature" of bodies, that is, their composition and properties. I will elucidate this question by exploring the relationship between touch and the healthy body in physiology, and between touch and simple drugs in pharmacology. As will be argued, both medical fields brought into discussion the associated notions of "tactile qualities" and "contact senses" to clarify the epistemic status of touch in the medical tradition. I will consider two Renaissance Galenic physicians in order to trace their conceptions of touch, as well as their references to ancient sources in medicine and natural philosophy.

First, I will survey the views of Jean Fernel (1497–1558), who was a professor of medicine at the University of Paris and the court physician of Henri II of France. As expounded at the beginning of the introduction to this chapter, Fernel published a textbook of Galenic medicine, the *Universa medicina* (1567). Divided into *Physiologia* (physiology), *Pathologia* (pathology), and *Therapeutices* (therapy), this work was a major reference in early modern medicine and the sciences, and was edited many times throughout the seventeenth century.[8] Most of Fernel's explanation of touch and tactile qualities was expanded in *Physiologia*, which was first published as *On the Natural Part of Medicine* (*De naturali parte medicinae*) in 1542. Nonetheless, the French physician also considered this topic in relation to drugs in *Therapeutices*. In so doing, he provided a clear and didactic synthesis of the medical tradition concerning touch, which was addressed to a readership of humanist physicians, well-versed in scientific texts in Latin and Greek.

To Fernel's views on touch, I will couple those of his commentator, the French physician Jean Riolan the Elder (1539–1605), who was a professor of medicine at the University of Paris from 1576 to 1605.[9] A scholastic physician and a defender of the medical establishment, Riolan published polemical treatises against the recognition of surgery as a medical discipline, and against the introduction of Paracelsian alchemy into therapy and the medical curriculum. He published his commentaries on Fernel's *Physiologia* in 1579, which were later included in his posthumous *Opera omnia*.[10] In this chapter, Riolan's account will be particularly helpful for identifying the ancient and medieval sources related to Fernel's account of touch

[6] Klestinec (2016); Harvey (2003); Von Hoffmann (2023).

[7] Siraisi (1987, p. 27 and pp. 125–127).

[8] Deer Richardson (2018, pp. 153–170); Hirai (2011, pp. 46–79); Henry (2011).

[9] Benoit (1995).

[10] Riolan, *Opera Omnia* (ed. Paris, 1610).

and contact sense. Indeed, Fernel rarely stated his references in his works and, as a humanist physician, contented himself with broadly referring to Galen and Aristotle. In contrast, Riolan expanded on the diverging and converging views on physiology across the works of Galen and Aristotle, as well as their various interpretations by Arabic-Latin authorities such as Avicenna (Ibn Sīnā) and Averroes (Ibn Rushd). In doing so, Riolan attempted to reconcile the views of Aristotle and Galen on touch. Such an effort was anything but new in the medical tradition, and dated back to scholastic medicine, as shown by Pietro d'Abano's *Conciliator*.[11]

If Fernel and Riolan did not encapsulate all the medical interpretations of touch in Renaissance physiology and pharmacology, they synthesized the traditional views and debates on Galenic and Aristotelian conceptions of touch that grounded early modern medicine. In this chapter, I will show that their interpretations advanced an original conception of touch as a gateway to the knowledge of bodies at the microlevel of their matter and elemental particles. This reasoning was grounded on the Galenic understanding of tactile senses and tangible properties, which expanded on the Aristotelian approach to touch and qualities. In the first section, I will examine Fernel's and Riolan's physiology of touch as a sensory faculty, before looking into the epistemic prominence of touch for the identification of "temperament" and the primary qualities. I will argue that touch as a sense and a sensation raised broader considerations on the notion of contact (*tactus*) producing knowledge, and related tactile (*tractabilis*) qualities at the level of body parts and their core elements. In the second section, I will investigate additional tactile qualities in pharmacology that involved a broader approach to touch as a contact sense, through the assessment of drug powers.[12] This will lead me to consider the commonality of touch with other senses, in particular, taste, in Galenic medicine.

5.2 The Physiology of Touch: Temperament and Primary Qualities

Celebrated for their prominent role in the knowledge of nature, the senses were examined in relation to their physiological functioning in the body. This was, at the very least, Fernel's rationale in his *Universa medicina*.[13] In expanded on this topic, he followed Galen's viewpoint on the physiology of the senses. As part of the animal realm, the senses stemmed from the "animal faculties" of the sentient soul,

[11] Ottosson (1984); Jacquart (1998); Siraisi (2019).

[12] I have previously discussed this subject in relation to taste, color and odor in Fernel's pharmacology, in Moreau (2023).

[13] Fernel, *Physiologia* (ed. Paris, 1567, reprinted in Forrester, 2003, p. 230): "Sublatis sensibus, qui quasi fundamenta quaedam sunt scientiarum, nihil iam quicquam relinquetur, unde sumi possit demonstrationis exordium".

which resided in the brain and provided the knowledge of external things.[14] In this regard, the sense of touch (*tactio*, *tactus*, *vis tangendi*) was part of these animal faculties and sensory functions.[15] Its instrument was one of the physiological spirits, the "animal spirit" (*spiritus animalis*), which flowed in the nervous system to bring the impulse that provided sensation and motion.[16] Such an animal spirit stemmed from the "vital spirit", a vaporous substance coming from respiration and processed in the heart. Sent to the brain through the arteries, it was refined into the animal spirit, which circulated in the spinal cord, the nerves, and their fibers.[17]

Anchored in the physiological framework of the animal faculties, the sense of touch was fundamental for the knowledge of external things and, more broadly, for animal survival.[18] Fernel drew this claim from Aristotle's *On the Soul* (*De anima*), which was part of the medical curriculum in the Renaissance for its discussion of the relationship between life, soul, and the senses.[19] As we will see in this section, this treatise was a recurring reference in Fernel's and Riolan's physiological works to enhance the epistemic role of touch in the knowledge of bodies through their tactile qualities. Their views were closely connected to the anatomical location of touch in the human body, such as skin and hands.

Distributed across the entire skin (*cutis*), the sense of touch was fundamental to physicians for detecting the physical properties of bodies.[20] As Fernel argued, human touch made it possible to identify the primary qualities—hot, cold, dry, and moist—which were central to both Aristotelian natural philosophy and Galenic medicine.[21] Indeed, it was through touch that physicians discerned "temperament"

---

[14] Fernel, *Physiologia* (ed. Paris, 1567, reprinted in Forrester, 2003, p. 330): "Facultatum animalium quae non utique stirpibus, sed solis animantibus obtigerunt, [...] duo prima summaque genera existunt". See Galen, *On the Usefulness of the Parts of the Body* (transl. Tallmadge May, 1968, pp. 400–407); Aristotle, *De anima*, II.2, 413b4–13.

[15] See Siegel (1970, pp. 174–193).

[16] See also Fernel, *Physiologia* (ed. Paris, 1567, reprinted in Forrester, 2003, pp. 488–494).

[17] Fernel, *Physiologia* (ed. Paris, 1567, reprinted in Forrester, 2003, p. 466): "Omnium primum spiritus est animalis appellatus, qui ex vitali per cervicis arterias in contextum admirabilem sublato, primum elaboratur, deinde in interiore altero cerebri contextu, qui a secundis nomen habet, vi cerebri ingenita in animalem convertitur".

[18] Fernel, *Physiologia* (ed. Paris, 1567, reprinted in Forrester, 2003, pp. 354–355): "Quanquam igitur tactu omnia praedita sunt animantia, quod unus ad salutem et conservationem maxime sit necessarius, eoque dempto animal effugere non possit tractabilium qualitatum exuperantias, quarum occursu mors statim irruit".

[19] Aristotle, *De anima*, III.12, 434b10–15 (ed. Barnes, 1984, vol. I, p. 691): "An animal is a body with soul in it: every body is tangible, i.e., perceptible by touch; hence necessarily, if an animal is to survive, its body must have tactual sensation". See also Aristotle, *De anima*, II.2, 413b4–5 (ed. Barnes, 1984, vol. I, p. 658): "The primary form of sense is touch, which belongs to all animals".

[20] Fernel, *Physiologia* (ed. Paris, 1567, reprinted in Forrester, 2003): "Nam tactionis sensus quem omnis ex aequo animantes participant, nervis nervorumve fibris in omne corpus diffusis indigebat: quibus susceptae tractabiles qualitates cum spinae medulla et cerebro, quae [...] sensus et motus principia sunt, continuatione communicantur".

[21] Fernel, *Physiologia* (ed. Paris, 1567, reprinted in Forrester, 2003, p. 230): "Ergo quod ἐνέργεια, id est re ipsa, aut quod dicere solent actu, calidum, frigidum, humidum, siccumve est, id tactu liceat

(*temperamentum, temperies, temperatura*).[22] Also called "complexion" in medieval medicine, the temperament determined the state of health and bodily constitution. According to the Galenic tradition, it resulted from the mixture of the primary qualities. Patients had their temperament characterized by the prominence of one or two of these qualities, which needed to be restored to a balanced state through adequate treatment.[23] To diagnose this temperament, physicians relied on their senses by touching the patient with the palm of the hand.[24]

The emphasis on the hand (*manus*) as the physician's main instrument has raised longstanding considerations in the medical tradition.[25] As Fernel claimed, the hand was the instrument of knowledge par excellence, which nature had only attributed to humankind alone. Such a statement drew on Aristotle's conception of touch, which Fernel's commentator expounded as well.[26] Indeed, for Riolan, the physician diagnosed the temperament of beings with the skin of the hand in the same way as an architect checked angles with a square.[27] As Fernel recalled, it was the skin of the fingers that was the most temperate, hence more adapted to be used in medical practice to identify the nature of bodies.[28] In his turn, Riolan specified that the softer part of the skin had more nerves, most notably on the fingertips and the palm of the hand, which felt touch more accurately.[29]

internoscere, non leonis quidem aut equi aut canis, sed solius hominis tactu, qui tum dignitate tum temperamenti mediocritate caeteris antecellit".

[22] Fernel, *Physiologia* (ed. Paris, 1567, reprinted in Forrester, 2003, pp. 194–197): "Unde cum sit tactus sensuum primus, isque plurimum corporeus, quae hunc feriunt qualitates omnium primae sunt et maxime corporis participes. [...] Sed quae solum tactum concitant, qualitates omnium sunt potentissimae. Ex harum autem genere aliae sunt principes, aliae ex his ortae: quae in molli, duro, crasso, tenui, laevi, aspero sunt corpore, quae variae dicuntur et quasi multiformes".

[23] Siraisi (1987, pp. 101–104).

[24] Stolberg (2021, pp. 56–58).

[25] Siraisi (1987, p. 103); Ragland (2022, p. 124 and p. 135); Harvey (2003); Von Hoffmann (2023, pp. 504–505).

[26] Aristotle, *De anima*, II.9, 421a20–24 (ed. Barnes, 1984, vol. I, p. 670): "While in respect of all the other senses we fall below many species of animals, in respect of touch we far excel all other species in exactness of discrimination. That is why man is the most intelligent of all animals".

[27] Riolan, *Opera Omnia* (ed. Paris, 1610, p. 20): "Manus inquam, est instrumentum instrumentorum. [...] Nullum est quod illius opera homo sibi non affabrefaciat, manu tangit et apprehendit, ingenio comprehendit omnia. [...] Cute, ut [...] architectus amussi, obliquum a recto, (est enim rectum iudex sui et obliqui) homo rerum omnium temperamenta dignoscit".

[28] Fernel, *Physiologia* (ed. Paris, 1567, reprinted in Forrester, 2003, p. 232): "Is autem cum per totam cutem aequabiliter fusus sit, quae rerum externarum appulsus sentiat, palam fit hanc quoque omnium extremorum vere mediam haberi, et omnis superlationis expertem, eam praesertim quae manum integit, maximeque summos digitos".

[29] Riolan, *Opera Omnia* (ed. Paris, 1610, p. 59): "Cutis, ministerio spiritus pellucidi excipit obiecti qualitatem, receptam ad primum sensorium transmittit nervorum ministerio. Ubi autem mollior est et temperatior, vel plures, aut acutior sensus nervos excipit, ut in extremis digitis, et in vola manus [...] perfectius sentit." See Galen, *On the Usefulness of the Parts of the Body* (transl. Tallmadge May, 1968, p. 125).

Powered by the innervation, softness, and temperate nature of the fingers, hand touch was essential, in medical practice, to appraise temperament through the tactile qualities of bodies. Such qualities were often called "tractable" (*tractabilis*), likely because these were the main properties that physicians could handle in their practice. On this point, Riolan gave insight into the relationship between touch, qualities, and the Aristotelian physics of matter and form. Such an appeal to the Aristotelian natural philosophy was developed in Galen's treatises such as *Mixtures* (*De temperamentis*) and *On Simple Drugs* (*De simplicium medicamentorum facultatibus*), and shaped the medical account of touch. Accordingly, Riolan, as a scholastic physician, defined the tactile or "tractable" qualities as tangible properties that were "accidental" to the extent that they pertained to the matter of bodies instead of their essence or "form". Whereas these sensory qualities were not the ontological features of beings, physicians considered them as the first qualities in the context of their practice.[30] This also entailed that the physician's sensory appraisal was limited to the identification of qualities that were operative or "in actuality" in the body.[31] As we will see in the second section, the knowledge of qualities that were latent or "in potentiality" was central in pharmacology and implied a different approach to touch.

To adjust their diagnosis of temperament, Galenic physicians trained their tactile skills by combining reason and experience. This rational–empirical approach to touch was anchored in Galen's exposition of the relationship between touch and qualities at the scale of the body parts and humors.[32] As Fernel noted, physicians recognized the hotness or coldness of the whole body, as well as its moistness or dryness, according to its soft or hard touch. However, when it came to assessing the temperament of body parts, the epistemic status of touch was more complex. Fernel and Riolan raised this question by classifying the temperament of body parts on a scale. This scheme spanned from body parts with a hot temperament, such as blood and the heart, to cold parts, such as fat, phlegm, and the brain. Moreover, Fernel enumerated the temperament of each organ and body part according to touch and reason, through the knowledge of their substance and composition.[33] Within this scheme, the temperament of some body parts could be inferred by touch only, whereas some theoretical knowledge and judgment was required in the case of other body parts, as touch alone could not always be trusted.

---

[30] Riolan, *Opera Omnia* (ed. Paris, 1610, p. 7): "Prima repugnantia ad sensum perspicua, illa est, quae primo et communi sensu, nempe, percipitur: tractabiles primae qualitates sunt: […] illae igitur medico prima erunt contraria, non absolute, inquam, sed ad sensum prima".

[31] Galen, *Mixtures* (transl. Singer, 1997, p. 239): "I therefore use the sense of touch to establish whether an object is hot in actuality; the case of an object which is not hot in that sense, but is apt to become so, and may thus be termed potentially hot, is one which I would attempt to investigate by logical methods".

[32] Galen, *Mixtures* (transl. Singer, 1997, pp. 240–246); Von Hoffmann (2023).

[33] Fernel, *Physiologia* (ed. Paris, 1567, reprinted in Forrester, 2003, p. 232): "Ut primum tactus aestimatione, sic rursum substantia et compositione, partium corporis humani temperamenta noscuntur".

Interestingly, this classification highlighted the uncertainty of touch and the need for a rational method in relying on sensory experience. Indeed, the immediate information provided by touch had often to be mitigated by the knowledge of organs and tissues, as well as fallacious sensations brought by surrounding organs. For instance, lungs might feel hotter than their actual temperament because of their proximity to the heart.[34] This reasoning followed Galen's considerations on the use of touch for diagnosing temperament. As expounded in *Mixtures*, the rational appraisal by touch applied to the heat and coldness of body parts. Physicians could still infer their dryness or moistness from their immediate sensation.[35] Nonetheless, Fernel noted some discrepancies between Galen and Aristotle about the temperament of the brain, the lungs, and fat. As for Riolan, he added the divergences between Avicenna's *Canon* and Averroes's *Colliget* on the order of "hot" and "cold" body parts, by discussing to what extent their temperament could be inferred by touch.[36] In agreement with Fernel, he recalled that touch always deceived if not governed by reason.[37]

Despite its ambivalent status in appraising the temperament of body parts, touch played a pivotal role in the Galenic exposition of the body's microstructure. Indeed, the medical approach to reason and experience opened onto the relationship between touch, tactile qualities, and the elements (*elementa*) in reference to Galen's *On the Elements According to Hippocrates* (*De elementis secundum Hippocratem*). In the medical tradition, the primary qualities were related to the four elements, which were the main components of the body. As Fernel explained, touch contributed to proving their existence because of the sensation brought by their opposite qualities.[38] Such a claim relied on Galen's *On the Elements* which stated, following Hippocrates' *Nature of Man* (*De natura hominis*), that common sensations such as pleasure and pain could not occur without the existence of qualities with some contrary nature.[39] As Galen added, these opposite qualities were overall haptic (ἁπταὶ) and subsumed all the other sensory properties of bodies.[40] Furthermore, it was from the primary qualities that Galen derived the four elements as the main constituents

[34] Fernel, *Physiologia* (ed. Paris, 1567, reprinted in Forrester, 2003, p. 234): "Porro autem pulmonis sive compositionem, sive calidum biliosimque nutrimentum, sive cordis vicinitatem et adhaesionem respicies, illum sane iecore tum calidiorem tum multo sicciorem iudicabis". See also Fernel, *Physiologia* (ed. Paris, 1567, reprinted in Forrester, 2003, pp. 240–242).

[35] Fernel, *Physiologia* (ed. Paris, 1567, reprinted in Forrester, 2003, p. 234): "Haec posterior humidorum siccorumque series tactus iudicio confirmata est, cui et substantiae compositio prorsus consentit".

[36] Riolan, *Opera Omnia* (ed. Paris, 1610, p. 21).

[37] Riolan, *Opera Omnia* (ed. Paris, 1610, p. 23): "Fallit saepe tactus nisi regatur ratione".

[38] Fernel, *Physiologia* (ed. Paris, 1567, reprinted in Forrester, 2003, pp. 200–201).

[39] Fernel, *Physiologia* (ed. Paris, 1567, reprinted in Forrester, 2003, p. 194): "Unde cum sit tactus sensuum primus, isque plurimum corporeus, quae hunc feriunt qualitates omnium primae sunt et maxime corporis participes: quod certe efficit ut ab his summis doloribus afficiamur aut voluptatibus". See Galen, *On the Elements According to Hippocrates* (transl. De Lacy, 1996, pp. 63–77).

[40] Galen, *On the Elements According to Hippocrates* (transl. De Lacy, 1996, p. 133): "And it is obvious to everyone that the changes of matter do not follow on visual, auditory, gustatory, or olfactory qualities; these too are generically tactile".

of bodies. At first, this reasoning was grounded in Aristotle's conception of touch in *On the Soul*: a sense stemming from compound bodies, i.e., bodies made of qualitatively distinct elements.[41] More broadly, it enhanced tactile experience as an epistemic gateway to the material structure of bodies at a micro-level that was not empirically accessible.

Conversely, the four elements pointed to the peculiarity of each sense and the materiality of the sensory object. As Fernel explained, the sensitive organs had a particular temperament through their connection with one of the four elements. In this regard, touch and taste both had a terrestrial nature.[42] This correlation between elements and the senses had a long philosophical tradition. The description of touch as an earthy sense referred to Galen's *On the Doctrines of Hippocrates and Plato* (*De placitis Hippocratis et Platonis*), especially his discussion of Plato's epistemology of the senses in *Timaeus*.[43] Whereas Plato had oriented the philosophical tradition towards the primacy of sight as the noblest form of perception, Galen implicitly followed Aristotle's accent on touch by taking up his attribution of the element earth to both taste and touch.[44] In turn, Fernel and Riolan associated the broader notion of tangibility, which pertained to both touch and taste, with earth as a cold and moist body. As Riolan explained, the linkage between touch and earth stemmed from the sensory requirement for solidity and consistency, while denoting a moist nature that was predominant in the case of taste.[45] For its commonality with touch, the sense of taste is further examined in the next section.

5.3 From Touch to Contact Sense: Texture and Taste in Pharmacology

To understand Fernel's conception of tactile senses in pharmacology, it is important to recall the commonality of touch with taste in the medical tradition. This theme was part of the medieval debates on the multisensory nature of touch, in the

[41] Aristotle, *De anima*, III.12, 434b10–12 (ed. Barnes, 1984, vol. I, p. 691): "But if a body has sensation, it must be either simple or compound. And simple it cannot be; for then it could not have touch, which is indispensable".

[42] Fernel, *Physiologia* (ed. Paris, 1567, reprinted in Forrester, 2003, p. 334): "Tactui instrumentum datum est plane terrenum, gustandi (quia ceu tactus quidam est) id etiam terrenum, sed humoris quoque alicuius particeps".

[43] Galen, *On the Doctrines of Hippocrates and Plato* (transl. De Lacy, 1978, p. 469): "[Plato] says that the body of the whole was made from fire and earth, so that it might be visible and tangible. Therefore in animal bodies also the organ of touch is earthy, the organ of sight fiery".

[44] Aristotle, *De sensu*, 2, 438b30–439a4.

[45] Riolan, *Opera Omnia* (ed. Paris, 1610, p. 59): "Respondetur, sensorium dici […] vel terrestre, ut tactus: non quia elementis omnibus non sit temperatum, sed quia unum praedominatur, ut humor in gustu: quandoquidem sola humida sapida, nec sine humore sapor gustatur: terra in tactu, quia soliditatem aliquam et consistentiam habere debent, quorum differentiae contactu discernuntur".

footsteps of Aristotle's discussion of touch as a single or a composite sense.[46] Among the scholastic issues at stake in Renaissance medicine, I will envisage the multisensory nature of touch as part of the broader category of "contact senses", which included taste, as well as the perception of tactile properties.[47] As we will see in this section, taste was considered a certain type of touch. Its instrument, the tongue, played a similar role to the hand in assessing the properties of drugs. Moreover, the commonality between touch and taste involved the status of primary and secondary qualities, both as tangible properties. As a reminder, the "secondary qualities", in Aristotelian natural philosophy, were ancillary to the primary qualities, and included textural features such as softness and hardness, smoothness and roughness.[48] In this section, I argue that the medical conception of taste offered an alternative viewpoint on these qualities. In this regard, the stances of Fernel and Riolan on touch, taste, and tangible qualities applied Aristotle's natural philosophy to Galenic medicine, with important consequences in pharmacology.

If Fernel did not dwell on the medieval discussion on the multisensory nature of touch, this topic was underlying his exposition of the commonality between touch and taste as for their anatomical location. Along with Riolan, he considered both touch and taste as tactile senses, which did not need an external "medium". In contrast, the other sensations needed to reach the external sensory organs—nose, ears, eyes—through some vapor, air, or aether.[49] However, by touch, the body could directly feel the primary qualities, which were processed by the brain through the nerves in the flesh. On this point, Riolan adopted the anatomical location of touch in the heart following Aristotle's related conception. Yet, following Galen, he considered the brain and nerves as the physiological core of external sensation, which expanded to flesh (*carnem*) as an internal organ.[50] Such a conception of touch also applied to taste (*gustus*): Fernel and Riolan agreed that it was a certain type of touch

---

[46] For the medieval debate, see Beneduce (2022). For Aristotle's discussion on touch as a single or plural sense, see Aristotle, *De anima*, II.11, 422b20 and III.1, 424b22–425a13.

[47] Riolan summarized the scholastic debate on the unicity or plurality of touch according to some stances he attributed to Aristotle, Alexander of Aphrodisias, Simplicius, and Avicenna—see Riolan, *Opera Omnia* (ed. Paris, 1610, p. 58).

[48] Fernel, *Physiologia* (ed. Paris, 1567, reprinted in Forrester, 2003, p. 332): "Caeterum vis tangendi nullo proprio conclusa organo coercetur, sed quum tactus ad animantis vitam maxime videtur necessarius, toto corpore aequabiliter fusus est. In his autem nervis potissimum collocatur, qui duriores iam facti intro forasque commeant: per quos corporum omnium tractabiles qualitates animadvertit [...] quarum extrema sunt calor et frigus, humor et siccitas, mollitudo durities, asperitas laevitas, gravitas et levitas". See Pasnau (2011).

[49] Fernel, *Physiologia* (ed. Paris, 1567, reprinted in Forrester, 2003, pp. 472–473): "Gustus autem et tactus medio corpore non indigent, sed coniunctione quadam et contrectatione functiones suas obeunt. Etenim si vel vapor in gustandi nervum, vel calor pungensve stylus in nudum apertumque tangendi nervum incidit, eximie et acriter sentitur, quanquam nulla caro media intercedit". See Galen, *On the Usefulness of the Parts of the Body* (transl. Tallmadge May, 1968, p. 405).

[50] Riolan, *Opera Omnia* (ed. Paris, 1610, p. 21): "Aristoteles ipse videtur suis interpretibus dedisse occasionem disputandi, modo enim ait cor esse primarium, atque, ut ita dicam, originale sive radicale principium, quia nervos inde oriri existimavit et sentiendi vim per nervos ad cutem influere, quod de cerebro affirmant medici: caro est medium internum, quando enim Aristoteles scribit contactum fieri absque medio [...] sentit enim nervi ministerio vi sensifica influente a cerebro ad interiectam carnem".

(*quidam tactus*), which did not require a medium because it used the tongue (*lingua*) as an internal organ of sensory perception.[51]

The Galenic approach to taste as a contact sense, with the tongue as its main instrument, was built on Aristotle's account in *On the Soul* as well as *On the Senses* (*De sensu*) in the so-called *Parva Naturalia*.[52] Both treatises were centered on sensory functions and sensations as physiological operations of the body and the soul.[53] In the Renaissance, they were particularly relevant for the integration of Aristotelian scholarship within the medical curriculum. From these treatises, Fernel and Riolan enhanced several commonalities between touch and taste. First, in the same way as touch, taste was a "necessary" sense indispensable to animal survival.[54] According to this view, animals that could not distinguish food from poison were doomed to death.[55] In addition, taste, like touch, was related to the element earth as a tangible body, although it also required moisture.[56] As Fernel noted from a Galenic viewpoint, some humor was indispensable to the correct perception of taste, thanks to the spongious flesh of the tongue.[57] What emerged from these different statements was the medical understanding of taste as a contact sense providing tangible, hence reliable sensations, which were conveyed by the tongue and the saliva for the perception of flavors. Remarkably, this approach to taste went beyond physiology and extended to the discussion of drug powers.

Building on the conception of taste as a contact sense and an important tool for differentiating food from poison, Galenic physicians elevated taste in one particular medical field: pharmacology. Indeed, tasting was fundamental for assessing the

---

[51] Riolan, *Opera Omnia* (ed. Paris, 1610, p. 58): "Lingua est instrumentum gustus […]. Non videtur medium esse necessarium: gustus enim est quidam tactus, contactus autem fit absque medio".

[52] Aristotle, *De anima*, II.4, 414b10–12 (ed. Barnes, 1984, vol. I, p. 660): "Flavours fall within the field of tangible qualities". Aristotle, *De anima*, II.9, 421a15–20 (ed. Barnes, 1984, vol. I, p. 670): "Our sense of taste is more discriminating because it is a sort of touch, which reaches in man the maximum of discriminative accuracy". See Aristotle, *De sensu*, 4, 441a1–5.

[53] Polansky (2007, pp. 313–320); Polansky (2024).

[54] Aristotle, *De anima*, III.12, 434b15–22 (ed. Barnes, 1984, vol. I, p. 691): "All the other senses […] apprehend through media; but where there is immediate contact the animal, if it has no sensation, will be unable to avoid some things and take others, and so will find it impossible to survive. That is why taste also is a sort of touch […] because it is the sense for what is tangible and nutritious".

[55] Riolan, *Opera Omnia* (ed. Paris, 1610, p. 21): "Gustus et tactus sunt sensus necessarii ad vitam, quibus nullum animal sine manifesta vitae iactura carere potest, utroque enim utilia ab inutilibus, ut alimenta a venenis, discernit". On the same line of argument, see Monika Mansfeld's chapter in this volume.

[56] Aristotle, *De anima*, II.10, 422a9–11 (ed. Barnes, 1984, vol. I, p. 671): "What can be tasted is always something that can be touched, and just for that reason it cannot be perceived through an interposed foreign body, for no more is it so with touch. Further, the flavoured and tasteable body is suspended in a liquid matter, and this is tangible".

[57] See Fernel, *Physiologia* (ed. Paris, 1567, reprinted in Forrester, 2003, p. 98): "Substantia illi [linguae] rara atque laxa, ut humore saporum vectore facile imbuatur". Fernel, *Physiologia* (ed. Paris, 1567, reprinted in Forrester, 2003, p. 332): "Gustandi facultas consedit, rerum sapores internoscens, spongiosae eius carnis adminiculo qua lingua praedita est".

powers of "simple" drugs (*simplicia medicamenta*) used in therapy. Physicians needed to identify their temperament, that is, their predominant qualities, in order to prescribe efficient remedies. As a reminder, the notion of "temperament" was proper to all natural beings, not only the human body, and was key for the identification of the active powers of drugs. For instance, a simple drug with a cold and moist temperament was ideal to cure a hot and dry ailment. Following this scheme, the pharmacological tradition discussed drug powers in reference to Galen's *On Simple Drugs*.[58] Among these powers, two types were perceptible by the senses. The first type of drug power included the temperament of drugs, as a mixture of the primary qualities, and the second type included their "matter" stemming from their secondary qualities, especially their texture. In the Renaissance, Galenic physicians named these types of active powers first and second "faculties" (*facultates*), respectively.

From Galen's pharmacology, the medical tradition put forward the role of taste to assess the sensory qualities of drugs, and hence their active powers. A key reference was the fourth book of *On Simple Drugs*, which examined the role of sensations, overall taste, in recognizing drug powers. As Galen explained, physicians should train their sense of taste by carefully masticating and memorizing the flavors of ingredients. In so doing, they would be able to distinguish the different flavors of simple drugs to apply a rational method of knowing their qualities.[59] Other sensations, such as odor and color, also played a role in the treatise, but were considered less pertinent for the identification of drug temperament, implicitly, because of their lack of tactility.[60] Indeed, in gustation, all the particles of the tasted bodies similarly touched the tongue and moved the sense according to their nature.[61] Following this view, Galen emphasized the role of taste for building a rational pharmacological method to assay drug powers. Interestingly, this claim expanded on his conception of haptic knowledge by hand touch, which was developed for the identification of temperament. This statement, in turn, was developed in late medieval pharmacology, through Arabic-Latin sources such as Avicenna's *Canon*, which stated taste as the safest means to infer the active powers of simple drugs.[62]

Taste was thus central in the pharmacological method to systematically identify the active powers of drugs, especially their first and second faculties. This method relied on a correspondence between flavor, temperament, and texture. In

---

[58] Martelli et al. (2020); De Vos (2021); Galen, *De simplicium medicamentorum temperamentis ac facultatibus I–VI* (ed. Kühn, 1826).

[59] Galen, *De simplicium medicamentorum temperamentis ac facultatibus I–VI* (ed. Kühn, 1826, p. 632): "Sed etiam gustandi exercere sensum in saporibus oporteat, principioque in iis exquirendis solicitos esse quae unam quampiam habeant qualitatem manifestissimam. Ut si acrimoniam cogitatione complecti clare velis, [...] efficies, assidue gustando, plurimumque dentibus mandendo, tum ita longo usu passionis existentis sensum memoriae diligenter mandando atque infigendo".

[60] Galen, *De simplicium medicamentorum temperamentis ac facultatibus I–VI* (ed. Kühn, 1826, pp. 700–703).

[61] Galen, *De simplicium medicamentorum temperamentis ac facultatibus I–VI* (ed. Kühn, 1826, p. 702): "In gustu autem omnes similiter gustabilium corporum particulae in linguam incidunt, sensumque movent pro sua natura singulae".

[62] McVaugh (2017).

expounding this scheme in his *Therapeutices*, Fernel explained drug assay through taste since their very ingestion in the mouth.[63] Indeed, drug faculties were activated by the heat and saliva of the mouth, which enabled their detection through their flavor.[64] On this point, Fernel took up Avicenna's types of flavors (*sapores*)—acrid, acidic, fatty, salty, austere, sweet, bitter, acerbic, in addition to insipid—each with a corresponding temperament and texture.[65] For instance, fatty drugs had a moderate temperament and thin matter, which displayed the faculty of loosening, softening, and emolliating the body. Bitter drugs had a hot and dry temperament with a thick matter, which pointed to the faculty of cleansing wounds and ulcers. In this regard, Fernel followed Galen's advice to physicians to train their sense of taste by specifying how flavors touched the tongue. For instance, an acrid taste struck and pricked the mouth and tongue with pungency, while a sweet taste coated them with a certain viscosity.[66] In consequence, taste was understood as a type of touch that not only allowed to detect the temperament of drugs through their primary qualities, but also their material texture through their secondary qualities, most notably their texture. As Riolan noted, such gustatory and textural features were also part of the "tractable qualities" that physicians detected through touch.[67]

Moreover, the Galenic conception of taste revealed a different approach to touch in pharmacology, as it implied an alternative perspective on the notion of qualities. Whereas the primary qualities in actuality could be assessed by hand touch, those of simple drugs were only in potentiality and needed to be appraised by taste.[68] As Galen put forward, it was the phase of ingestion, especially the processing of these bodies by the heat and moisture (*saliva*) of the mouth, which activated their properties and allowed them to be recognized by taste.[69] In the Renaissance, these properties included the first and second types of drug faculties, at the exclusion of hidden or "occult" powers that escaped the senses and could only be understood with hindsight through multiple trials. Hence, flavors were deemed key tactile sensations that

---

[63] Fernel, *Universa Medicina* (ed. Paris, 1567, pp. 344–557).

[64] Fernel, *Universa Medicina* (ed. Paris, 1567, p. 411): "Ut autem facultates secundae, ita et medicamentorum sapores ex eorum materia primis qualitatibus instructa prodeunt: quumque horum magna sit ex origine affinitas, optima ratione sapores indices erunt et internuncii primarum atque secundarum facultatum". See also Fernel, *Universa Medicina* (ed. Paris, 1567, pp. 411–414).

[65] Moreau (2023).

[66] Fernel, *Universa Medicina* (ed. Paris, 1567, p. 412): "Sapor acer est qui linguam et os acrimonia ferit atque compungit. […] Pinguis sapor […] linguam orisque partes lentore quodam oblinit".

[67] Riolan, *Opera Omnia* (ed. Paris, 1610, p. 327): "Gustabiles sunt, sapores ingrati, qui palatum, et linguam inficiunt […]. Ad qualitates tractabiles pertinent, affectus simplices cutis, ut laxitas, durities, asperitas, quae duplici nomine tractabiles dicuntur, primo, quia cutem immutant, et afficiunt tactus sensorium, secundo quia a medicis contactu percipiuntur".

[68] Aristotle, *De sensu*, 4, 441b20–24 (ed. Barnes, 1984, vol. I, p. 701): "Savour brings into actual exercise the perceptive faculty which pre-existed only in potency".

[69] Fernel, *Physiologia* (ed. Paris, 1567, reprinted in Forrester, 2003, p. 472): "Itaque sapor succo humorique permistus, aut […] in humore constitutus (eius enim qualitas quaedam est et affectio) simulac adhibetur admoveturque linguae, diffusus cum humore gustum incitat". See Galen, *Mixtures* (transl. Singer, 1997, p. 268).

signaled primary and secondary qualities. This reasoning stood in contrast to the Aristotelian classification of flavor and texture as "secondary qualities". By comparison, Galenic pharmacology granted flavors a prominent epistemic status, viewing them as indicators of a drug's active powers.

Remarkably, the pharmacological focus on taste put forward texture as a tactile quality that reflected the microstructure of drugs. The texture and density of drugs as solid, liquid, or aerial bodies pointed to the level of thinness or thickness of their portions. As Fernel showed in his *Therapeutices*, texture reflected the ability of elemental parts to flow as thin, mediocre, or thick portions, following Galen's description of thin parts (λεπτομερής) and thick parts (παχυμερής) in *On Simple Drugs*.[70] Because of their minute structure in particles, drugs could transform the body by condensing or coagulating its humors and, conversely, by moistening, penetrating, or dissolving these very humors.[71] Galen's reasoning was centered on the ability of simple drugs to take effect or to mix into efficacious compounds if they were reduced to their smallest parts. In suggesting the ability of drugs to penetrate the tissues of the organism, this scheme unveiled the natural divisibility of bodies into elemental particles, as well as their porosity. This added up to the insight into the composition of bodies in the four elements, which hand touch revealed in the case of the body's temperament. With taste, too, physicians could gain knowledge about the elements through the materiality and tactile properties of bodies.

5.4 Conclusion

As has been argued, Fernel and Riolan followed Galen's emphasis on touch and tactile senses in their conception of physiology and pharmacology. In the context of Renaissance medicine, their interpretation built on the role of touch in the diagnosis of the patients' constitution, and on taste in the assay of drug powers. Conveyed by the hand and the tongue as the physician's main instruments, the sense of touch raised further considerations on the method of diagnosing temperament through reason and experience, and on the analysis of drugs through tangible properties. From Fernel's and Riolan's explanations, it emerges that haptic experience, including touch and taste, was understood as an epistemic gateway to the microstructure of bodies at the level of their elements and particles. Besides, Fernel's and Riolan's views shed new light on the epistemic status of secondary qualities, in particular, textural features. Indeed, thickness and viscosity were

[70] Debru (1997); Galen, *De simplicium medicamentorum temperamentis ac facultatibus I–VI* (ed. Kühn, 1826, p. 385 et passim).

[71] See for instance Fernel (ed. Paris, 1567, pp. 412–413): "In mediocri materia [salsus sapor] consistit cum calore et siccitate. Aqueae enim materiae, (quae non exquisite simplex est) partes quasdam terrenas permistas calor externus tandem praeassans et urens atque siccans, salsum saporem inducit. [...] Quocirca salsum moderate penetrat et incidit, corrandendo stimulat et expurgat, valide siccando tenues ichores absorbet, corpora conservat et a putredine tuetur".

granted a central role in Galenic pharmacology for inferring the material composition of drugs.

As shown through our examination of Renaissance physiology, the emphasis on touch was grounded in the conception of temperament as a mixture of the four primary qualities. The latter, as tactile properties, revealed the composition of bodies into the four elements. Such a claim referred to Galen's works, overall *Mixtures* and, to a lesser extent, *On the Elements According to Hippocrates*, following Aristotle's conception of touch in *On the Soul* and *On the Senses*. From these Galenic and Aristotelian works, Renaissance physicians like Fernel and Riolan took up a series of *topoi* concerning the epistemic role of touch. Both put forward the superiority of human touch through the hand as an instrument of knowledge par excellence, as well as the essentially tactile nature of the primary and secondary qualities. Moreover, their anatomical description of touch as a medium-less sense, since skin provided direct access to the object of sensation, opened onto a broader understanding of touch as a contact sense that shared common grounds with the sense of taste.

Indeed, our exploration of Galenic pharmacology has highlighted a conception of haptic knowledge that encompassed Aristotle's notion of contact sense and the broader notion of tactility. This perspective positioned taste and texture as key tactile objects in drug analysis. Grounded in Galen's *On Simple Drugs*, this reasoning gave the sense of taste a prominent role in assaying the active powers of drugs. Its main instrument, the tongue, also pertained to the physician's expertise by touch. Thus, in Renaissance Galenic medicine, the pharmacological method of tasting functioned as a reliable epistemic tool for assessing bodily conditions through tactile qualities. Texture, in particular, pointed to the materiality of bodies and their structure into particles of diverse size. This line of reasoning reinforced the connection between tactile experience and the microstructure of bodies into elements.

To develop a haptic expertise, Renaissance Galenic physicians trained their senses, mostly through hand touch and taste, by combining their perception with a systematic knowledge of nature. In this respect, Fernel and Riolan suggested the need to use knowledge and experience in assessing temperament by hand touch. Moreover, they showed that taste offered the advantage of detecting qualities that were in potentiality within bodies, since they were activated during their ingestion in the mouth. In consequence, physicians also needed to hone their sense of taste to determine the active powers and material composition of drugs through their flavors. This conception of taste supported the development of a pharmacological method of treatment aligned with the tenets of Aristotelian natural philosophy concerning touch and the notion of contact sense. As a result, the epistemic role of taste and the tongue was integrated with that of hand touch, to form a tactile approach to bodily diagnosis and therapeutic practice in the late Renaissance.

Acknowledgements This chapter was written with the support of an FWA Postdoctoral Research Fellowship in the Department of History and Philosophy of Science (HPS) at the University of

Cambridge. I am grateful to Chiara Beneduce and Mattia Mantovani for their insightful comments and for the opportunity of presenting the initial version of this research at *The Sense of Touch* conference at Radboud University (Nijmegen) in 2023.

References

Primary Sources

Aristotle. (1984). In J. Barnes (Ed.), *The Complete Works of Aristotle: The Revised Oxford Translation*. Princeton University Press.

Fernel, J. (1567). *Universa Medicina*. Paris.

Fernel, J. (1567). *Physiologia*. Reprinted and Translated in J. M. Forrester (2003), *The Physiologia of Jean Fernel (1567)*. American Philosophical Society Press.

Galen. (1826). De simplicium medicamentorum temperamentis ac facultatibus I–VI. In K. G. Kühn (Ed.), *Claudii Galeni Opera Omnia* (Vol. 11–12, pp. 619–703). Knobloch.

Galen. (1968). M. Tallmadge May (transl.), *On the Usefulness of the Parts of the Body: Peri Chreias Morion: De Usu Partium. [Book 1–8]*, Vol. I. Cornell University Press.

Galen. (1978). P. De Lacy (transl.), *On the Doctrines of Hippocrates and Plato*. Akademie Verlag.

Galen. (1996). P. De Lacy (transl.), *On the Elements According to Hippocrates*. Akademie Verlag.

Galen. (1997). *Mixtures*. In P. N. Singer (transl.), *Selected Works* (pp. 202–289). Oxford University Press.

Riolan, J. (1610). *Opera Omnia*. Paris.

Secondary Sources

Beneduce, C. (2022). "Is Touch One Sense or Several?": A Late Medieval Scientific Question. In M. Gensler, M. Mansfeld, & M. Michałowska (Eds.), *The Embodied Soul: Aristotelian Psychology and Physiology in Medieval Europe between 1200 and 1420* (pp. 253–276). Springer.

Benoit, R. (1995). Conceptions médicales à l'Université de Paris d'après les cours de Jean Riolan à la fin du XVIe siècle. *Histoire, Économie & Société, 14*, 25–50.

Bynum, W. F., & Porter, R. (Eds.). (2004). *Medicine and the Five Senses*. Cambridge University Press.

De Vos, P. S. (2021). *Compound Remedies: Galenic Pharmacy from the Ancient Mediterranean to New Spain*. University of Pittsburgh Press.

Debru, A. (1997). Philosophie et pharmacologie: La dynamique des substances leptomères chez Galien. In A. Debru (Ed.), *Galen on Pharmacology: Philosophy, History, and Medicine* (pp. 85–102). Brill.

Deer Richardson, L. (2018). *Academic Theories of Generation in the Renaissance: The Contemporaries and Successors of Jean Fernel (1497–1558)*. Springer.

Harvey, E. D. (2003). The Touching Organ: Allegory, Anatomy, and the Renaissance Skin Envelope. In E. D. Harvey (Ed.), *Sensible Flesh: On Touch in Early Modern Culture* (pp. 81–102). University of Pennsylvania Press.

Henry, J. (2011). "Mathematics Made No Contribution to the Public Weal": Why Jean Fernel (1497–1558) Became a Physician. *Centaurus, 53*, 193–220.

Hirai, H. (2011). *Medical Humanism and Natural Philosophy: Renaissance Debates on Matter, Life and the Soul*. Brill.

Jacquart, D. (1998). *La médecine médiévale dans le cadre parisien (XIVᵉ–XVᵉ siècle)*. Fayard.

Kambaskovic-Sawers, D., & Wolfe, C. T. (2014). The Senses in Philosophy and Science: From the Nobility of Sight to the Materialism of Touch. In H. Roodenburg (Ed.), *A Cultural History of the Senses in the Renaissance* (pp. 107–126). Bloomsbury.

Klestinec, C. (2016). Touch, Trust and Compliance in Early Modern Medical Practice. In A. Whitehead, A. Woods, & S. J. Atkinson (Eds.), *The Edinburgh Companion to the Critical Medical Humanities* (pp. 209–224). Edinburgh University Press.

Martelli, M., Petit, C., & Raggetti, L. (2020). Introduction. New Perspectives on Galen's Treatise *On Simple Drugs* and the History of Pharmacology. *Archives Internationales d'Histoire Des Sciences, 70*, 6–15.

Maurette, P. (2014). *De rerum textura*: Lucretius, Fracastoro, and the Sense of Touch. *The Sixteenth Century Journal, 45*, 309–330.

McVaugh, M. (2017). Determining a Drug's Properties: Medieval Experimental Protocols. *Bulletin of the History of Medicine, 91*(2), 183–209.

Moreau, E. (2023). Temperament and the Senses: The Taste, Odor and Color of Drugs in Late-Renaissance Galenism. *Early Science and Medicine, 28*, 526–552.

Moshenska, J. (2014). *Feeling Pleasures: The Sense of Touch in Renaissance England*. Oxford University Press.

Ottosson, P.-G. (1984). *Scholastic Medicine and Philosophy: A Study of Commentaries on Galen's* Tegni *(ca. 1300–1450)*. Bibliopolis.

Pasnau, R. (2011). Scholastic Qualities, Primary and Secondary. In L. Nolan (Ed.), *Primary and Secondary Qualities: The Historical and Ongoing Debate* (pp. 41–61). Oxford University Press.

Polansky, R. (2007). *Aristotle's* De Anima*: A Critical Commentary*. Cambridge University Press.

Polansky, R. (2024). *Aristotle's* Parva Naturalia*: Text, Translation, and Commentary*. Walter de Gruyter.

Ragland, E. R. (2022). *Making Physicians: Tradition, Teaching, and Trials at Leiden University, 1575–1639*. Brill.

Siegel, R. E. (1970). *Galen on Sense Perception: His Doctrines, Observations and Experiments on Vision, Hearing, Smell, Taste, Touch and Pain, and their Historical Sources*. Karger.

Siraisi, N. G. (1987). *Avicenna in Renaissance Italy: The* Canon *and Medical Teaching in Italian Universities after 1500*. Princeton University Press.

Siraisi, N. G. (2019). *Taddeo Alderotti and His Pupils: Two Generations of Italian Medical Learning*. Princeton University Press.

Stolberg, M. (2021). *Learned Physicians and Everyday Medical Practice in the Renaissance*. Walter de Gruyter.

Van der Eijk, P. J. (2008). Therapeutics. In R. J. Hankinson (Ed.), *The Cambridge Companion to Galen* (pp. 283–303). Cambridge University Press.

Von Hoffmann, V. (2023). Can Mixtures be Identified by Touch? The Reception of Galen's *De complexionibus* in Italian Renaissance Medicine. *Early Science and Medicine, 28*, 498–525.

Chapter 6
Touch in Second Scholasticism: Francisco Suárez, John Poinsot, Bartolomeo Mastri and Bonaventura Belluto

Daniel Heider

Abstract The chapter deals with three theories of four eminent Second Scholastics, namely the Jesuit Francisco Suárez, the Thomist John Poinsot, and the Baroque Scotists Bartolomeo Mastri and Bonaventura Belluto regarding the following four issues: the unicity or plurality of touch; the necessity or the dispensability of tactile species; the organ of touch; the necessity or the dispensability of the tactile medium. It shows that, while with respect to the first two queries all the authors agree, they significantly differ with respect to the last two. While Poinsot endorses the necessity of the medium in order to ensure that touch can produce sensation, Suárez and Mastri and Belluto deny it. Accordingly, unlike Poinsot who restricts the organ of touch to the nerves, the latter embrace the more permissivist account according to which the tactile *sensorium* are all animate bodily parts. Contrary to Poinsot, they substantially reinterpret the Aristotelian principle according to which "a sensible placed on a sense impedes sensation" by saying that it concerns only the inherence (and not the contiguity) of the sensible in the tactile organ.

Keywords Francisco Suárez · John Poinsot · Bartolomeo Mastri · Bonaventura Belluto · Touch · Unicity of touch · Tactile species · Organ of touch · Tactile medium

6.1 Introduction

A standard discussion on the sensory modality of touch in Second Scholasticism commonly covers topics such as those of the organ, the medium, the object, and the number of tactile power(s). There exists a considerable diversity of views related to these queries in sixteenth and seventeenth century Scholasticism. I would like to illustrate this diversity using the examples of three theories proposed by four

D. Heider (✉)
Faculty of Theology, University of South Bohemia, České Budějovice, Czech Republic
e-mail: heiderd@tf.jcu.cz

C. Beneduce et al. (eds.), *The Sense of Touch*, Studies in the History of Philosophy of Mind 32, https://doi.org/10.1007/978-3-032-16308-0_6

eminent early modern scholastics. I shall present the views of the Spanish Jesuit Francisco Suárez (1548–1617), the Portuguese Dominican and Thomist John Poinsot, known also as John of St. Thomas (1589–1644), and two Italian Conventual Franciscans and Scotists, Bonaventura Belluto (1600–1676) and Bartolomeo Mastri (1602–1673), all expressed in their commentaries and disputations on Aristotle's *De anima*. All these authors stand in a clear line of succession. On the one hand, Poinsot reacts (largely critically) to Suárez; on the other hand, Mastri and Belluto frequently criticize Poinsot's view, and are significantly inspired by Suárez's philosophy, even to such a degree that their philosophical system is labeled by some as a combination of Scotism and Suarezianism.[1]

My aim is to present two sorts of comparisons. In the first part, I shall point out that our authors share similar views on the issues of the number of tactile powers and the necessity of sensible (tangible) species; in the second, I shall focus, in more detail, on disagreements among the authors concerning the controversial issue of the organ of touch, and a related question about the (non-)existence of the medium of touch. I shall leave aside several other specific issues that are treated by our scholastics in their disputations and commentaries on Aristotle's *De anima* and elsewhere. First of all, in considering only *De anima* texts, I shall not deal with topics on touch discussed in theological contexts, such as the issue concerning the kind of tactile perception that the damned after the Last Judgment have in hell.[2] Second, I shall set aside the issue of the various comparisons and rankings of human touch, both related to other human senses and the tactile power in other (non-rational) animals. Third, I shall not discuss the topic of the seat of bodily pain (according to many, falling under the ambit of the object of touch as well) and its bodily causes. Accordingly, I shall not analyze the issue of whether bodily pain is to be taken as an act of the cognitive or the appetitive power.[3] Fourth, I shall set aside general issues, such as the nature and causality of sensible impressed species, which are treated by Second Scholastics in a more general context of the sensory powers as such, rather than in discussions of particular external sensory modalities. My focus will therefore be on touch in its specificity.

6.2 Doctrinal Agreement

6.2.1 One Touch

Based on Aristotle's criterion of one-to-one correspondence between a sense and a sensory object, the case of touch results as a kind of anomaly because touch is actually said–by Aristotle himself–to correspond to a plurality of sensory modalities.

[1] See Forlivesi (2014).

[2] For this issue, see, among others, Duns Scotus (2013), *Ordinatio*, IV, dist. 44, q. 2 pp. 125–34: "Utrum homines damnati post iudicium cruciabuntur igne infernali".

[3] For this issue, see Heider (2018; and especially 2024).

Rather than a single sensory modality, it seems to be a collective noun that signifies a plurality of sensory modalities differing in kind. As is well-known, this view was suggested by Aristotle in *De anima* II.11 (henceforth: *DA* II.11):

> […] if touch is not a single sense but a group of senses, there must be several kinds of what is tangible. It is a problem whether touch is a single sense or a group of senses. […] Every sense seems to be concerned with a single pair of contraries, white and black for sight, sharp and flat for hearing, bitter and sweet for taste, but in the field of what is tangible we find several such pairs, hot cold, dry moist, hard soft, etc.[4]

Every sense is concerned with a single pair of contraries through which it is individuated. Sight is related to black and white, and to all the intermediate colors mixed from these two; taste is concerned with bitter and sweet, and with flavors resulting from the blending of these two extreme secondary qualities, etc. The case of touch seems to be less clear. A tangible object of the sublunary realm is composed of four elements that correspond to the different mixtures of the four basic contrary qualities. These qualities are hot and cold, and wet and dry. All other tangibles, hard and soft, heavy and lightweight, viscous and brittle, rough and smooth, and coarse and fine, are derived from these basic qualities.[5] Given the existence of these two pairs of qualities, however, it seems that at least two tactile powers must exist: the first apprehending the hot and the cold, and the second sensing the wet and the dry. Accordingly, common sense talk about five external senses, also accepted by Aristotle,[6] has to be rejected in favor of a revisionist account admitting a greater number of external senses.

In their interpretation of Aristotle's formulations, Suárez, Poinsot, Mastri and Belluto agree that such a revisionist account is false.[7] In doing so, they make it clear that the generally accepted "object criterion" of the external senses' individuation must be qualified. Following Thomas Aquinas's deduction of five external senses from the sensibles' fivefold effect on the sense powers (*STh.* 1, q. 78, art. 3),[8] in his early treatise *Commentaria una cum quaestionibus in libros Aristotelis De anima* (written in the early 1570s), disputation 7 "On the External Senses Individually", question 15 "On the Number of the External Senses in Themselves, and in the

---

[4] Aristotle (1995), *De anima*, II.11, 422b18–26 p. 672.

[5] Aristotle (1995), *De generatione et corruptione*, II.2, 329b6–330a29 pp. 539–40.

[6] See Aristotle (1995), *De anima*, III.1, 424b20–2, p. 675: "That there is no sense in addition to the five—sight, hearing, smell, taste, touch".

[7] As Beneduce (2022) shows, the tension in Aristotle's formulations stimulated the intense debate about the issue of the unity and the plurality of tactile sensory modality in many medieval authors running from the twelfth to the fifteenth century.

[8] Aquinas (1889), *Summa Theologiae* (henceforth: *STh.*), I, q. 78, art. 3, c.1889 p. 254: "Sed in quibusdam sensibus invenitur immutatio spiritualis tantum, sicut in visu. In quibusdam autem, cum immutatione spirituali, etiam naturalis; vel ex parte obiecti tantum, vel etiam ex parte organi. Ex parte autem obiecti, invenitur transmutatio naturalis, secundum locum quidem, in sono, qui est obiectum auditus: nam sonus ex percussione causatur et aeris commotione. Secundum alterationem vero, in odore, qui est obiectum olfactus: oportet enim per calidum alterari aliquo modo corpus, ad hoc quod spiret odorem. Ex parte autem organi, est immutatio naturalis in tactu et gustu: nam et manus tangens calida calefit, et lingua humectatur per humiditatem saporum. Organum vero olfactus aut auditus nulla naturali immutatione immutatur in sentiendo, nisi per accidens".

Subjects in which They Exist", Suárez unambiguously endorses the existence of five senses, and thus the existence of a single touch. Importantly, for Aquinas and Suárez (and for all our Second Scholastics), no external sense's operation can occur without the causal concurrence of a sensible species or likeness (Aristotelian forms without matter). According to all, a sensible species is the necessary co-principle of the cognitive assimilation of sensation.[9] Considering the distinction between an object's natural alteration and its intentional action there are, however, important differences related to the concomitance of these two as far as particular sensory modalities are concerned. While a colored object affects sight purely intentionally by imprinting its species of color onto a sense organ and does not naturally alter the eye (an eye does not turn red upon the reception of the species of a red color), an intentional effect on the other senses is always accompanied by a natural action. An audible object's production of a species requires the local motion of air; the emission of an odorous species from an odorous object is conditioned by qualitative changes, i.e., by the heating of the odoriferous object. Both require a natural change on the part of the sensible object. In addition to this change on the part of the object, the proximal senses of taste and touch also require a natural change in the sense organ. While in taste this modification operates through the moistening of the tongue, that is, the tongue is affected through the quality of wetness, and not through its proper sensible, namely a flavor, in the sensory modality of touch this alteration *is* caused by the tangible itself. Accordingly, in sensing a wet object a hand will become wet; in sensing a hot object it will turn out to be hot. And since there are no other modes of affecting than these five as described above, there are no more than the five proposed senses either. This is also why in the case of the sense of touch the basic qualities must agree in a common proximate genus–even though really and naturally distinct, they agree in a single "*ratio sensibilis*."[10]

Besides the argument from the different modes of natural affecting accompanying intentional modification, Suárez introduces other, supporting, arguments as well. The first concerns a biological criterion of individuation. Since there is a single organ (which turns out to be the whole animate body, as we shall see below) of touch, there must be also a single power of touch. Second, Aristotle's statement that one sense is concerned only with a single pair of contraries is false; there are a

---

[9] Regarding Aquinas, this position is clear already from what immediately follows in the quote introduced in the previous note. He says that "Ad operationem autem sensus requiritur immutatio spiritualis, per quam intentio formae sensibilis fiat in organo sensus. Alioquin, si sola immutatio naturalis sufficeret ad sentiendum, omnia corpora naturalia sentirent dum alterantur." The same view is embraced by Suárez who in his *Commentaria una cum quaestionibus de Aristotelis De anima* (henceforth: *CDA*), disp. 5, q. 1, n. 3, vol. 2 p. 286 says that "Unio obiecti cognoscibilis cum potentia [through intelligible species] est necessaria in omni cognitione". For the identical view, see Bartolomeo Mastri and Bonaventura Belluto (1727), *Philosophiae ad mentem Scoti cursus integer* (henceforth: *CPS*), disp. 4, q. 4, nn. 38–9 (ed. Pezzana, vol. 3, f. 69a). For Poinsot's espousal of the sensible species, see Ioannes a S. Thomae (2008), *Cursus philosophicus Thomisticus* (henceforth: *CPT*), vol. III: *Naturalis philosophiae*, part IV: *De ente mobili animato*, q. 6, art. 2, pp. 177–83).

[10] Aquinas (1889), *STh*. I, q. 78, art. 3, ad 4, pp. 254–5.

number of counterexamples. Contrary to Aristotle, for a number of late scholastics including Suárez sight perceives not only colors, but also light. Smell perceives various kinds of odors; some are nutritive (connected to the sense of taste), others are aesthetic, such as the odor of flowers.[11] Finally, regarding the proper exegesis of the abovementioned Aristotle's passage in *DA* II.11, Suárez makes clear that Aristotle presents this view only as a tentative position that must be further investigated.[12]

In his Thomistic *Cursus philosophicus*, part 4 "On Mobile Animated Being" (first published in 1635), question 5 "On the External Senses Individually", article 6 "What Is the Power, Act and Medium of Touch?" Poinsot expresses his agreement with the conclusion that there is a single sensory modality of touch. Given the objects' conformity in their effect on the tactile organ, all the aforementioned tangible qualities must belong under a single (unnamed) intentional genus. Poinsot spells out this conclusion in terms of the contrariety of qualities insofar as their convenience or inconvenience to the constitution (*temperamentum*) or complexion (*complexio*) of the tactile organ is concerned.[13] All these binary qualities agree in being either convenient or inconvenient to the material composition of the tactile organ. Poinsot stresses that this beneficial or harmful character of any pair of contrary qualities affects the tactile organ's *temperamentum*, and that these pairs of

[11] For this distinction between nutritive and non-nutritive (aesthetic) odors, see Suárez, *CDA* disp. 7, q. 10, n. 7, pp. 696–8.

[12] Suárez (1981), *CDA*, disp. 7, q. 15, n. 5, pp. 756–8: "[...] nam omnes qualitates primae tangibiles habent eumdem modum immutandi sensum, et organum huius est unum et eiusdem requirens temperamentum; ergo superfluum est fingere distinctionem. Neque argumentum urget, quia non cur [unus] sensus tantum sit unius contrarietatis. Visus enim percipit unam contrarietatem, nempe albedinem et nigredinem, et praeterea lucem [...] illae quattuor qualitates primae conveniunt in uno genere communi; et illud est adaequatum obiectum tactus, continens sub se duas secundas contrarietates; quod nihil refert ad unitatem obiecti sensibilis, quia in omnibus illis speciebus est eadem ratio sensibilis". For Suárez's view of five sensory modalities and the single power of touch, see Heider (2021, pp. 140–4).

[13] The notions of "*complexio*" and "*temperamentum*" are ubiquitous in Second scholastics' views of touch. Their frequent occurrence confirms that these concepts have the "central role in early modern psychological thought" (Bos [2022]). As Suárez suggests (see note 30), the concept "*temperamentum*" seems to be inherited from Galen, especially from his treatise *De temperamentis*. This notion refers to the fact that corporeal living beings are complexionate in the way that each has its characteristic mixture of ingredients. This complexion concerns not only the elemental and the humoral mixtures, but above all the mixture of the four basic qualities. Since the tactile organ always includes the qualities of the perceived external objects, in order to be able to sense them it cannot be deprived of them, but it must be well-tempered with respect to them. Accordingly, the sense of touch, as said by Aristotle (see *De anima* II.11, 424a3–5, p. 674) can apprehend only excess, that is, touch can sense only through being deflected from its medial state. If the organ is not well-balanced with respect to these qualities, the sense of touch either will not sense at all, or its apprehension will be unreliable and misrepresenting. Besides, if this deflection from the state occurs abruptly and to an excessive degree, as Poinsot says, something inconvenient to the body and pain will follow. For all our authors, the mental state of pain, however, is not to be identified with tactile sensation. Tactile sensation is only the *cause* of the mental states of pain and pleasure, which are the acts an appetitive power. For a crucial role of Galen in the dissemination of the notion of *temperamentum* in the Middle Ages, see Siraisi (1990, pp. 101–4).

contrary qualities are related to the central contrariety "beneficial/harmful" only accidentally.[14]

In their five-volume Scotistic *Cursus philosophicus*, the section on *De anima* (first published in 1643) of volume 3, disputation 5 "On the Sensory Powers Individually", question 6 "On the Sense of Touch", article 1 "What Is the Organ of Touch, and How Many Kinds of Touch Are There?" Mastri and Belluto agree that there is only one kind of touch.[15] Like our other Second Scholastics, they claim that this feature makes all the tangible contrary qualities univocal in "*esse sensibili*." They also introduce the biological criterion based on the unity of the organ. Like Poinsot, they also refer to the effect on the tactile organ by tangibles due to the contrariety "convenient and inconvenient" regarding bodily constitution. They claim that this feature makes all the tangible contrary qualities univocal in "*esse sensibili*." In addition, they formulate the following objection to their own theory: the existence of a single organ does not necessarily mean the existence of one power, since there are two powers in many bodily parts. In fact, there are two powers in every one of the bodily parts of four external senses. In the tongue we experience both touch and taste; in the pupil we have both sight and touch; etc. So, why should not two (or even more) kinds of touch exist in the entire body as well?[16] The Scotists reject this analogy. Touch is the universal sense, not only in the way that it occurs in all animals, but also in the way that it is distributed over all animate bodily parts. As the *universal* sense it can well co-exist with the other external senses. However, this compatibility does not extend to two or more senses of touch.[17]

---

[14] Ioannes a S. Thoma (2008), *CPT*, vol. III: *Naturalis philosophiae*, part IV: *De ente mobili animato*, q. 5, art. 6, pp. 164–5: "Nihilominus probabilius videtur, tactum esse unum sensum secundum speciem atomam [...]. Et ratio sumitur ex eo, quod omnes illae contrarietates, quae dicuntur esse primae, conveniunt in uno genere proximo, quod est obiectum tactus, licet genus illud sit innominatus. Quod sic explico, quia licet calidum et frigidum, siccum et humidum sint primae contrarietates in genere naturali, possunt tamen convenire in genere intentionali, quatenus habent eundem modum immutandi sensum, scilicet cum contactu rei tangibilis et alteratione sentientis. Reducuntur etiam ad unam contrarietatem obiectivam, quia sentiuntur secundum aliquem excessum, vel convenientiam ad temperamentum sensus, id est secundum convenientiam vel disconvenientiam ad ipsum sentientem: quod autem hic excessus sit in calore et frigore, sicco et humido, materialiter et per accidens se habet".

[15] Mastri and Belluto (1727), *CPS*, vol. 3, disp. 4, q. 6, art. 2, n. 181, p. 109: "Dicimus sensum tactus esse unum specie specialissima".

[16] *Ibidem*, n. 182, p. 109: "[I]dem organum materialiter est perfectibile a duabus potentiis animae, ergo ex unitate organi corporei non recte infertur unitas potentiae, ant. Prob. nam eadem lingua, & nervi, imo quodlibet organum alterius sensitivae potentiae est etiam organum tactus, hic enim per totum corpus diffunditur".

[17] *Ibidem*, n. 183, p. 109: "[...] esto omnes illae qualitates quoad esse reale non videantur ad unam genericam reduci, tamen in esse sensibili possunt univocari sub ratione tangibilis qualitatis, quatenus omnes eodem modo immutant sensum tactus, sub qua ratione generica potest reperiri, una prima contrarietas immutandi sensum convenienter, vel inconvenienter [...] sensus tactus propter suam communitatem, & universalitatem est ubique per totum corpus diffusus, ideoque est cum aliis sensibus compossibilis; at duo sensus tactus in eodem organo sunt superflui, quia unus sufficit erga omnem tangibilem qualitatem, sicut una potentia visiva, quamvis color, & lux sint duo objecta, nam conveniunt in una rationes visibilis qualitatis".

6.2.2 The Necessity of Tactile Species

While at the end of the sixteenth and the first half of the seventeenth century the existence of sensible species of the so-called distal senses of sight, hearing, and smell was accepted by a number of scholastics, the existence of species of the so-called proximal or contact senses of taste and touch remained more controversial. For many Jesuits and philosophizing doctors, the sensible likenesses in taste and touch were far from necessary.[18] Nevertheless, the authors treated in this chapter differ from this progressive Jesuit stream; for all considered here, these likenesses are necessary for the explanation of the assimilative nature of cognition in all the external senses.

Suárez's arguments for the existence of tactile species are introduced in the context of his rejection of the view according to which we can cognize tangible qualities *inherent* (not tangent) in our body. For him, if this inherence were the case, all tangible species would turn out to be redundant. But since we do not sense these qualities inherent in our bodies, a tactile species must be posited. But how does Suárez argue for the claim that we do not sense the qualities inherent in the tactile organ? He adduces the following (phenomenological) experience: we often perceive heat of a temperature greater than the one manifested in our hand. Thus, we do not feel a heat inherent in our hand but only that of an external object, such as from fire. The antecedent is proved thusly: if someone approaches fire, she feels a more intense heat than the one produced in her hand. Suárez argues that, in fact, it is not possible that the intense heat experienced could be produced in her hand that quickly. Therefore, we must be sensing the heat that comes from outside, that is, from the fire. Moreover, as soon as we pull our hand away from the fire, we immediately stop feeling the same intense heat we felt before. Again, if the heat we feel were inherent in our hand, it could not disappear that quickly. This non-perception of the inherent qualities, however, does not mean that any inherence of heat in the tactile organ is not a necessary condition in touch; the opposite is true. As mentioned, the tactile (as well as gustatory) organ must be affected not only *intentionaliter*, but also *realiter* in order for a tactile sensation to be produced. Following the *litera* of Aristotle's *DA* II.11,[19] all the authors agree that touch can sense an object

[18] According to Pedro Hurtado de Mendoza (1578–1641) Suárez contributed to this process of the elimination of sensible species from the philosophical scene. In this context, Hurtado refers to the criterion of the existence of intelligible species from Suárez (1981), *Disputationes metaphysicae*, disp. 35, sect. 4, n. 18 (ed. Berton, *Opera omnia*, vol. 26, 1861, p. 464): "[S]pecies intelligibilis poni solet, vel ut objectum intelligibile in potentia fiat intelligibile in actu, vel ut objectum quod erat separatum aut distans et improportionatum, conjungatur vel proportionetur potentiae". If an object is in contact with a cognitive power, and if it is proportionate to it, the sensible species are redundant. For Hurtado's theory of the external and internal senses, cf. Heider (2026 forthcoming). For Hurtado (1624), see his *Universa philosophia*, *De anima*, d. 12, s. 1, §10 f. 611b. For a denial by the Jesuits of the first generations of the sensible species in some external senses, cf. Leinsle (2006, p. 388; 2016, pp. 107–10).

[19] Aristotle (1995), *De anima*, II.11, 424a3–5, p. 674.

only if its organ becomes deflected from its resting mid-point by the object's excessive quality. In this, touch differs from the other senses, whose organs do not contain the qualities they sense. This is the reason why the sense of touch cannot experience a heat which is the same as the organ's. It can sense only those qualities that exceed the qualities that are constitutive of the organ's *temperamentum*.[20]

The same argument is presented by Mastri and Belluto too. After rejecting several arguments for the existence of sensible species, above all by referring to the possibility of intellectual intuition operating without any mediation of an intentional likeness, they formulate the "Suarezian" argument that can be called "reasoning from the existence of the external medium in touch." As the example of fire shows, touch can sense the heat of an external object through an *external* medium, namely air. As they note, for many of their contemporaries sensible species have to be posited in the distal senses due to the absence of immediate contact between the powers and the objects. However, if an object is contiguous to the organ, the species seems to be unnecessary. The two Scotists show that the existence of an external medium in touch implies more: once we admit this intentional action in a case where an external medium exists, we must do the same for all other cases, that is, also for cases where objects are in immediate contact with the organ.[21]

Poinsot also concurs with the claim that sensible species are necessary co-principles for a tactile operation. He considers the following objection against the necessity of a tangible species: species emitted by sensible objects are necessary to determine the powers of a sense where objects cannot do that. However, if objects are in direct contact with them, the powers of a sense are sufficiently determined by the objects; in such a case, no species will be required. In his reply, Poinsot affirms that while material contact is necessary for touch (and taste), it is only a necessary, and not sufficient,

[20] Suárez (1981), *CDA* disp. 7, q. 13, n. 4, pp. 724–6: "[…] principaliter [arguitur], nam tactus saepe sentit maiorem calorem quam [sit] sibi inhaerens; ergo non sentit [sibi] inhaerentem, sed [calorem] rei calefacientis. Si enim sentiret inhaerentem, non maiorem [sentiret]; sed […] Antecedens patet, nam dum quis est igni applicatus, sentit vehementem calorem; qui tamen non est in illo productus totus. Quod patet, [tum] quia cito sentit, et non posset tam cito produci, tum quia si separetur ab igne, statim non sentit tam intensum."; *Ibidem*, n. 5, p. 728: "Secunda conclusio sequitur ex his: Hi duo sensus [the senses of taste and touch; D.H.] immutantur [a suis obiectis] per species intentionales. Probatur in communi. 1° Quoniam sentiunt qualitates extrinsecas; ergo recipiunt ab illis aliquas similitudines per quas sentiunt". For the sensible likenesses in Suárez's account of touch (and taste), cf. also Heider (2021, pp. 208–11).

[21] Mastri and Belluto (1727), *CPS*, vol. 3, disp. 5, q. 6, art. 2, n. 204, p. 112: "Vera, & communior sententia species quoque admittit in sensu tactus, & consequenter etiam in gustu, olfactu; quae opinio quanto est communior, tanto est difficilior probationis […]. Solet probari, quia objectum cum sit nimis crassum, nequit immediate concurrere ad sensationem, quae est immaterialior"; *ibidem*, n. 205, p. 112: "Sed certe nulla ex his rationibus concludit, vel aeque demonstrat qualibet etiam in cognitione intuitiva intellectiva necessario speciem intervenire, quod esse falsum ostendemus […]"; *ibidem*, n. 207, p. 113: "Probatur modo conclusio, quod tactus fiat immediate per species, contingit aliquando tactum fieri per speciem, ergo semper fit per speciem; antec. Prob. contingit aliquando objectum tactus habere medium extrinsecum […] ergo quandoque fit per species, ista conseq. Prob. quia ideo per adversarios negatur tactum fieri per species, quia objectum est valde approximatum potentiae, itaut sit contiguum cum illa, & eadem ratione requirunt species in visu, & auditu, quia objectum est distans, ergo si objectum tactus quandoque est distans, causabit species ad sui tactionem".

condition. Even though fire touches a log, the underlying piece of wood is not sensed. Sensation requires a vital and intentional contact that is different from and more than a physical contact, since physical contact is characteristic of non-cognitive beings, too.[22]

6.3 Doctrinal Differences: The Organ and the Medium

While our scholastics agree about the number of tactile powers and the necessity of species, they substantially differ in the issues related to the topics of the seat of the sensory organ and the necessity of a medium in tactile sensation. In the following subsections, I shall present the authors' views on these two controversial issues. I shall show that a doctrinal division runs between Poinsot on the one side, and Suárez and Mastri and Belluto on the other.

6.3.1 Suárez

In question 14 "What Is the Organ of Taste and Touch?" disputation 7 "On the Particular External Senses" of his *CDA*, Suárez lays out four views related to the issue of the organ of touch. He leaves no doubt that this issue is "more dubious" than that concerning the organ of taste. According to the first opinion, the organ of touch is the *heart*; all other parts, such as nerves, skin, and especially flesh are only the internal *media* standing between the organ and external tangibles. In this connection Suárez refers to *De sensu et sensato* where Aristotle says:

> The organ of touch consists of earth, and the faculty of taste is a particular form of touch. This explains why the sensory organ of both touch and taste is closely related to the heart.[23]

Like all our scholastics, and despite the seeming advocacy by Aristotle, Suárez rejects this view: we feel heat in our hand and on the surfaces of our body, and not in our heart.[24] Moreover, as he argues elsewhere, the heart is not even the inner gen-

---

[22] Ioannes a S. Thoma (2008), *CPT*, part 4, q. 6, art. 2, p. 180: "[...] communi Philosophorum consensu receptum est, dandas esse species intentionales impressas pro potentiis cognoscitivis [...]"; *Ibidem*, p. 182: "Ratione eadem veritas stabilitur, quia potentia quantumcumque habeat virtutem efficiendi vitaliter actum cognitionis, dependet ab obiecto, ut simul cum ipso cognitionem eliciat. Et hoc ideo est, quia potentia per cognitionem debet esse assimilativa vitaliter, non solum communi ratione, qua omne agens intendit assimilare sibi effectum eo modo, quo potest, sed speciali ratione, quatenus tendit ad uniendum et coniungendum sibi obiectum et trahendum ad se [...]. Et quod ista actuatio et determinatio intrinseca potentiae debeat fieri per speciem, constat [...] non ordinatur ad constituendum unum esse physicum vel naturam ex potentia et objecto, quia repraesentatio sola obiecti non constituit naturam tertiam".

[23] Aristotle (1995), *De sensu et sensibili*, 2, 439a1–2 (ed. Barnes, vol. 1, p. 697).

[24] Suárez (1981), *CDA*, disp. 7, q. 14, n. 1, p. 734. For Poinsot's, and Mastri and Belluto's critique of this view, see below. Unlike Second scholastics, the thirteenth century scholastics authors regarded this view as "default." For this view, see Monika Mansfeld's chapter in this volume.

eral principle of sensory cognition; this is the brain. The heart is merely the organ of a sensory appetite.[25]

According to the second opinion, the organ of touch is the *nerves* that are diffused throughout the whole body.[26] These nerves, being a sort of fascicle of ropes, lie inward, under the skin and flesh. Again, flesh and skin are only the internal media. Famously, Aristotle compares the function in touch of flesh with that of air and water in sight, hearing, and smell, and formulates his famous "axiom" according to which "a sensible placed on a sense impedes sensation":

> In general, flesh and the tongue are related to the organs of touch and taste, as air and water are those of sight, hearing, and smell. Hence in neither the one case nor the other can there be any perception of an object if it is placed immediately upon the organ, e.g. if a white object is placed on the surface of the eye. This again shows that what has the power of perceiving the tangible is seated inside. Only so would there be a complete analogy with all the other senses. In their case if you place the object on the organ it is not perceived, here if you place it on the flesh it is perceived; therefore, flesh is the medium of touch.[27]

Whatever Aristotle knew about nerves, he makes clear that despite all appearances that, from the fact that when an object comes into contact with flesh we seem to feel the pain in that place of that contact, we are not allowed to infer that the organ of touch is flesh, i.e., the place of this immediate contact.[28] Here, he introduces his famous experiment to confirm this description:

> […] no indication can be drawn from the fact that if the object comes into contact with the flesh it is at once perceived. For even under present conditions if the experiment is made of making a sort of membrane and stretching it tight over the flesh, as soon as this web is touched the sensation is reported in the same manner as before, yet it is clear that the organ is not in this membrane. If the membrane could be *grown* on to the flesh, the report would travel still quicker.[29]

According to the third opinion, which is attributed by Suárez to Galen, the organ of touch should be located on the surface of the body, namely in the skin.[30] Neither

---

[25] Suárez (1991), *CDA*, disp. 11, q. 1, n. 7 (ed. Castellote, vol. 3, 1991, p. 332).

[26] Suárez (1981) in *CDA*, disp. 7, q. 14, n. 1, vol. 2, pp. 734–6, subsumes under this view also Avicenna's opinion according to which "hoc organum non esse solam carnem, nec tantum nervum, sed carnem nervosam".

[27] Aristotle (1995), *De anima*, II.11, 423b17–26, p. 674.

[28] Friedrich Solmsen (1961) notes that it was not until post-Aristotelian medicine when *pneuma* currents identified as *neura* were discovered. Interestingly, although Aristotle did not identify nerves through the sensory evidence, in his commentary on his *De anima* Averroes notes that Aristotle did anticipate them by locating the organ of touch within the body. For this historical attribution, which also explains why a number of Aristotelian-scholastic authors ascribed this view to Aristotle himself, see Hasse (2000, pp. 98–9) and Beneduce (2023, p. 172). For Averroes, see Averroes (1953), *Commentarium magnum in Aristotelis de anima libros,* § 108 and § 116, p. 298 and p. 312.

[29] Aristotle (1995), *De anima*, II.11, 423a1–6, p. 673.

[30] *CDA* disp. 7, q. 14, n. 1, p. 736: "Aliter opinatur Galenus, lib. [1] *De temperamentis*, cap. <9>, et <5 *De simplicium medicamentorum facultate*, cap. 4>, dicens organum tactus esse cutem, quia est magis temperata, caro vero est nimium sanguinea et calida, nervus nimium frigidus et exsanguius,

flesh nor nerves can be the organ of touch. While the first is too hot and sanguine, the latter is too cold and bloodless. In brief, they both fall short of the well-balanced elemental *temperamentum* that is necessary for the delicate tactile organ. Since the organ of touch, unlike that of the other senses, cannot be deprived of the qualities it senses, it must be well-tuned with respect to all of them. But the best elemental constitution as regards touch can be found only in the skin, not in flesh or nerves.

> Lastly, according to the most permissive view, the tactile *sensorium* exists in all bodily parts including flesh, skin, and nerves, except for the "earthy" parts, such bones, fingernails, teeth, hair, etc., which are not animated.[31]

Despite Suárez's general conclusion that the tactile organ is flesh, he is far from accepting it as simply flesh, that is, as distinct from the skin and nerves.[32] When claiming that the tactile organ is flesh he actually subscribes to the final view according to which *all* parts of the body comprise the organ of touch. In other words, he accepts 'flesh' synecdochically. Suárez presents four arguments for his all-embracing view. The first is the argument from experience, which Aristotle already had found inconclusive.[33] We feel pain both in the skin and exposed flesh (in ulcers, e.g.) as well. Second, there are parts (stomach, veins) in which no nerves exist, and yet we feel sensation in them. Third, he agrees with Galen that the best *temperamentum* is to be found in skin, followed by flesh, and lastly by nerves. He notes that human skin is of such an exquisite complexion that it is the most perfect *sensorium* in the entire animal kingdom. Fourth, the tactile organ must be located in all the sentient peripheral parts of the body. The soft tissues of the human body by nature are designed to protect the inner, hard parts, such as bones, that, being inanimate, do not sense.[34]

Two main objections to this view are recognized by him. First, this view is at odds with Aristotle's *De anima* II.11. Second, and more importantly, the skin cannot be the tactile organ but only the medium, because "a sensible placed on a sense impedes sensation." The organ must lie inward, under the skin and flesh. While Suárez briefly replies to the first objection by referring especially to some parts of Aristotle's later treatise *De partibus animalium* in which he explicitly says that flesh is the organ of touch,[35] much more space is devoted by him to the rejection of the second argument. Not only does he reject (i) the view that sight, hearing, and smell

---

cutis vero se medio habet". As mentioned, this quote shows that Galen can be regarded as the authority, at least for Suárez, with respect to the notion of *temperamentum*.

[31] *Ibidem*: "Alia opinio sit omnes partes corporis esse sensorium tactus, tantum exceptis quae terrestres sunt et durae, ut ossa et dentes, ungues, etc. [...] Quae mihi videtur vera."

[32] For these views in Suárez (1981), see *CDA* disp. 7, q. 14, n. 1, pp. 734–6.

[33] See note 29.

[34] Suárez (1981), *CDA*, disp. 7, q. 14, n. 2, p. 736: "Organum tactus esse carnem ipsam."; *ibidem*, n. 3, p. 740: "[...] in illis animalibus [those that do not have flesh; D.H.] aliquid proportionale carni est organum tactus".

[35] Aristotle (1995), *De partibus animalium*, II.8, 653b22–5, p. 1018: "For an animal is by our definition something that has sensibility and chief of all the primary sensibility, which is that of touch; and it is the flesh, or analogous substance, which is the organ of this sense [...]".

differ from taste and touch in that the former three require only an external medium and the latter two only an internal medium, but Suárez also substantially revises (ii) Aristotle's axiom that "a sensible placed on a sense impedes sensation."

First of all, he denies that all the first three senses necessarily require an external medium. This is true only for the sight of colors, which requires an illuminated medium, since light is a quality that can be actualized only through the diaphanous. However, this is not true for the light found in a medium, which can be contiguous to the eye and yet can be seen. The same applies to hearing and smell too: a sound can penetrate to the internal ear, which is the organ of hearing; an odoriferous evaporation can permeate the brain and thus attain the interior part of the nose, which is the organ of smell. Equally, as we have seen, Suárez denies that an internal medium is necessary for taste and touch. The skin, and not a bodily part under it, experiences the heat of fire or water when it is immersed in it. Unlike Aristotle, who obscurely states in his hypothetical claim that two bodies cannot immediately touch each other either in water or in air since there are always some medial particles of air or water between them, Suárez is clear that two bodies can be in direct contact.[36] Chewed and moistened food penetrates through the pores of the tongue to the lingual nerve, which is the organ of taste. At the same time, as has been shown, Suárez does not agree that an external medium is to be found only in the first three (distal) senses. Skin also feels the heat of a distant object (fire), and is affected by this heat through the medium of air. So, how does Suárez then assess the Aristotelian principle "a sensible placed on a sense impedes sensation"? He distinguishes two different meanings of this axiom. While the first is related to *inherence*, the second concerns the *contiguity* of the object and the organ. While in the first sense the axiom is true, in the second it is false. Suárez is confident that the argument repeatedly employed, especially by Thomists, according to which a medium (being sort of an agent sense) must be posited in order to explain the "spiritualization" of sensible species is entirely false.[37]

[36] Aristotle (1995), *De anima*, II.11, 423a29–b1, p. 673: "[I]n water two bodies cannot be in contact with one another. The same holds of two bodies in air–air being to bodies in air precisely what water is to bodies in water–but the facts are not so evident to our observation, because we live in air, just as animals that live in water would not notice that the things which touch one another in water have wet surfaces".

[37] Suárez (1981), *CDA*, disp. 7, q. 14, n. 4, p. 742: "[F]alsum est quod inter obiectum sensibile et organum debet semper esse aliquod medium; quin potest, quamvis sit illi contiguum, sentire potest, ut patet in tactu [...]. In olfactu similiter evaporatio odorifera aliquando ascendit ad cerebrum, et consequenter usque ad organum odoratus [...] et tunc etiam sentitur odor; ergo sensibile contiguum sensui causat sensationem. Sonus etiam, quando est vehemens et fit prope aures, saepe intrat usque ad membranam, et forte per poros illius usque ad sensorium auditus, et tamen audit illum. De gustu patet etiam."; *Ibidem*, n. 5, p. 744: "[...] tactus potest sentire calorem rei absentis, et immutari ab igne per aerem, v.g."; *Ibidem*, n. 6, p. 746: "[...] lumen autem medii, quod est contiguum visui, forte videtur per modum medii [...] Species enim non redduntur spirituales propter medium, sed quia obiectum habet virtutem ad faciendas illas, et quia sensus est capax illarum". See also Heider (2021, pp. 211–5).

6.3.2 Poinsot

In article 6 "What Are the Power, Act, and Medium of Touch?" of question 5 "On the External Senses Individually" in the *De anima* section of his *CPT*, Poinsot agrees that the tactile organ is extended over the whole body, and cannot be restricted to the heart. In order that an animal could protect itself from harm and seek what is beneficial, the organ must be spread over the body. A difficulty lies in the question of whether this organ exists only in the nerves, or both in nerves and flesh. Probably following Averroes's note, Poinsot is sure that Aristotle locates the organ of touch only in the nerves that underlie flesh. Flesh is not the organ but only the medium. But at the same time, probably listening to the critique of Suárez and others, he considers Aristotle's axiom as rather "difficult." As noted, Aristotle's argument runs as follows: what is placed on an organ cannot be sensed; however, we sense anything placed on flesh; therefore, flesh cannot be the organ but only the medium. Poinsot makes clear that this argument is not sufficient. It proves even the opposite claim, namely that flesh is the organ of touch, because flesh has a medium, namely skin. Moreover, it is obvious that we feel an object with an exposed nerve, where no medium exists. In order to explain Aristotle's principle, he hastens to say that in this case the object is not felt by the part of the nerve that is in immediate contact with the object touched, but is sensed only by the nerve's *proximate* part. No doubt, with this claim Poinsot is able to argue for the necessary existence of the medium. This rejoinder, however, as he rightly notes, raises the question why we could not feel an object that is contiguous to a certain part of the flesh through *its* proximate part. In his reply, not entirely satisfactorily, Poinsot rejects this analogy by saying that, unlike nerves, flesh has no natural medium. This is why only nerves, and not flesh, can sense what is placed upon them through the medium of a proximate bodily part.[38] Following Aristotle, Poinsot is thus able to reject Suárez's argument from experience according to which we feel a tangible object both in the flesh and the

[38] Ioannes a S. Thoma (2008), *CPT*, part IV, q. 5, art. 6, pp. 166–7: "Sententia Aristotelis collocat tactum in nervis sub carne […] solum carni attribuit esse medium, non organum. Haec auctoritas rationem insinuat de se satis difficilem, quia si ideo organum tactus attribuitur nervo, quia non potest sentiri, quod ponitur supra sensum, sentimus autem per tactum aliquid positum supra carnem, et sic caro non est organum, potius hac ratione deberet attribui carni sensorium tactus, quia caro licet sentiat, etiam habet medium, quo sentiat, scilicet cutem, et ita tangit rem non immediate positam supra se, sed supra aliquid distinctum, internum tamen […] Quodsi dicatur carnem etiam detracta cute sentire, hoc idem instatur in nervo, quia denudata carne adhuc sentit, et quidem vehementissime, et tunc dicimur, quod sentit nervus non in ea. parte, cui superponitur sensibile, sed ratione partis proximioris […] Cur autem similiter non dicemus, quod caro sit organum tactus sentiendi non id, quod immediate superponitur, sed quod proxime parti? Respondetur esse disparem rationem, quia caro cum non habeat aliquid sibi suprapositum sicut nervus, quod deservire possit, ut connaturaliter et per directam emissionem aliquid tangat, ideo non potest connaturaliter esse organum tactivum, id tamen potest convenire nervo, quia connaturali modo habet carnem supra se tamquam medium, quo sentiat, eo vero deficiente per accidens et indirecte et minus connaturaliter potest sentire id, quod proximiori parti apponitur, caro vero quia non habet quomodo directe tangat per aliquid superpositum, non mirum, quod etiam indirecte non possit tangere".

skin. For Poinsot, such an experience is not enough to show that flesh is the organ of touch. Moreover, if Suárez and others observe that there are a number of bodily organs that do not possess nerves, such as the stomach and liver, and yet these organs sense, this is only due to the subtle and hardly visible nerves that exist in them.[39]

But what are Poinsot's reasons for the endorsement of the crucial principle "a sensible placed on a sense impedes sensation"? In article 3 "Whether a Sensible Placed on a Sense Is Sensed, and Whether It Harms the Sense?" of question 4 "On Sentient Soul, and Its Powers Universally", in his *De anima* section of the *CPT* he, at first, introduces two reasons. The first reads as follows: since a visible object placed on the eye cannot be seen, the same should be true of all the external senses. If we are to provide a *unified* account of the external senses, we have to assume the existence of a medium in touch, and in other external senses as well. The second reason, which has already been mentioned, is that some medium must be posited due to the necessity of the spiritualization of sensible species. Poinsot finds both arguments inconclusive. Like Suárez, he replies that although this placement upon the organ impedes the vision of a colored (opaque) object, this need not be necessarily the case when a luminous object is placed on the eye. Moreover, there are many other differences between the external senses, so why not assume a special case also in the issue of the medium? The second argument is rejected because it commits a fallacy of equivocation. A medium is not an "instrumental cause" that would spiritualize a species. It does not make a species more perfect with respect to its representation; rather the opposite is true–the more distant an object is, the less perfect its sensible species will be, and the less perfectly it will be sensed.[40]

Having dismissed these arguments, Poinsot formulates two main reasons for the validity of the Aristotelian axiom. First, if a sensible placed on an organ were sensed, it would be possible to see something without a species, and to see reflectively the sensory power itself. If Aristotle's principle were denied, an external sense would become a reflective power, not only with respect to its own act–regarding imperfect or implicit cognition, this is conceded by a number of authors–, but also with respect to the power itself. However, this is something entirely incompatible with the materiality of the external senses. Poinsot justifies his consequence as follows: a species just emitted from an object is no more immaterial than the object itself. But if such

[39] *Ibidem*, p. 168: "Ducit [Suárez] autem argumentum ab experienti: Videmus enim non solum nervum, sed etiam carnem sentire, ut si caro pungatur vel vena; ergo non solum nervus sentit, sed etiam aliaepartes [...] experientia debet in hac parte praeponderare, videmus autem nos in cute et carne actu sentire. Cur ergo negabimus ibi elici sensationem et consequenter esse organum tactus, et non solum medium? Tum quia in pulmone, in hepate, in corde, non dantur nervi, et sentitur dolor. Ergo non solum nervus est organum tactus [...] quod si caro pungatur sit sensus, non tamen monstrat, quod fiat ab carne, et non in nervo coniuncto, sicut posita specie in pupilla statim sentimus visionem, non tam constat, an fiat a pupilla, an a nervo visorio. Ad aliam replicam dicitur, etiam in illis partibus dari nervos, aut aliqua principia nervorum, qui ob suam subtilitatem satis vehementer sentiunt; vel etiam in partibus illis propinquis vel contiguis dantur nervi, a quibus fiat ipsa sensatio".

[40] Ioannes a S. Thoma (2008), *CPT*, part IV, q. 4, art. 3, p. 122.

a species could serve as a principle for the co-causation of vision, there would be no reason to deny that also a luminous object could directly concur in this vision. But if that were possible, the pupil itself, being luminous, could be seen by itself, too.[41] Poinsot also introduces a second argument. Any sensory operation is an act that is more perfect than any natural operation of inanimate beings, such as heating. The natural alteration of heating requires a propinquity between an agent and a patient. However, cognition and cognitive beings, in order to be differentiated from and elevated above this kind of natural efficient causality, must be independent from this condition of propinquity–they must be regarded as powers that operate essentially through a medium.[42]

6.3.3 Mastri and Belluto

In article 1 "What Is the Organ of Touch, and How Many Kinds of Touch Are There?" in question 6 "On the Sense of Touch" in the *De anima* section of their *Cursus philosophicus*, Mastri and Belluto, like Suárez and Poinsot before, do not spend much time on the position that the heart is the tactile organ. This view is clearly false, since there are many other bodily parts (especially surface parts) that also generate a tactile sensation. In the next step, they present three views, of which two are regarded by them as extreme. According to the first view, the organ of touch is only the nerves diffused over the whole body, in the same way as fibers run throughout a tree leaf. This view was advocated, among others, by Poinsot and Scotus.[43] According to the second view, the organ of touch is flesh. The third (inter-

---

[41] *Ibidem*: "[S]i sensibile immediate positum supra sensum posset sentiri, non repugnaret videri aliquid sine specie, imo et ipsum sensum sentire seipsum, atque adeo reflectere supra se, quod repugnat eius materialitati et limitationi [...] Sequela vero probatur, quia species, ut immediate exit ab objecto, non magis immaterialis exit, quam sit ipsa immaterialitas obiecti in se, v.g. quam sit ipsa lux [...]. Ergo si talis species prout immediate exit ab obiecto servire posset sensui, non est cur ipsum obiectum lucidum factum per seipsum non posset etiam servire sensui. Quodsi potest, ergo ipsa pupillam, quae in se lucida est et ad alios oculos mittit species, cum videatur ab eis, poterit a seipsa videri seu ab oculo in ea existente".

[42] *Ibidem*, p. 123: "Denique hoc est conveniens operationi sensitivae, quae elevatior et immaterialior est quam naturalis. Ad alterationem autem et motum naturalem requiritur indistantia applicationis inter agens et passum, quia agens naturale determinatur a sua materia, circa quam operatur. Ergo conveniens est, quod natura cognoscitiva sit aliquo modo independens a tali indistantia et contactu naturali, et sic operetur non existente sensibili supra se, sed aliquo modo distanti et per aliquod medium".

[43] Duns Scotus (2006), *Quaestiones super secundum et tertium De anima*, q. 2, n. 6, p. 14: "Ad istam quaestionem dicendum quod caro non est organum sensus tactus, sed aliquid infra carnem, scilicet nervus, vel aliquid loco nervi coextensum corpori."; *Ibidem*, n. 13, p. 17: "Ad illud 'ubicumque est tactus in animali, ibi est nervus vel organum': non autem requiritur quod ibi sit nervus formaliter in qualibet parte carnis, sed sufficit quod sit prope vel iuxta realiter vel formaliter, et sic in qualibet parte virtualiter, quia virtute nervi prope exsistentis quaelibet pars potest sentire tangibile".

mediate) position regards both flesh and nerves as the organ of touch. Suárez is mentioned as one of the representatives of this middle view.[44]

Surprisingly, Mastri and Belluto do not defend the second view of their master, but rather the third position one that they attribute to Suárez. Not only nerves, but also skin and flesh, are the tactile *sensoria*.[45] They are convinced of this, despite Scotus's explicit conclusion in the body of his Question 2 "Whether Flesh Is the Organ of Touch?" in his *Quaestiones super secundum De anima*. They note that in his reply to a third objection to his view, the Subtle Doctor after all qualifies his view about the nerves as the organ of touch. The well-known objection goes like this: touch can be experienced in many parts of the body in which no nerves but only flesh exists. Therefore, flesh must be taken as the tactile organ, rather than nerves. In his reply, Scotus says that this identification of the tactile organ with nerves ought not to be understood in the sense that nerves exist formally in all fleshy parts; it is sufficient that they exist proximately to them, and due to this proximity these fleshy parts acquire a power (virtue) to sense a tangible object. Due to this, as Scotus says, virtual existence of nerves in flesh, flesh is sufficiently capable of sensation. As does Suárez, Mastri and Belluto prove this all-inclusive view of the tactile organ by an argument from experience, one rejected by Aristotle: we feel pain in some fleshy part where we have been stabbed. They also claim that several fleshy parts, such as the lungs, liver, and heart, have no nerves, and yet we feel stimuli in them. The same holds for nerves. If we are pricked in a nerve, we feel an intense pain there. Since nerves transport spirits throughout the whole body, they have to be regarded as a more primary organ than flesh.[46]

Like Suárez, Mastri and Belluto agree that the main argument against their view (that flesh, including skin, is the organ of touch and not only the medium) is based on Aristotle's axiom "a sensible placed on a sense impedes sensation." How do the two Scotists address this principle? Can a sensible that is adherent (not inherent) to a sense organ according to them impede sensation? In the explicit treatment of this issue, they start with the presentation of two traditional views. The first, Thomistic, states that the reply to the question must be affirmative. As seen in Poinsot, the contiguity of an object with an organ *does* impede sensation. On the contrary, Jesuits and Scotus are convinced of the opposite. They maintain that this does not hold necessarily for all the senses, and that a unified account of the external senses according to this position is nothing more than an illusion. Within the deniers of this Aristotelian principle, however, they find doctrinal differences. There are the differences as to which external sense(s) is to be exempted from this axiom. Mastri and Belluto note that Suárez exempts the sensory modalities of touch and taste, while Scotus exempts also the senses of smell, hearing, and even sight with respect to the perception of light. The only exception is the sight of colors–that requires an

[44] Mastri and Belluto (1727), *CPS*, vol. 3, disp. 5, q. 6, art. 1, n. 175, p. 108.

[45] *Ibidem*, n. 176: "Dicimus non solum nervus, sed etiam cutem, carnem, &c. esse sensoria tactus, ita tamen quod caro a nervis dependeat".

[46] *Ibidem*, nn. 177–8.

illuminated medium.[47] In fact, we have seen that Suárez actually denies the validity of the principle in the same (in fact, most extensive) manner, as did Scotus.[48]

Unsurprisingly, Mastri and Belluto endorse the second view, according to the more extended version that was defended by Scotus and Suárez. They proceed by destroying the Thomistic foundation, according to which the principle "a sensible placed on a sense impedes sensation" is true because every sensation requires a *proportion* between the species and the sense. On this foundation, since the sensory power (as mentioned) in a way approximates immateriality (of course, it does not have a literal spirituality because this holds only for intelligible species), so its principle, that is, the sensible species, must operate as well. However, since the species emitted from objects are too coarse-grained and insufficient for the immateriality of the sense powers, they require spiritualization; this is executed by the medium.[49] Mastri and Belluto produce no less than five objections to this argument, of which I shall mention only the first two. First, as we have seen, the claim that a material object cannot concur with a power in the production of sensation is simply false. Clearly, a material object *can* concur immediately in the production of an intellectual intuition. Accordingly, if such a concurrence between these two is possible, *a fortiori* it is not necessary in the case of senses and sensible. Thus, there is no universal need for this spiritual proportionalization of an object in order to accord with the immateriality of cognitive powers, of whatever sort. Second, it is unclear what is actually meant by the species being too "coarse-grained and distant", and what is understood by the required "subtilization" that seems to be required. In line with Poinsot, they say that the only possible interpretation is the following: an agent (an object) produces its species more strongly when an object is nearer the power than when it acts over a long distance. While at the outset of its emission a species acts strongly, while progressing through the medium it gradually weakens. But if this weakening is what Thomists mean by the terms "spiritual" and "immaterial," then, certainly, this "spirituality" is entirely irrelevant for their argument, or it even contradicts their argument. It is not true that an attenuated species is more perfect and immaterial than one that is not attenuated; on the contrary, an attenuated species is less perfect. It causes a sensation that is less perfect than one produced by a species nearer the object. A species weaker in its intensity and emitted from a more distant object does not become more proportional to the power, nor does it become more

---

[47] Mastri and Belluto (1727), *CPS*, vol. 3, disp. 5, q. 7, n. 213, p. 113.

[48] See note 37.

[49] Mastri and Belluto (1727), *CPS*, vol. 3, disp. 5, q. 7, n. 214, pp. 113–4: "Sed praecipue ostenditur evertendo fundamenta Thomistarum; nam primo probant hanc propositionem esse universaliter veram, eo quia sensibile positum supra sensum impediat sensationem, quia inter species, & sensum, ut fiat sensatio, debet esse proportio, itaut sicut potentia valde accedit ad immaterialitatem, sic etiam species debent esse immateriales, quae statim ab objecto prodeunt, licet sint immaterialiores, & minus crassae ipso objecto, attamen non sunt ita immateriales, sicut potentia sensitiva [...] ergo indigent medio, in quo species veluti subtilientur, & attenuentur [...]".

apt to produce a sensation. If such were the case, we would see an object at a great distance more perfectly than one nearby.[50]

Mastri and Belluto also reject Poinsot's argument that denying this principle would entail the dispensability of the species in general, and would result in the sensory power's self-sensation.[51] They are convinced that Poinsot's argument is false, and is guilty of *petitio principii*. On the one hand, Poinsot argues for the existence of the species by referring to the principle "a sensible placed on a sense impedes sensation"; on the other, he demonstrates this proposition by saying that the denial of this principle would entail the species' redundancy. They are also clear that Poinsot's claim that the rejection of the principle would result in the possibility of a power sensing itself is false, because species are generated through a transitive and not an immanent action. Lastly, they agree with Suárez's interpretation of Aristotle's *dictum*, according to which the Stagirite took this proposition in the sense of inherence, rather than in the sense of the object's and the organ's contiguity.[52]

Granting their arguments for the existence of a tactile species and against the above-mentioned Aristotelian axiom, it is not difficult to anticipate Mastri and Belluto's view of the role of the medium in touch. In line with Suárez and Scotus, they affirm that a tactile act *per se* requires neither an external nor an internal medium; both media intervene only accidentally.[53] What Poinsot says, namely that the organ in the case of a naked nerve is a part that is *proximate* to the part that comes into contact with an object is assessed by them as entirely arbitrary. While "to be the sensorium of touch" means to receive a species and produce a sensation,

[50] *Ibidem*, n. 215, p. 114: "Haec ratio proprie consistit in quodam verborum apparatu, & multa continet falsa, & sine fundamento asserta; nam in primis urgetur ex hoc, quod objectum valde materiale potest immediate concurrere ad sui intiuitivam cognitionem cum intellectu, quae est potentia vere immaterialis, ergo non requiritur haec proportio inter speciem, & potentiam in immaterialitate. Tum, quia nequit intelligi, in quo consistat haec major crassities, quam habet species, dum statim exit ab objecto, & deinde amittit per medium, nisi dicatur [...] quod agens fortius agendo in propinquum, quam in distans producit speciem intensiorem in principio, remissiorem in progressu; sed ista immaterialitas potius officit sensationi; quam juvet, ergo non est ad propositum".

[51] *Ibidem*, n. 216, p. 114: "Deinde adhuc arguunt si sensibile positum supra sensum posset causare sensationem, posset quoque sine specie sentiri. Prob. seq. Quia species, ut immediate exit ab objecto, non est magis immaterialis, quam sit ipsa immaterialitas objecti in se, ergo posset quoque sensationi servire, sicut species, immo & sensus ipse posset se ipsum sentire causando in se speciem sui".

[52] *Ibidem*, n. 217, p. 114: "Hae quoque rationes nihil concludunt; prima enim videtur committere circulum, nam existentiam specierum Thomistae ut plurimum probant per istam propositionem, quod sensibile positum supra sensum non faciat sensationem, modo propositionem istam ostendunt per hoc, quod si sentiretur, frustra essent species, in quod circulum committunt [...] Neque tandem sequitur, posse sensum seipsum sentire, quia species fit actione transeunte, non immanenti [...]. Suarez explicat [Aristotle's principle; D.H.] de sensibili inhaerente organo, non de contiguo." For Suárez's explication of the principle in the sense of inherence, see Suárez, *CDA* disp. 7, q. 14, n. 6, t. 2, pp. 744–6. Cf. also Heider (2021, p. 215).

[53] Mastri and Bellut (1727), *CPS*, disp. 5, q. 6 art. 2, n. 186, p. 109: "Dicendum est ad actum nullum medium neque externum, neque internum per se requiri, sed tantum quandoque per accidens intervenire".

the phrase "to be the medium of touch" indicates the reception of a species and its communication to a sensorium. But if nerves as such have the capacity to cause sensation, as natural agents they must exercise this capacity already in the part that comes into direct contact with the object, and not only in the proximate part.[54] Moreover, like Suárez, they hold that the rejection of the necessity of an external or internal medium does not mean the dismissal of the accidental intervention of such media. They mention Aristotle's example of a membrane (conceived, perhaps, as an internal medium) grown over the flesh through which an object tangent to the membrane can be felt; likewise, following Suárez, they repeat the example of the fire whose heat is perceived through the medium of air (a case of an external medium).[55]

6.4 Conclusions

We have seen that our representatives of three different scholastic traditions in early modern scholasticism are in substantial agreement regarding the unicity of the sense of touch (accordingly, they are in line with the view that there are only five sensory modalities), and in their opinions about the necessary existence of tactile species. Despite the sense of touch being the sense of contact, they agree that we need to assume the tactile species to explain tactile sensation and haptic intentionality. At the same time, however, it has been shown that there is a doctrinal divide between Poinsot on one side, and Suárez and Mastri/Belluto on the other concerning the issues of the tactile organ and the necessity of medium.[56] While the Thomist located the organ of touch in the subsurface nerves and in their parts that are proximate to those in contact with tangible objects, holding so in order to uphold the Aristotelian principle "a sensible placed on a sense impedes sensation," the Jesuit and the two Scotists formulated what can be called a "permissivist account," according to which

---

[54] *Ibidem*, n. 187, pp. 109–10.

[55] *Ibidem*, n. 188, p. 110.

[56] Like in the case of the issue about the unicity and the plurality of the tactile power, these controversies over the organ and the medium of touch were, of course, far from being limited to Second scholasticism. The dispute concerning the plausibility and the interpretation of the principle according to which "a sensible placed on a sense impedes sensation" mirrors the doctrinal ambiguities in formulations that can be found in Aristotle's textual *corpus*. While in his *On the Soul* II.11 and some sections of *Parts of Animals* Aristotle espouses the view that flesh is the medium of touch, and thus embraces the validity of the abovementioned axiom; in the other passages of *Parts of Animals*, he is inclined to the view according to which flesh is the tactile organ. The above-presented Second Scholastic controversy over these issues thus can be seen as a late continuation of the exegetical struggles concerning Aristotle's position. For these doctrinal tensions in Aristotle, see Redmond (2019). For *Parts of Animals*, in which he endorses the view that flesh is the organ, see (except of the reference in note 35), see *De partibus animalium*, II.1, 647a19–21, p. 1007. For a different view, according to which the tactile organ is something internal to animals' bodies, see Aristotle (1995), *De partibus animalium*, II.10, 656b36–657a1, p. 1023. For the intense debates about these issues in medieval philosophy, within the so-called "controversy between philosophers and physicians", see Hasse (2000, pp. 98–106).

all bodily parts, with the exception of inanimate ones, constitute the proper organ of touch. Not only were their theories not guided by this axiom, but they also explicitly denied its validity. To put it more charitably, they reinterpreted the principle concerning contiguity by appealing to the object's inherence. As we have seen, from Poinsot's theoretical point of view, this Suarezian-Scotistic reinterpretation significantly contributed to the gradual elimination of the sensible species in early modern philosophy.

Acknowledgements I would like to thank Chiara Beneduce and Mattia Mantovani for a number of helpful comments on the draft of this chapter. My thanks are due also to Robert Andrews and Joshua Keatley for their proofreading.

References

Primary Literature

Aquinas, Thomas (1889). *Pars prima Summae Theologiae*. In *Opera omnia* (vol. 5). Rome: ed. Leonina.

Aristotle (1995). *The Complete Works of Aristotle*, edited by J. Barnes. The Revised Oxford Translation, Vol. 1, Bollingen Series 71.2. Princeton: Princeton University Press.

Averroes (1953). *Commentarium magnum in Aristotelis de anima libros*, edited by F. S. Crawford. Cambridge, Mass.: The Mediaeval Academy of America.

Hurtado de Mendoza, Pedro (1624). *Universa philosophia, De anima*. Lyon: Ludovici Prost.

Mastri, Bartolomeo and Belluto, Bonaventura (1727). *Libros de anima, Libros de generatione & corruptione, Libros de coelo & metheoris*. In *Philosophiae ad mentem Scoti cursus integer* (vol. 3). Venice: Pezzana.

Ioannes a S. Thoma (John Poinsot) (2008). *Cursus philosophicus Thomisticus*. Vol. 3: *Naturalis philosophiae pars IV et Indices*, edited by B. Reiser. Hildesheim, New York: Olms Verlag (reprint of Rome edition, 1929).

Duns Scotus, Ioannes (2006). *Quaestiones super secundum et tertium De anima*, edited by T. B. Noone et al. The Franciscan Institute, St. Bonaventure University: The Catholic University of America Press.

Duns Scotus, Ioannes (2013). *Ordinatio*, in *Opera omnia* (vol. 14), ed. Commissio Scotistica. Civitas Vaticana: Typis Polyglottis Vaticanis.

Suárez, Francisco (1861). *Disputationes metaphysicae*. Tome 1. In *Opera omnia* (vol. 25), edited by C. Berton. Paris: Ludovicus Vivès.

Suárez, Francisco (1981). *Commentaria una cum quaestionibus in libros Aristotelis De anima* (vol. 2), edited by S. Castellote. Madrid: Sociedad de estudios y publicaciones.

Suárez, Francisco (1991). *Commentaria una cum quaestionibus in libros Aristotelis De anima* (vol. 3), edited by S. Castellote. Madrid: Sociedad de estudios y publicaciones.

Secondary Literature

Beneduce, C. (2022). "Is Touch One Sense or Several?" A Late Medieval Scientific Question. In M. Gensler, M. Mansfeld, & M. Michałowska (Eds.), *The Embodied Soul. Aristotelian Psychology and Physiology in Medieval Europe between 1200 and 1420* (pp. 253–271). Springer.

Beneduce, C. (2023). John Buridan: The Human Body at the Intersection of Natural Philosophy and Medicine. *American Catholic Philosophical Quarterly, 97*(2), 161–182.

Bos, J. (2022). Complexion (cf. temperament). In D. Jalobeanu & C. T. Wolfe (Eds.), *Encyclopedia of Early Modern Philosophy and the Sciences* (pp. 358–359). Springer.

Forlivesi, M. (2014). Mastri, Bartolomeo. In M. Sgarbi (Ed.), *Encyclopedia of Renaissance Philosophy*. Springer.

Hasse, D. N. (2000). *Avicenna's De anima in the Latin West* (Vol. 1). The Warburg Institute.

Heider, D. (2018). Suárez on Pain and Touch. *Pensamiento: Revista de Investigación e Información Filosófica, 74*(1), Núm. Extraordinario, 75–90.

Heider, D. (2021). *Aristotelian Subjectivism: Francisco Suárez's Philosophy of Perception*. Springer.

Heider, D. (2024). Pain in Second Scholasticism: Cognitive or Affective State? In D. Svoboda, P. Sousedík, & L. Novák (Eds.), *Second Scholasticism, Analytical Metaphysics, Christian Apologetics: Essays in Honour of Stanislav Sousedík* (pp. 151–189). Studia Neoaristotelica: Supplementum III). Editiones scholasticae.

Heider, D. (2026). Hurtado on External and Internal Senses. In D. D. Novotný & L. Novák (Eds.), *Pedro Hurtado de Mendoza (1578–1641): System, Sources and Influence* (pp. 153–175). Brill.

Leinsle, U. G. (2006). Dilinganae Disputationes. *Der Lehrinhalt der gedruckten Disputationen an der Philosophischen Fakultät der Universität Dillingen 1555–1648*. Schnell & Steiner.

Leinsle, U. G. (2016). Ontology and Experience–Rodrigo de Arriaga and Christoph Haunold on the 'Species Sensibiles'. In P. Dvořák & T. Machula (Eds.), *Special issue of Filosofický časopis: Explorations in late scholasticism* (pp. 103–118).

Redmond, D. (2019). Aristotle on the Organ and Medium of Touch. *Méthexis*, 31, 103–121.

Siraisi, N. G. (1990). *Medieval and Early Renaissance Medicine: An Introduction to Knowledge and Practice*. The University of Chicago Press.

Solmsen, F. W. (1961). Greek Philosophy and the Discovery of the Nerves. *Museum Helveticum, 18*(4), 169–197.

Chapter 7
The Sense of (Interoceptive) Touch in Descartes

Alison Simmons

Abstract Touch has two sides to it: an outward-facing "exteroceptive" side and an inward-facing "interoceptive" side. Through the former we perceive the many bodies with which we make contact in our environment. Through the latter we perceive our own body "from the inside." What, if anything, does Descartes have to say about the curious perceptual experience of feeling one's own body interoceptively from the inside? How does it contribute to the Cartesian function of the senses, including touch itself, as guardians of the body? And what philosophical and interpretive puzzles concerning it do Descartes' text give rise to?

Keywords Touch · Interoception · Internal senses · Bodily awareness · Bodily ownership · Descartes · Pain

7.1 Interoception: The Neglected Side of Touch

Descartes famously casts the senses in the role of bodyguards: they are given to the human being to protect and preserve its body in its travels through the world.[1] Of all the senses, touch clearly has a particularly important role to play in this project. In touch, after all, the human body comes into direct contact with other bodies poised to benefit or harm it: my hot summer foot discovers the refreshing cool water of the

[1] See Meditation 6, AT VII 83 and 89. References for Descartes' works are to Descartes (1964-1974), citing volume and page numbers. Translations are guided by Descartes (1984-1991), with some alterations of my own. Short titles are as follows: *Discourse* = *Discourse on Method*; Meditation = a particular Meditation in *Meditations on First Philosophy*; *Passions* = *Passions of the Soul*; *Principles* = *Principles of Philosophy*, citing Part and Article; *Treatise* = *Treatise, on Man*; *World* = *The World.*

A. Simmons (✉)
Harvard University, Cambridge, MA, USA
e-mail: asimmons@fas.harvard.edu

C. Beneduce et al. (eds.), *The Sense of Touch*, Studies in the History of Philosophy of Mind 32, https://doi.org/10.1007/978-3-032-16308-0_7

ocean or the barbed tentacles of an angry jellyfish. But touch has a quieter, and more neglected, side. Whether in direct contact with an external body or not, we also feel our own bodies "from the inside": I feel the internal heat of my foot, its flexion as I walk, its cooling when it hits the water and, of course, the pain of the jellyfish's sting. Unlike the other sense modalities, touch faces outward and inward at once. What, if anything, does Descartes have to say about the curious experience of feeling one's own body from the inside? And how, if at all, does it contribute to the function of the senses as bodyguards?

Let's start by focusing in for a moment on the complexity of the experience of touch. Indulge me. Place your right hand on your left. Now run your right hand over and around the left and pay attention to what you feel. This one intentional action of touching yields a multitude of sensory experiences, navigable by a mere shift in attention. First, your right hand actively observes many properties of the left: its temperature (it's warm!), its texture (ah, nice and smooth…wait, what's that little bump?), and (if you move your right hand around enough) its size and shape. Second, your left hand, while not doing any active exploring, passively feels itself touched by the right hand and observes some of its properties (those chilly fingertips!). Meanwhile, and third, each hand also feels itself: the right hand feels its own movement through space; each feels its own shape or configuration; each feels itself to be warm or cold, achy, itchy, or maybe tingly. In this exercise, each hand is observing subject one moment, observed object the next. And sometimes it is subject and object at once, as in the third case, when the hand feels itself. In this exercise, the "duality" of touch is on display: it is both exteroceptive or outward-facing, observing the many bodies surrounding us, and interoceptive or inward-facing, able to feel one body in particular from the inside.[2]

The distinction I'm invoking here between exteroceptive and interoceptive touch has a complex history, and the meanings of the terms are still being negotiated in the scientific, philosophical, and clinical literature today, so a word of clarification is in order. Exteroception is relatively uncontroversial. It is sensory perception caused by a stimulus outside the human body; experientially, it is the observation of an external object and its properties. Vision, audition, olfaction, gustation, and touch are exteroceptive senses. Interoception is the more controversial notion. I will use the term broadly to refer to what Erik Ceunen and his colleagues call the "subjective perception of the body state" or what Bud Craig calls "the material me."[3] It is sensory perception caused by a stimulus internal to the human body; experientially, it is the internal feeling of one's own body and its properties. As such, it includes the perception of bodily states involved in homeostatic regulation (e.g., hunger and thirst), proprioception (sense of bodily position), kinesthesis (sense of bodily motion), nociception (sense of bodily pain), hedonoception (sense of bodily pleasure), and thermoception (sense of bodily heat).[4]

[2] On the "duality" of touch, see Fulkerson (2014), Batty (2016), Longo (2017), and Matthen (2021).

[3] Ceunen et al. (2016, p. 12) and Craig (2002, 2015).

[4] This is an expansive use of the term. Charles Sherrington introduced the adjective "interoceptive" in 1906 more narrowly to target a set of physiological receptors stimulated by the internal function-

As described so far, the terms "exteroception" and "interoception" primarily target *what* we are perceiving, viz., things going on outside our body in the "external" world or things going on inside our own body.[5] But the terms also suggest two different *ways of perceiving* or *perspectives on* a body: "from the outside" and "from the inside." In exteroception, we encounter bodies "from the outside" in the sense that we *observe* them. What we observe is publicly available to all perceivers. (Note that exteroception, so understood, would include seeing or touching one's own foot.) In interoception, by contrast, we experience one particular body alone, our own body, "from the inside" insofar as we have a peculiarly intimate or private access to it of a sort that others do not have. We tend to capture this phenomenon in English by saying that we "feel" rather than "perceive" our body. Interoception and exteroception are thus different sorts of sensory encounter with bodies, including our own.[6]

Some will argue that touch proper should be restricted to exteroceptive touch. As such, touch includes the (typically active) haptic perception of things like the shape and location of bodies through grasping and muscular exploration (think of grasping a water bottle) and the (typically passive) tactile perception by the skin of things like the texture, temperature, motion (and possibly shape[7]) of bodies that are brought into contact with it (think of sensing a spider crawl up your arm).[8] But it does not

ing of organs of the body, such as those involved in feeling one's heartbeat, feeling the lungs fill with air, and feeling bloated (see Sherrington [1906]). When the noun "interoception" came on the scene later in the century, the concept had evolved to cover a more wide–ranging set of sensory responses to the things going on inside the body, including deep tissue pain, hunger and thirst, muscular sensations, hot flashes, and the like (see Ceunen [2016]).

[5] I will frequently put "internal" and "external" in scare quotes because the distinction is ambiguous: sometimes it refers to the distinction between mind and world (e.g., ideas of bodies versus actual bodies); other times, as here, it refers to the distinction between one's own human body and bodies outside of its boundaries. For the most part I will be using the distinction in the latter sense.

[6] I will assume for present purposes that both exteroception and interoception are genuinely perceptual phenomena. There was a tendency in twentieth century analytic philosophy to treat many of what I'm calling interoceptive sensations, such as those of pain, as non-perceptual mental states, i.e., as "mere sensations" devoid of representationality or intentionality. On this view, the genitive in "sensation of pain" is a descriptive genitive while the genitive in, say, "sensation of color" is an objective genitive; the verb "sense" in "to sense pain" is non-accusative while the verb "sense" in "to sense a color" is accusative. My focus here will be on Descartes, who, *pace* many of his twentieth–century commentators, does not make this distinction between mere sensation and perception. As a matter of terminology, the terms "sensation" and "sensory perception" are interchangeable for Descartes (*Principles* IV.189, AT VIII-A 316). As a matter of substance, I've argued elsewhere that all sensory perceptions (and so all sensations) are representational or intentional for Descartes; all the genitives are objective genitives and all the verbs of sense are accusative. See Simmons (2016); see also Baker and Morris (1996, pp. 128–138) and Clarke (2003, pp. 194–195).

[7] For a recent defense of spatial perception through (passive) contact with the skin, see Haggard et al. (2017).

[8] Fulkerson (2014) discusses notable exceptions to these generalities: haptic perception might be passive (e.g., when I forcibly place your hand on a piece of ice and you perceive its coldness); tactile perception might be active (e.g., you actively attend to the motion of the spider crawling up your arm).

include such interoceptive experiences as hunger, thirst, proprioception, kinesthesis, nociception, hedonoception, thermoception, and so on. These, my imagined interlocutor insists, are not cases of touch but something else. Call it "bodily awareness" if you like. I'm inclined to be more liberal, at least at the start of inquiry, and include interoceptive bodily awareness under the umbrella of touch. Whether we officially declare interoceptive experiences cases of touch or not, all parties must agree that they are intimately linked with exteroceptive touch (causally, functionally, and phenomenologically).[9] Indeed we are increasingly learning that the boundaries between them are far from clear.[10]

It is interoceptive touch that I am chiefly interested in here. It is receiving an enormous amount of attention today among both scientists and philosophers.[11] It is even in the general press and making Nobel news (with a reference to Descartes included).[12] It is therefore a good moment to go back to some of our sources to interrogate the history of this curious sensory phenomenon. My target is Descartes: what, if anything, does Descartes have to say about interoceptive touch? Some, like Antonio Damasio, say: nothing, Descartes failed to recognize that the mind is embodied in an organism![13] Others, such as Frédérique de Vignemont, have gone so far as to say, "Descartes might […] have been the first advocate of the significance of interoception for bodily awareness."[14] Spoiler alert: I'm on de Vignemont's side. I aim to show that Descartes was, in fact, fully attuned to the significance of interoception in our cognitive economy and that his observations raise important questions about it.[15]

7.2 Descartes on Touch

Before turning to interoception in particular, it is worth reviewing what Descartes has to say about touch more generally. Touch is, of course, part of the human sensory system. As such, Descartes will insist that touch sensations are among the obscure and confused perceptions of the mind that arise from its union with body.

[9] For discussion, see, e.g., O'Shaughnessy (1995), Fulkerson (2014), and Batty (2016).

[10] See Bertnson and Khalsa (2021).

[11] See, e.g., Fulkerson (2014), Craig (2015), de Vignemont and Alsmith (2017), de Vignemont (2018), Tsakiris and De Preester (2018), and Alsmith and Longo (2023).

[12] See Robson (2021) and https://www.nobelprize.org/prizes/medicine/2021/press–release/. Last accessed on December 26, 2024.

[13] Damasio (1994, pp. 245–252).

[14] De Vignemont (2020); see also Martin (1995).

[15] Interoceptive touch has not been entirely overlooked in Descartes scholarship. French scholarship has been most attentive to it: Martial Gueroult's *Descartes selon l'ordre des raisons* (1953) is an important foundational text for this topic; Jean-Luc Marion's *Sur la pensée passive* (2013) also explores it in some detail. In English-language scholarship, monographs such as Baker and Morris (1996), Carriero (2009), and Clarke (2003) have given it some attention. More recently, Curley and Koivuniemi (2015) and Chamberlain (2018, 2023) have made it the subject of scholarly articles.

As such, they fail to resemble (or clearly and distinctly represent) the distal properties that cause them: sensations of warmth are caused by bodies that increase the motion in individual nerve fibers terminating in the skin; sensations of smoothness are caused by bodies that tug evenly on nerve fibers terminating in the skin; sensations of pain are caused by bodies that produce violent motions or ruptures in the nerve fibers.[16] Until we engage in proper epistemic meditation, this lack of resemblance between physical cause and sensational effect leads us into epistemic trouble: we tend to judge, e.g., that bodies are warm in just that way they sensorily appear to be and that where we feel no bodies there are no bodies.[17] Touch, in fact, comes in for extra vigorous epistemic finger wagging. Having argued that a vessel full of gold or lead in fact contains no more matter, metaphysically, than a so-called "empty" vessel, Descartes writes: "This may seem very strange to many people whose reason extends no further than their fingertips and who suppose that there is nothing in the world except what they touch."[18]

But, like all other sensations, touch sensations are useful for keeping our own bodies healthy and navigating them around a world of other objects. That's their job.[19] As such, they are "sufficiently clear and distinct" (*satis clarae & distinctae*) and "more frequently report the truth than the false" (*multo frequentius verum indicare quàm falsum*).[20] So far, Descartes' discussion of touch is in keeping with his discussion of the other four so-called external senses: sight, hearing, taste, and smell.

Touch does have some distinctive features. Unlike the other sense modalities, touch does not have a special dedicated sense organ. It is a function of the nerves (or, more precisely, the filaments or fibers inside the nerves) that terminate in the skin all over the body. As for the ancient view that the skin is the organ of touch, Descartes writes: "although external objects do not ordinarily touch [the nerve filaments] directly, but rather touch the skin surrounding them, there is no more reason to think of the skin as the sense organ than there is to think of gloves as the sense organ when we feel something while wearing gloves."[21] Nor does touch have a single special object or (in Aristotelian lingo) proper sensible, as sight has color and smell odor. The qualities that Descartes recognizes as accessible through touch are

---

[16] See *World* AT XI 5–6, *Treatise*, AT XI 144, and *Principles* IV.191, AT VIII-A 318.

[17] See, e.g., Meditation 6, AT VII 83.

[18] *World* AT X 21: "ce qui peut sembler bien étrange à plusieurs dont la raison ne s'étend pas plus loin que les doigts, et qui pensent qu'il n'y ait rien au Monde que ce qu'ils touchent."

[19] Meditation 6, AT VII 83.

[20] Meditation 6, AT VII 83 and AT VII 89. Even considered in its role of self–preservation Descartes seems to have a low regard for touch, deeming it one of the least sensitive of the senses—trailed only by taste (*Principles* IV.192, AT VIII-A 218). He offers as an explanation that the nerves of touch are moved only by the third (and coarsest) element of matter, while the nerves of sight are moved by matter of the more subtle second element (*Treatise,* AT VI 151).

[21] *Treatise,* AT XI 142–14: "Car encore que, pour l'ordinaire, ce ne soit pas eux qui soient immédiatement touches par les object extérieurs, mais les peaux qui les environment, il n'y a pas toutefois plus d'apparence de penser que ce sons ces peaux qui sont les organs du sense, que de penser, lorsqu'on manie quelque corps, étant ganté, qu ce sont les gants qui servent opur le sentir." Descartes was wrong. We now think that the mechanoreceptors for touch are, after all, in the skin.

quite varied: hotness and coldness, wetness and dryness, hardness and softness, heaviness and lightness, smoothness and roughness.[22] They extend to the Aristotelian common sensibles: motion, rest, shape, and size. And Descartes sometimes includes pain and tickles (*chatoüillement*) among the qualities sensed by touch.[23]

With the possible exception of sensing pain and tickles, Descartes' treatment of touch so far does not suggest much recognition of interoceptive touch. He seems to be focused on how "external" objects, objects beyond the limits of our own bodies, are perceived by touch when they impact our body. It is even possible that the reason Descartes includes pain and tickles among the tactile qualities is that, spontaneous foot cramps aside, they are often the result of something else impacting our own body, like knives and feathers (though, of course, we do not attribute the pain and tickle qualities to knives and feathers the way we do colors and shapes).

Now one might interject: Descartes elsewhere draws a distinction between the "external" and "internal" senses. Perhaps what I'm calling "interoceptive touch" is just Descartes' internal senses. If so, then I should be looking for interoception not in his discussion of touch, but in his discussion of the internal senses. In fact, most Descartes interpreters who are attentive to interoceptive phenomena (or bodily awareness) discuss it under the label "internal sense."[24] While there is indeed some overlap between the exteroception-interoception distinction and Descartes' external-internal sense distinction, I'll argue below that the two distinctions do not line up neatly. For the moment, I'll just note that in Descartes' most sustained discussion of the external-internal sense distinction in *Principles* IV.191, sensations of pain and tickle explicitly fall under the external sense of touch rather than the internal senses, which makes for a mismatch between the distinctions. (Though, as we'll see, sensations of pain and tickle are a bit of a moving target in Descartes' shifting perceptual taxonomies.)

7.3 Descartes on Interoception

7.3.1 The Meditations

The fact that interoceptive touch doesn't obviously appear in Descartes' official catalogues of the sense modalities doesn't mean that he's not aware of it or that he takes no theoretical interest in it. Although you might miss it if you aren't looking for it, interoception figures prominently in his 1641 *Meditations on First Philosophy*. It makes an appearance early on in Meditation 1. As Descartes' meditator begins to raise doubts about her sense-based beliefs, she targets the following: the beliefs

[22] *Treatise*, AT XI 144 and *Principles* IV.191, AT VIII-A 318.

[23] *Treatise*, AT XI 144 and *Principles* IV.191, AT VIII-A 318.

[24] See Baker and Morris (1996), Carriero (2009), Clarke (2003), and Chamberlain (2018, 2023). Curley and Koivuniemi (2015) are a rare exception.

"that my eyes are open, that I am moving my head and stretching out my hands" and, further, "whether I even have such hands or such a body at all."[25] By the end of the Meditation 1, the meditator announces, "I shall consider myself as not having hands or eyes, or flesh, or blood or senses, but as falsely believing that I have all these things."[26]

Now, it's *possible* that Descartes' meditator came to believe that her eyes are open, that she is shaking her head, that she is stretching her hands out and that she has a body by means of the exteroceptive senses of sight and touch—by looking at and touching her body.[27] But that's unlikely. First, we typically *feel* that our eyes are open, that we are shaking our heads, that we are stretching our hands from the inside. These are interoceptive sensory acts. Second, as we'll see later, it is not through exteroceptive sensing but only through interoceptive sensing that Descartes thinks we come to experience a body (or part of a body) *as our own*. Exteroceptively seeing my open eyes in a mirror, seeing or holding my shaking head, and seeing or touching my outstretched hand would not lead me to experience these things *as mine* unless they were associated with simultaneous interoceptive perceptions. So interoceptive touch seems to be the source of one of the most basic sense-based beliefs that Descartes targets for doubt in Meditation 1, viz., that we have bodies.

Having successfully doubted the senses into submission in Meditation 1, and having subsequently used her pure intellect to establish some key metaphysical truths (I exist, God exists, extension is the essence of body), the meditator returns to the senses in Meditation 6 to re-appraise them. Her self-appointed task here is to "consider what I should now believe about [the senses]."[28] To that end, she reviews the things she took herself to perceive by the senses; she then reviews the doubts she raised about them in Meditation 1; and, finally, she re-instates a number of sense-based beliefs with some important caveats. Interestingly, this entire exercise centers around interoceptive touch.[29]

In the meditator's review of her former sense-based beliefs, the very first belief she considers is the one I called attention to above: "I sensed myself to have a head, hands, feet, and other limbs making up the body that constitute that body that I regard as part of myself or even as my whole self."[30] As I suggested previously, the

---

[25] Meditation 1, AT VII 19: "Age ergo somniemus, nec particularia ista vera sint, nos oculos aperire, caput movere, manus extendere, nec forte etiam nos habere tales manus, nec tale totum corpus."

[26] Meditation 1, AT VII 22–23: "considerabo meipsum tanquam manus non habentem, non oculos, non carnem, non sanguinem, non aliquem sensum, sed hæc omnia me habere falso opinantem."

[27] Weirdly, a related passage from the *Principles* suggests as much: "They were inclined instead to take 'themselves' to mean only their bodies—the bodies which they *saw with their eyes and touched with their hands*" (AT VIII-A 9, italics mine): "sed contra potius intellexerunt sola sua corpora, quæ oculis videbant, et manibus palpabant."

[28] Meditation 6, AT VII 74: "considerabo quid mihi nunc de iisdem sit credendum."

[29] Carriero (2009) and Chamberlain (2018) note this as well.

[30] Meditation 6, AT VII 74: "Primo itaque sensi me habere caput, manus, pedes, et membra cætera ex quibus constat illud corpus, quod tanquam mei partem, vel forte etiam tanquam me totum spectabam."

belief that I have a body that I consider to be part of myself (or my whole self) is grounded in interoception. (I'll look more closely at it below.)

The second thing that the meditator recounts herself to have sensed is that "this body [viz., her body] was situated among many other bodies that could affect it in various favorable or unfavorable ways."[31] This fact was gauged by the interoceptive feelings of pleasure and pain that external bodies cause her to feel in her body. So even our beliefs about the benefit or harm of other, "external" bodies are grounded, at least initially, in interoceptive experiences of pain and tickling.

The third thing the meditator recalls is that "I also had sensations within me of hunger, thirst, and other such appetites, and also of physical propensities toward cheerfulness, sadness, anger and similar emotions."[32] This is another list of interoceptive sensory experiences put forward for examination, doubt, and re-examination.

Only after all this does the meditator turn to what she seemed to sense in "external" bodies, that is, bodies outside the limits of her own: "And outside me, besides the extension, shapes and movements of bodies, I also sensed their hardness and heat, and other tactile qualities, and further light and colors, and odors, and flavors, and sounds, the variety of which enabled me to distinguish the sky, the earth, the seas, an all other bodies, one from another."[33] Only now do the exteroceptive senses turn up, reporting on bodies "outside me." And note that the meditator begins the list with exteroceptive touch. Touch seems to provide a natural transition from inner to outer.

Now of course all of these sense-based beliefs about bodies, one's own and "external" ones, fell into doubt in Meditation 1. When the reasons for doubting them are reviewed in Meditation 6, something interesting happens: in addition to reviewing the dream argument from Meditation 1 that casts all sense-based beliefs about bodies into doubt, the meditator introduces a new illusion into the mix, viz., phantom limb pain.[34] The meditator seems particularly interested in the fact that even this sort of interoceptive sensory experience is subject to illusion: "So even in my own case it was apparently not quite certain that a particular limb was hurting, even if I felt pain in it."[35]

[31] *Ibidem*: "sensique hoc corpus inter alia multa corpora versari, a quibus variis commodis vel incommodes affici potest."

[32] *Ibidem*: "Atque, præter dolorem et voluptatem, sentiebam etiam in me famem, sitim, aliosque ejusmodi appetitus; itemque corporeas quasdam propensiones ad hilaritatem, ad tristitiam, ad iram, similesque alios affectus."

[33] *Ibidem*, AT VII 74–75: "foris verò, præter corporum extensionem, &t figuras, & motus, sentiebam etiam in illis duritiem, & calorem, aliasque tactiles qualitates; ac præterea lumen, & colores, & odores, & sapores, & sonos, ex quorum varietate cælum, terram, maria, & reliqua corpore ab invicem distinguebam."

[34] *Ibidem*, AT VII 77.

[35] *Ibidem*, AT VII 77: "ideoque etiam in me non plane certum esse videbatur membrum aliquod mihi dolere, quamvis sentirem in eo dolorem." Descartes' early unpublished work, *The World*, explores a different pain illusion that mixes interoceptive and exteroceptive touch: a solider feeling pain takes himself to have been wounded in battle; when the surgeon removes his armor he must

The meditator is now ready to remove the doubts of Meditation 1: "I do not think that everything I seem to have acquired from the senses should be admitted, but neither do I think that everything should be called into doubt."[36] After establishing that bodies exist, the meditator enquires into the particulars. The first set of sense-based beliefs that she considers include: (a) that she has a body, (b) that when she's hungry and thirsty her body needs food and drink; and (c) that she is united to or intermingled with that body.[37] And, once again, it is interoceptive touch sensations of pain, hunger, and thirst that the meditator invokes when she reinstates these beliefs in the famous "sailor in his ship" passage:

> There is nothing that my own nature teaches me more expressly than that I have a body that is unwell when I sense pain, that needs food and drink when I suffer hunger or thirst, and so on. So I should not doubt that there is some truth in this.[38]

Of course, the reinstated beliefs have changed a bit: I am no longer misled into thinking that hunger and thirst exist in my stomach and throat, or pain in my foot, in quite the way that they appear to. But I'm nevertheless secure in my belief that *I have a stomach and throat and foot* and (if all is working well) that *they are in need of food and drink and medical attention.* Interoceptive touch thus grounds the initial set of cleaned up sense-based beliefs that the meditator re-instates concerning her embodied nature and about the needs of her body.

The meditator goes on to consider what counts as a "true error of nature" (*verus erro natura*) in which we have a sensation that not only misrepresents the metaphysical nature of body (something we are now epistemically alert to) but also misleads us about our bodily well-being. The discussion centers on thirst. The dropsy patient feels thirst at a time when drink would not be beneficial for him.[39] Of course, true errors of nature like this can also turn up in exteroceptive perception: a color-blind person might have trouble picking out the ripe from unripe berries, resulting in indigestion. But Descartes focuses our attention on an interoceptive experience of our own body.

Finally, toward the end of Meditation 6, the meditator introduces the idea, developed earlier in the *Treatise* and *Dioptrics,* that God has set up an "institution of nature" between types of brain state (in particular, types of pineal gland state) and (entirely dissimilar) types of sensation in the mind such that the occurrence of one is hard-wired to give rise to an occurrence of the other.[40] The examples that the

---

"inform [the soldier] of what he was feeling [*l'avertir de ce qu'il sentoit*]," viz., a buckle or strap caught under the armor rather than a battle wound (AT XI 6).

[36] Meditation 6, AT VII 77–78: "non quidem omnia, quæ habere videor a sensibus, puto esse temere admittenda; sed neque etiam omnia in dubium revocanda."

[37] *Ibidem*, AT VII 80–81.

[38] *Ibidem*, AT VII 80: "Nihil autem est quod me ista natura magis expresse doceat, quam quod habeam corpus, cui male est cùm dolorem sentio, quod cibo vel potu indiget, cùm famem aut sitim patior, & similia; nec proinde dubitare debeo, quin aliquid in eo sit veritatis."

[39] Meditation 6, AT VII 85.

[40] For its introduction in the *Treatise,* and *Dioptrics,* see AT XI 143 and AT VI 130. For a recent discussion of the institution of nature, see Mantovani (2022).

meditator invokes to illustrate this institution of nature are, yet again, cases of interoceptive touch, viz., pain and thirst:

> […] when the nerves in the foot are set in motion in a violent and unusual manner, this motion, by way of the spinal cord, reaches the inner parts of the brain, and there gives the mind a sign for sensing something, namely the pain as if it existed in the foot. This stimulates the mind to do its best to get rid of the cause of the pain, which it takes to be harmful to the foot […] In the same way, when we need drink, there arises a certain dryness in the throat; this sets in motion the nerves of the throat, which in turn move the inner parts of the brain. This motion produces in the mind a sensation of thirst, because the most useful thing for us to know about the whole business is that we need drink in order to stay healthy.[41]

The critical examination of the senses in Meditation 6 is thus shot through with interoceptive touch. And for good reason: one of the chief concerns of Meditation 6 is to get clear about the mind's relation not to any old body, but to its own body: it is really distinct from it insofar as it can exist apart from it (once the body dies), but it is also united to it in this life in such a way that it forms a single thing with it. The sensory experience that most needs close scrutiny in the *Meditations* is therefore interoceptive touch, which testifies to our having a body (and to its wellbeing or lack thereof) in the first place and which also can mislead us into thinking we can have no existence without it. (We are under no illusion that we cannot exist without most external bodies.)

7.3.2 Interoception outside the Meditations

Outside the *Meditations*, too, Descartes has more to say about interoceptive touch than you might think. In his early *Treatise on Man* (written in the 1630s, but not published until after Descartes' death) and *Dioptrics* (1637), he shows an interest in its underlying physiology. He explores the physiology of one critical form of interoception, viz., proprioception. What, Descartes asks, occasions "the soul to sense that the arm is turned toward object B"?[42] His answer: the effluence of animal spirits from the pineal gland into the nerves down toward the muscles in the arm. More generally, the position of those nerves in the brain that extend to the muscles of our limbs are "ordained by nature to enable the soul […] to know the place occupied by each part of the body it animates relative to all others (Fig. 7.1)."[43]

---

[41] Meditation 6, AT VII 88: "cùm nervi qui sunt in pede vehementer & præter consuetudinem moventur, ille eorum motus, per spinæ dorsi medullam ad intima cerebri pertingens, ibi menti signum dat ad aliquid sentiendum, nempe dolorem tanquam in pede existentem, a quo illa excitatur ad ejus causam, ut pedi infestam, quantum in se est, amovendam […] Eodem modo, cùm potu indigemus, quædam inde oritur siccitas in gutture, nervos ejus movens & illorum ope cerebri interiora; hicque motus mentem afficit sensu sitis, quia nihil in toto hoc negotio nobis utilius est scire, quam quod potu ad conservationem valetudinis egeamus."

[42] *Treatise,* AT XI 181: "cela même donnerait occasion à l'âme, de sentir que le bras se tourne vers l'objet B."

[43] *Dioptrics* 6, AT VI 135.

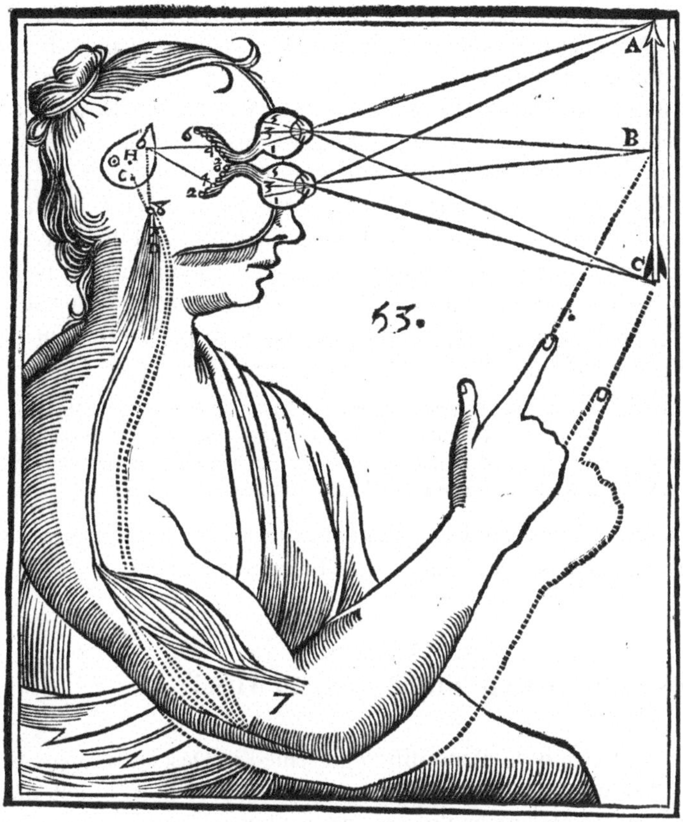

Fig. 7.1 Gerard van Gutschoven. René Descartes, *Traité de l'homme* (Paris, 1664), 79. Source: *Traité de l'Homme* (Paris, 1677), 74. The Huntington Library, San Marino, California, RB 701994. Open domain

Descartes describes another critical form of interoception, the kinesthetic experience in the soul as it registers the changing motions of the arm, which change the position of nerves in the brain.[44] In his famous discussion of perceiving distance "as if by natural geometry" (*comme par une géométrie naturelle*) in the *Treatise on Man*, Descartes sometimes suggests that a blind man is aware of the distance between two hands holding sticks as they triangulate on a distant object and that a sighted man is similarly aware of the distance between his eyes and the direction in which he is looking when he focuses on objects near or far.[45] He twice in the *Treatise on Man* discusses proprioceptive and kinesthetic illusions caused by external

[44] *Treatise,* AT XI 182.

[45] *Treatise,* AT XI 160. But see Hatfield (2015) for an alternative account.

constraint on our muscles, and the object illusions that result from them.[46] And in a letter to the Marquess of Newcastle of October 1645 he details the physiology underlying hunger and thirst.[47]

In his last published work, the *Passions of the Soul* (1649), Descartes chronicles in some detail the various interoceptive bodily sensations that accompany the passions. Experience teaches us what's happening in our body when we undergo the passions: in love we feel that "the pulse has a regular beat, but is much greater and stronger than normal; we feel a gentle heat in the chest; and the digestion of food takes place very quickly in the stomach."[48] In hatred, by contrast, we feel "the pulse is irregular, weaker and often quicker; we feel chills mingled with a sort of sharp, piercing heat in the chest; and the stomach ceases to perform its function, being inclined to vomit and reject the food we have eaten."[49] In joy, we feel

> the pulse is regular and faster than normal, but not so strong or great as in the case of love; we feel a pleasant heat not only in the chest but also spreading into all the external parts of the body along with the blood which is seen to flow copiously to these parts; and yet we sometimes lose our appetites because our digestion is less active than usual.[50]

Again, these are not just things happening in the body, but things we experience sensorily as happening in our bodies as we undergo a passion.

7.4 Why Is Interoception So Easy to Miss?

If interoceptive touch is so plentifully represented in Descartes' texts, why is it so easy to miss? Why hasn't there been more attention paid to it? I think there are two reasons, one physio-phenomenological and one textual.

First, Descartes himself gives us a good physio-phenomenological reason why we might tend to overlook it: it falls victim to habituation. In all sensory perception, we stop noticing things that are there all the time. Sensory perception requires change and novelty: "we cannot sense any body unless it is the cause of some change in our sense organs—that is, unless it somehow moves the minute parts of

---

[46] *Treatise*, AT XI 160–61 and AT XI 181–182.

[47] Descartes to the Marquess of Newcastle, October 1645 AT IV 326–328.

[48] *Passions* 97, AT XI 401–402: "le battement du pouls est égal et beaucoup plus grand et plus fort que de coutume; qu'on sent une douce chaleur dans la poitrine, et que la digestion des viandes se fait fort promptement dans l'estomac."

[49] *Passions* 98, AT XI 402: "le pouls est inégal et plus petit, et souvent plus vite; qu'on sent des froideurs entremêlées de je ne sais quelle chaleur âpre et piquante dans la poitrine; que l'estomac cesse de faire son office et est enclin à vomir et rejeter les viands qu'on a mangées."

[50] *Passions* 99, AT XI 402–403: "le pouls est égal et plus vite qu'à l'ordinaire, mais qu'il n'est pas si fort ou si grand qu'en l'amour; et qu'on sent une chaleur agréable qui n'est pas seulement en la poitrine, mais qui se répand aussi en toutes les parties extérieures du corps avec le sang qu'on voit y venire n abondance; et que cependant on perd quelquefois l'appétit, à cause qu la digestion se fair moins que de coutume."

the matter of which these organs are composed."[51] And so, "bodies that are usually there are the least perceivable, and those that are always there can never be perceived at all."[52] This is a bit of an exaggeration but the general point is well taken. In the case of exteroceptive touch, you don't much notice the ground under your feet until you step on a pebble or in a hole. Interoceptive touch is particularly susceptible to habituation. After all, as Descartes' meditator points out in Meditation 6, you are never separated from your own body in the way that you are from other bodies; it's always there.[53] So long as all is going well, you barely notice it at all: "the heat of the heart is very great, but we do not feel it because it is usually there. The weight of our body is not small, but it does not discomfort us."[54] Under these circumstances, you have to make an effort to bring the body (or body part) into view, as you may do in a yoga or mediation class.[55] It's only when things change, usually for the worse, that we take notice: the pain the pebble or hole causes in the foot, or the ache from walking up an incline, is something we feel. Interceptive touch hovers in the background of our sensory experience most of the time. And so it tends to be overlooked. That doesn't mean it's not there or that it's not important (in our lives *or* in the text). In fact, Descartes suggests to Arnauld that we are and must be aware of our embodiment to some extent all the time: if we weren't, we would not bother to engage in intentional action at all.[56]

Second, perhaps precisely because it is phenomenologically recessive in our lives, interoceptive touch rarely gets explicitly theorized or even taxonomized. This is certainly true in Descartes' work. "Interoception" and "exteroception" are not Cartesian terms. Nor does the distinction map neatly onto Descartes' own taxonomies. The natural place to look for a mapping, as I noted earlier, is Descartes' distinction between external and internal senses. His external senses include the usual exteroceptive modalities: sight, hearing, taste, smell and touch. His internal senses, however, include a strange pair: natural appetites (e.g., hunger and thirst) and passions (e.g., love and sadness).[57] This distinction does not map onto historical precedents of the internal-external sense distinction.[58] Nor does it map onto the distinction

[51] *World*, AT XI 21–22: "nous ne saurions sentir aucun corps, s'il n'est cause de quelque changement dans les organs de nos sense, c'est à dire s'il ne remuë en quelque façon les petites parties de la matière dont ces organs sont composez."

[52] *World*, AT XI 21: "ce sont celles qui y sont le plus ordinairement, qui le peuvent être [senties] le moin, & qu celles qui y sont toujours, ne le peuvent étre [senties] jamais."

[53] Meditation 6, AT VII 76.

[54] *World*, AT XI 21: "La chaleur de nôtre coeur est bein grande, mais nous ne la sentons pas, à cause qu'ell est ordinaire. La pesanteur de nôtre corps n'est pas petite, mais elle ne nous incommode point."

[55] Chamberlain (2017) makes this point.

[56] Descartes to Arnauld, 29 July 1648, AT V 222; see also Descartes to Elisabeth, 28 June 1643, AT III 694.

[57] See Meditation 6, AT VII 76–77 and *Principles* IV.190, AT VIII-A 316–317.

[58] Traditionally, the distinction is drawn between sight, smell, hearing, taste, and touch on the one hand (external senses) and common sense, imagination, memory, *phantasia*, and the estimative senses on the other (internal senses). See e.g. Knuutila and Kärkkäinen (2013) and Casini (2014).

I'm making between exteroception and interoception. Hunger and thirst are interoceptive perceptions. And at least the somatic accompaniments of the passions are (e.g., the feeling of butterflies in the stomach that accompanies anxiety). But Descartes' internal senses extend beyond the domain of sensing one's own body: in the passions we are first and foremost sensing our soul, according to Descartes. Meanwhile, many paradigm cases of interceptive touch, such as proprioception, kinethesis and thermoception, are not included on Descartes' list of internal senses, though he clearly does recognize that they exist, as we've seen. Two other paradigm cases of interoception, pain and tickling, jump back and forth between the two categories in Descartes' texts. He seems not to know what to do with them. Descartes sometimes identifies pain and tickling as internal sensations: "what could be more internal than pain?" the meditator asks in Meditation 6.[59] But, as I noted earlier, pain is sometimes treated as a tactile quality of the external sense of touch alongside rough and smooth, wet and dry, hot and cold, hard and soft.[60] So the interoceptive-exteroceptive and internal-external distinction do not map neatly onto each other.

Descartes' last published work, the *Passions of the Soul*, gives us a more promising taxonomy. Here Descartes distinguishes *three* categories of perception that originate from nerves in the body: those we refer to external bodies, those we refer to our own body, and those we refer to the soul.[61] With this trichotomy, Descartes separates the two former internal senses (natural appetites and passions) and makes some adjustments: "The perceptions we refer to our body or to certain of its parts are those of hunger, thirst and our other natural appetites. To these we may add pain, heat *and the other states we feel as being in our limbs, and not as being in objects outside us*."[62] The list starts with familiar internal sensations (hunger, thirst), but then adds pain, heat "and other states we feel as being in our limbs." The passions, meanwhile, are peeled off from the internal senses and get their own category, perceptions referred to the soul. What seems to unite the new group of sensations that are referred to the body is that we "feel [them] as being in our limbs." Here, at last, have a category that looks like interoceptive touch. Descartes only comes to it as a category of its own in his last published work. (Even though it's been playing a critical role all along.)

Interestingly, we find in his discussion of "perceptions we refer to our body" Descartes' recognition of the fact that, in touch, exteroception and interoception are closely intertwined: "at the same time and by means of the same nerves we can feel

---

[59] Meditation 6, AT VII 77: "quid dolore intimius esse potest?" See also *Treatise*, AT XI 176.

[60] *World*, AT XI 5; *Treatise*, AT XI 142 and 162; and *Principles* IV.191, AT VIII-A 318. For evidence of the instability of Descartes' own internal–external sense distinction, in Meditation 6 he identifies the pleasant taste of some food as an *internal* rather than an external sensation (AT VII 83).

[61] *Passions* I.22, AT XI 345.

[62] *Passions* I.24, AT XI 346–347, italics mine: "Les perceptions que nous rapportons à notre corps ou à quelques-unes de ses parties sont celles que nous avons de la faim, de la soif et de nos autres appétits naturels, à quoi on peut joindre la douleur, la chaleur et les autres affections que nous sentons comme dans nos membres, et non pas comme dans les objets qui sont hors de nous."

the cold of our hand and the heat of a nearby flame, or on the other hand, the heat of our hand and the cold of the air to which it is exposed."[63] One and the same physiological system gives rise to two quite different of touch experiences that require only a shift in attention.

7.5 Why Does it Matter?

Why am I harping on all this? Why is interoception important to the Cartesian story? I suggested earlier that interoception is critical to Descartes' account of the human being. It's time to unpack and defend that claim. The human being is an embodied mind, i.e., a mind united to a body.[64] The embodied mind has different cognitive faculties tasked with different jobs: the intellect is for doing metaphysics; the senses are for taking care of ourselves as embodied minds. "[T]he conduct of life depends entirely on the senses," Descartes tells us in the opening of the *Dioptrics*.[65] More specifically, the senses help us to navigate through a world of bodies poised to benefit or harm us: "The proper purpose of the sensory perceptions given me by nature is simply to signify to the mind what is beneficial or harmful for the composite of which the mind is a part."[66] The senses carry out their job by representing or showing us (*exhibere*) bodies not as they are in themselves, but "as they are related to us and benefit or harm us" (*quatenus ad nos referuntur & prodesse possunt aut nocere*).[67] Now this is the job of all the senses, exteroceptive and interoceptive alike. But interoceptive touch has a unique role to play in the embodied mind's mission to preserve itself.

7.5.1 Interoception and Bodily Ownership

First, it is through interoception that we learn that we *have a body of our own* in the first place. Like all sensations, interoceptive sensations are represented as existing in (i.e., as modes of) body. Just as colors and thermal qualities are naturally represented as modes of bodies, so too pains and tickles are naturally represented to us as modes of body.[68] But when interoceptive sensations are represented as modes of

[63] *Passions* I.24, AT XI 347: "Ainsi nous pouvons sentir en même temps, et par l'entremise des mêmes nerfs, la froideur de notre main et la chaleur de la flamme dont elle s'approche, ou bien, au contraire, la chaleur de la main et le froid de l'air auquel elle est exposée."

[64] See, e.g., *Treatise*, AT XI 119 and Meditation 6, AT VII 87.

[65] *Dioptrics* 1, AT VI 81: "Toute la conduit de nôtre vie depend de nos sense."

[66] Meditation 6, AT VII 83: "sensuum perceptionibus, quæ proprie tantùm a naturâ datæ sunt ad menti significandum quænam composito, cujus pars est, commoda sint vel incommode."

[67] Descartes to More, 5 February 1649, AT V 271; see also *Principles* II.3, AT VIII-A 41.

[68] See, e.g., *Principles* I.67, AT VIII-A 33. One might wonder how any sensation winds up being represented as a mode of body at all, given that it is, according to Descartes, a mode of mind. Is the

body, they are represented as modes of *my* body or even as modes of a (partly) material *me*. And this in turn forges an important distinction, representationally or experientially, between my body (or my embodied self[69]) and other "external" bodies. Pains and tickles "we feel as being in *our* limbs, and not as being in objects outside us."[70] Again: "I also had sensations *within me* of hunger, thirst, and other such appetites" that are contrasted with sensations "*outside me*" of "extension, shapes, and movements of bodies [and] hardness and heat and of other tactile qualities."[71] So: interoceptive touch sensations are experienced as modes of my body; exteroceptive touch sensations are experienced as modes of other, external bodies. Distinguishing between my body and other external bodies is obviously critical to the project of

representation of a sensation as a mode of body merely doxastic, i.e., a belief that results from a (mis)judgment? Or is it built into sensory experience itself? If the latter, is it built in psychologically and naturally by what Descartes describes as a "teaching of nature" (Meditation 6, AT VII 76 and 80–83) or a "third-grade" (pre-doxastic) judgment (Sixth Replies, AT VII 437–438)? Or is it already built into the "second-grade" sensations that result from the "institution of nature" established by God between pineal states and sensations? This is a matter of considerable interpretive debate and the texts are multiply ambiguous. My own view is that while Descartes insists that our *belief* that bodies are colorful and painful is grounded (at least in part) in the (mis)judgement that the bodies that cause our sensations of color and pain also resemble them, it is *also* grounded in the fact that those bodies already *sensorily appear* to be colorful and painful; the judgment does not *make them so appear* but provides reason to affirm the appearance. In many passages the language of "judgment" is clearly that of the natural or psychological "third–grade" variety and not the doxastic variety that depends on the will. Moreover since doxastic judgments are, according to Descartes, formed by the will's affirmation or denial of *what is already represented to the mind by a perception* (sensory or otherwise), it seems that any doxastic judgment that colors and pains are in body *must* be based at least in part on a perceptual representation of them as such. There are also passages, however, that suggest that the representation of sensations as modes of body is not judgmental at all but are rather built directly into the second–grade sensations themselves. Meditation 6, for example, suggests that God's institution of nature pairs a certain type of pineal state with the sensation of "pain-in-the-foot" (AT VII 87).

[69] "Self" is another moving target in Descartes work. Descartes famously identifies the self (or I) with the mind in Meditation 6: "and so it is certain that I am really distinct from my body and can exist without it" (*certum est a corpore meo revera essence distinctum, et absque illo posse existere*) (AT VII 78). More accurately, he identifies the self with the pure intellect [and will?] here, since he goes on to insist that "I can clearly and distinctly understand myself as a whole without [the faculties of sensing and imagining]" (*facultates imaginandi et sentiendi, sine quibus totum me possum clare et distinct intelligere*) (AT VII 78). And yet he also identifies a more expansive self (or I): the embodied mind or mind-body union. He singles out "my nature" as what "God has bestowed on *me as a combination of mind and body*" (*mihi, ut composito ex mente et corpore, a Deo tributa sunt*) (AT VII 82, italics mine). And most of the cognitive activities of the *Meditations* are activities that only an embodied mind (a mind-body union) have, since they involve the use of the senses, imagination and memory. The "I" that engages in these activities, and so the "I" of the *Meditations*, can only be an embodied mind. For discussion see Chamberlain (2020).

[70] *Passions* 24, AT XI 346–347, italics mine: "nous sentons comme dans nos membres, et non pas comme dans les objets qui sont hors de nous."

[71] Meditation 6, AT VII 74, italics mine: "sentiebam etiam in me famem, sitim, aliosque ejusmodi appetitus" and "extensionem, et figuras, et motus, sentiebam etiam in illis duritiem et calorem, aliasque tactiles qualitates."

self-preservation. I need to know which part of *res extensa* I should be protecting and which parts are poised to affect it for good or ill.

Now one might well ask what the connection is between interoceptive sensations of pains, tickles, hunger, thirst, etc. and the experience of a particular body as *mine* (of "bodily ownership" as we now refer to it). The question raises a chicken-and-egg problem: which comes first, the distinction between my body and external bodies *or* the distinction between interoceptive and exteroceptive sensations? Are sensations like pain and hunger represented as modes of *my body* because they are represented as modes of a hunk of extension that is (somehow) already represented as my body? (Call this CHICKEN.) Or does a particular hunk of extension come to be represented as (part of) my body because sensations like pain and hunger that are represented as modes of it have a *this-belongs-to-me* character to them? (Call this EGG.)

While a number of philosophers working on bodily ownership today opt for some version of CHICKEN,[72] I think it's pretty clear that Descartes opts for EGG. The very distinction between my body and other bodies is always tied up, for Descartes, with sensations of pain, tickles, hunger, thirst, and other interoceptive sensations. In the next section I will look at Descartes' sailor-in-his-ship passage, which provides perhaps the central evidence that interoception is the key to distinguishing my body from other bodies, since it suggests that interoception is our cognitive route to knowing that I, as a mind, am *united* to a particular body. For now, suffice it to note that Descartes offers up no other cognitive mark of the important distinction between my body and other, external bodies. And it is not clear that anything else in his cognitive arsenal *could* do the work. The exteroceptive senses represent colors, sounds, flavors, odors, tactile qualities, textures, sizes, and shapes as modes of all bodies alike: there is nothing particularly special about my visual representation of (what is in fact) my foot, as compared my visual representation of (what is in fact) your foot, to suggest the first is mine and the second yours.[73] The intellect fares no better: it represents all bodies in the same way, as hunks of *res extensa* really distinct from myself as a mind. Only in interoception do we find a representation of a body as mine.

It is true, as Chamberlain notes, that Descartes cites a few possible "reasons" for identifying one body in particular as mine in the Meditation 6: (a) "I could never be separated from it, as I could from other bodies"; (b) "I felt all my appetites and emotions in, and on account of, this body"; and (c) "I was aware of pain and pleasurable ticklings in parts of this body, but not in other bodies external to it."[74] These are not,

[72] They are motivated in good part by the clinical phenomenon of depersonalization, i.e., feeling a bodily sensation as in (what is in fact) one's hand, but failing to experience the hand as one's own. See Bermudez (2017), de Vignemont (2007, 2018). If I read them correctly, Billon (2017) and Gallagher (2017) might be defenders of EGG.

[73] It is true, however, that my visual perspective on my own body is different from my perspective on yours. I'll discuss this difference in the next section.

[74] Meditation 6, AT VII 76: "neque enim ab illo poteram unquam sejungi, ut a reliquis; omnes appetitus et affectus in illo et pro illo sentiebam; ac denique dolorem et titillationem voluptatis in ejus partibus, non autem in aliis extra illud positis, advertebam."

I think, things that *confer* a sense of ownership on the body. They are reasons adduced for *believing* that one body in particular is mine. The reasons are themselves grounded in our interoceptive *experience* of embodiment. Take the second two: it is the fact that *these sorts of sensations* are represented in this hunk of *res extensa* and not others that makes this hunk and not others appear to be mine; and that gives me a *prima facie* reason to believe that it is mine.[75] As for the fact that I could never be separated from this body as I could others, what suggests that to us in the first place? It's not that I understand it to be inseparable from me; on the contrary, I understand it to be separable from me! Nor do I see, hear, smell, taste, or exteroceptively feel it to be inseparable from me; sometimes I don't see, hear, smell, taste or touch it at all. But I do feel it, interoceptively, all the time—at least to some extent. A sensory deprivation tank may eliminate my exteroceptive sense perceptions altogether, but it will not relieve me of my interoceptive experience of embodiment. It must therefore be something about interoception that grounds the sense of bodily ownership. (Just what that something is, I'll turn to below.)

It is worth noting that EGG faces a challenge that CHICKEN may avoid: it has to build up a whole and relatively stable representation of the body out of fleeting and fluxuating sensory pieces. According to EGG, where interoceptive sensations are felt to be, so my body is felt to be. But interoceptive sensations, like all sensations, are temporary and they occur in isolated locations: I feel pain in my toe now, a tingle in my finger a moment ago, an ache in my back all the time but it moves around. The body I regard as mine, however, is, and is experienced as, a persistent body with reasonably well-defined (if plastic) boundaries. What accounts for that? Descartes says nothing about this, but I see no reason to think he could not respond to the challenge in the same way he responds to the challenge that we perceive other, external bodies as persistent with reasonably well-defined (if mutable) boundaries: the answer is constancy. The color and perspectival shape sensations that an object produces in us are shifting all the time as the illumination changes and as we move. And yet we do not perceive bodies not as kaleidoscopic collections of changing colors shapes. Descartes proposes that the mind produces an experience of persistent objects with stable colors and shapes in changing conditions of illumination viewed from different perspectives through habitual judgments made by the imagination, memory, and intellect (together with input from other sensory modalities). These judgments transform a collection of fleeting and changing sensations into a perception of a stable object in the world.[76] The same may account for our experi-

---

[75] One might argue for a CHICKEN interpretation of these passages by shifting the emphasis: it is because these sensations are experienced as being in (and only in) *this* body, a body already identified as mine, that they are experientially appropriated into my body. But Descartes seems to be offering these sensations in particular as a means for distinguishing my body from other bodies and offers no other means for distinguishing them. Contemporary CHICKEN defenders, by contrast, offer other means for distinguishing my body from others, such as distinctive spatialities (Bermudez, 2017), agency and control (de Vignemont, 2020), and selective protection (de Vignemont, 2017).

[76] For more on how this works, see Simmons (2003).

ence of a stable and well-defined body in the face of every changing interoceptive sensations. This answer does not undermine the suggestion that the interoceptive sensations provide the basis for the representation of a body as mine.[77]

If I'm right, interoceptive sensations suggest to me that this bit of *res extensa*, this body, has a special relationship to me. It is mine. And so interoception marks (or perhaps creates?) a salient and important boundary between my body and other "external" bodies. This is an important first step toward preserving it.

7.5.2 Interoception and the Mind-Body Union

Second, it is through interoception that we learn that we not only have a body but are *united to it*, so that our mind and body form a single thing (*unam quid*), viz., a human being. I *am*, at least in part, a body. The claim that I am *united to* a body or *am,* at least in part, a body sounds stronger than the earlier claim that I *have* a body; or perhaps it simply fleshes out what it is to "have" a body. Either way, it is again interoception that provides our cognitive access. The intellect does not inform us that our mind is united to any body at all: at least when it's working properly with clear and distinct perceptions, it represents mind as really distinct from body, as noted earlier. Descartes is explicit in his correspondence with Elisabeth that the intellect is not helpful in representing the mind's union with body: "things that pertain to the union of the soul and the body are known only obscurely by the intellect alone."[78] It is through the senses, Descartes tells Elisabeth, that we know about it, indeed know it "very clearly" (*tres-clairement*).[79] Again: "everyone experiences that he is a single person with both body and thought so related by nature that the thought can move the body and sense the things which happen to it."[80] Descartes similarly writes to Arnauld in the Fourth Replies that the mind's union with the body is something "we experience constantly through the senses."[81]

But this is an overstatement. The exteroceptive senses do not inform me that I am united to my body. There is nothing about the experience of seeing my knees or even hearing them crunch when I lunge to suggest that I am united to them in a way that I am not united to your knees. Granted, my visual perspective on my knees is a little different from my visual perspective on yours; I can't view them from all angles as

[77] Thanks to Matt Fulkerson for pressing me on this issue.

[78] Descartes to Elisabeth, 28 June 1643, AT III 691: "les choses qui appartiennent à l'union de l'ame & du corps, ne se connoisssent qu'obscurement par l'entendement seul."

[79] *Ibidem.*

[80] *Ibidem*, AT III 694: "chacun éprouve toujours en soi-meme [...] qu'il est un seule personne, qui a ensible un corps & une pensée, lesquels sont de telle nature que cette pensée peut mouvoir le corps, & sentire les accidens qui lui arrivent."

[81] Fourth Replies, AT VII 228–229: "Verumtamen non inficior arctam illam mentis cum corpore conjunctionem, quam sensibus assidue experimur." For a more extensive treatment of our sensory apprehension of the mind-body union, see Simmons (2014) and Simmons (2017).

I can yours. And there are parts of my body, like the backs of my knees, that I can't very well see at all. But there is nothing in all that to suggest that these knees are part of *me.* (At most the visual experience suggests I'm viewing the world from the perspective of somewhere in this body that constrains what I see and how I see it.) What's more, I *do* feel united to parts of my body that I cannot see at all, as when I feel an itch in the center of my back. The sensory experience through which I feel I am united to my body is interoception. It's feeling my knees bend "from the inside" when I lunge and then feeling that searing pain in them (at the same time that I hear the crunch) that tells me they are part of me: that *I* am squatting and *I* am in pain.

The famous "sailor in his ship" passage from Meditation 6 provides textual evidence that Descartes thinks so too. It is "through these *sensations of pain, hunger, thirst, and so on*" that "nature teaches me that I am present in my body not merely as a sailor is present in his ship, but rather that I am very closely joined and as it were intermingled with it, so that I compose a single thing with it."[82] The critical difference between the sailor in his ship and embodied human mind is *not* that the sailor *understands* what's going on in his ship while the mind *senses* what's going on its body. The sailor's cognitive relationship to his ship is sensory too: he "perceives through vision if something in the ship is broken."[83] The difference is that the sailor *observes* through exteroceptive sense perception (vision) what's going on in his ship while the mind *feels* through interoceptive experiences of pain, hunger, thirst, and so on what's going on in its body. Interoception is thus the mind's cognitive access to its union with body.

All this raises a question: what is it about interoception that teaches me that I am united with my body—that I am, at least in part, a body? I keep saying that interoceptive sensations uniquely represent the body "from the inside." But what, exactly, does this mean? At least in part, it means that interoception elides subject and object, and does so in a particularly salient way: when my hand itches, I experience my hand as both perceiving subject and object perceived at once. In experiencing it as a perceiving subject, I experience it as (part of) me. Now all Cartesian sensory experiences involve some sort of elision between mind and body: in both seeing orange on the surface of a kumquat and feeling pain in my foot, qualities that are really modes of mind (phenomenal orange, felt pain) are represented as modes of body (of the kumquat, of the foot). As Malebranche colorfully puts it: the mind "spreads itself onto the objects it considers by clothing them with what it has stripped from itself."[84] But in the case of interoception, something more happens. It is not just the sensory qualities that turn up, representationally, in bodies. The sensing subject does too! Strictly speaking, the mind is the subject of sensory perception; it's the perceiver

[82] Meditation 6, AT VII 81, italics mine: "Docet etiam natura, per istos sensus doloris, famis, sitis, etc., me non tamtum adesse meo corpori ut nauta adest navigio, sed illi arctissime essen conjunctum et quasi permistum, aedo ut unum quid cum illo componam." See also *Discourse* 5, AT VI 59.

[83] Meditation 6, AT VII 81: "nauta visu percipit si quid in nave frangatur."

[84] Malebranche, *Search After Truth* I.12, V (1980, p. 58): "elle se plait, pour ainsi dire, à se répandre sur tous les objets qu'elle considére, en se dépoüillant de ce qu'elle a pour les en revêtir." (1958–78, p. 138).

doing the sensing.[85] In interoception, however, the body is experienced as the very subject of the perception; the foot in which the pain seems to be located also seems to be feeling the pain. I become one, phenomenologically, with the foot.[86] That does not happen in exteroceptive sensory experience. When I see an orange kumquat, or feel its smooth texture, the distinction between perceiving subject and perceived object is phenomenologically clear in the experience. I do not come to feel one with the kumquat, even if it is "clothed" in qualities "stripped from" myself.

Identifying my body as (part of) *me*, as subject of experience, is an important, second step toward preserving it. I may care what happens to your body. I may even care what happens to the orange kumquat. But I *really* care what happens to *me*. If I experience this body as part of *me*, I will be sure to give it my attention and concern.

7.5.3 Interoception and Health

Third, through interoception I'm aware not simply that I have a body, or that it is part of me, but whether it is in good or bad working order, and what it needs. It is also through interoception that I am motivated to take an active interest in these matters.

The human body has to maintain a certain functional unity if it is to stay alive. If the body loses its functional integrity through damage to any of its critical parts, it dies and so the human being goes out of existence: the soul departs the body.[87] In order for a human being to stay in existence, then, its mind needs a way to register the well-being, harms, and needs of its body, and to do so in a way that motivates it to care. The human intellect is, once again, hopeless for this task. It represents the body to which it is united, like all body, as a hunk of extension. If all the body's limbs were ripped off and its lungs stopped breathing, the intellect would represent it as *changed,* but it would be hard-pressed to consider it as *harmed* (except, perhaps, instrumentally as something it wants to use as a tool for some task). As far as the intellect is concerned, the dismembered and breathless body is not malfunctioning or harmed: it's simply a hunk of extension inevitably obeying the laws of motion.[88]

---

[85] *Dioptrics* 4, AT VI 109.

[86] Koivuniemi and Curley (2015, p. 7) gesture at this double confusion: "In judging the pain to be in my foot, I attribute something which is in fact a mental state to a physical object; and I also conceive a thinking thing (myself) as it if were something physical, spread throughout the body." Brewer (1995) may also be suggesting that this elision between subject and object is responsible for the experience of the body as mine or as me (see pp. 305–306).

[87] See *Passions* 5–6 and 30, AT XI 330–331 and 351. For discussion see Des Chene (2001), Brown (2011), Hutchins (2016), Hatfield (2017).

[88] See Meditation 6, AT VII 84–85. Carriero (2009, p. 397) reads the intellect's prospects more optimistically, based on an analogy between it and the sailor in his ship.

It is, once again, the senses, and in particular some of the interoceptive touch sensations, that inform the mind of the well-being, harm and needs of the body. How do they do that? Most of these sensations are, as Descartes sees it, intrinsically pleasant or unpleasant: a gentle squeeze of the hand and a modest healthy meal generally produce pleasant bodily sensations; a door slammed on the hand and the bloating from an overindulgent meal produce unpleasant bodily sensations. The body parts they are felt to modify are thereby represented as doing well or badly. Light tickles are "naturally agreeable to the mind because [they are] a sign of robust health in the body with which it is closely conjoined."[89] They are "instituted by nature to testify to the body's healthy condition and strength."[90] Pain, by contrast, tells me something is wrong with the body. It is "instituted by nature to indicate to the soul damage that the body receives from this [violent] action [in the nerves] and the body's feebleness in not being able to resist it."[91] Through their intrinsic valence as pleasant or unpleasant, interoceptive sensations both represent the body as well or harmed.

The exteroceptive senses play an important role as well, of course. They are, according to Descartes, universally pleasant or unpleasant too: if something smells or tastes bad, that is an indication that it's not going to digest well and you had better not eat it. The external senses are *harbingers* of benefit and harm, representing external bodies insofar as they may enhance or do damage to our body, and so the union.[92] But interoception is on the front line, telling us how things stand *right now* with our own body.[93]

Of course, reporting on the condition of the body is one thing. If it is to keep its body alive, the mind also needs to be motivated to take action, i.e., to tend to the damage the body suffers and to fulfill its needs by eating when it is hungry. Here too interoception comes to the rescue. For, as both John Carierro and Colin Chamberlain have pointed out, Descartes thinks that these interoceptive sensations not only represent the goods and harms of the body (through their pleasantness and unpleasantness); they represent them to the mind *as its own goods and harms*.[94] Again, it does this by eliding subject and object of perception. A pleasant tickling sensation, Descartes writes, "represents [the body's robust health] to the soul as a good *which belongs to it* insofar as it is united with the body" and a pain sensation "represents [the body's dam-

[89] *Principles* IV. 191, AT VIII-A 318: "sensus titillationis, menti naturaliter gratus, quia vires corporis, cui arcte conjuncta est, ei testatur."

[90] *Passions* 94, AT XI 399: "instituée de la nature pour témoigner cette bonne disposition et cette force."

[91] *Passions* 94, AT XI 400: "instituée de la nature pour signifier à l'âme le dommage que reçoit le corps par cette action, et sa faiblesse enc e qu'il ne lui a pu resister." See also *Treatise,* AT XI 143–144. For a recent treatment of the affective character of interoceptive touch (or bodily awareness), see de Vignemont (2023, ch. 2).

[92] See *Principles* II.3, AT VIII-A 41 and Descartes to More, 5 February 1649, AT V 271.

[93] For further discussion see Simmons (2008). Akins (1996) offers an especially helpful account of "narcissistic" character of thermoreception in particular.

[94] See Carriero (2009, p. 394 and p. 400) and Chamberlain (2016, pp. 227–228).

age and ill health] as evils which are always unpleasant to the soul."[95] If the body's benefits and harms are represented as *my* benefits and harm, I'm going to be naturally motivated to attend to them.[96] And that is a third way in which interoception plays an especially important role in the life of the embodied mind.

7.5.4 Interoception and Agency

Fourth, if I'm going to take care of my body, I have to move it in response to what's going on in and around it. I have to raise my hand to get food to my mouth. I need to run from danger. And I need to exercise just to keep the machine in good working order. Now Descartes does not seem to think it is necessary for a mind to feel its body interoceptively in order to move it. In *Discourse* 5, he suggests that a mind merely inhabiting a body (but not united to it) would have the power to move that body, even though it would not have "feelings and appetites like ours and so constitute a real man."[97] In such an arrangement, the mind would be like our sailor in his ship or, to quote a less familiar passage from the *Treatise on Man,* like a "fountain-keeper who must be stationed at the tanks to which the fountain's pipes return if he wants to produce, or prevent, or change their movements in some way."[98] The mind in this case has an instrumental relation the body, using it as a tool to achieve some end. And perhaps we embodied human minds sometimes use our bodies this way. But Descartes suggests that the interoceptive sensations of the embodied human mind are, as a matter of fact, connected to our own voluntary bodily movement. We've already found him saying to Arnauld, "if we weren't [aware of our embodiment], we would not bother to engage in intentional action at all."[99] Whether it's strictly speaking necessary or not, guiding my body around the world and making adjustments to it is usually easier with the help of interoception (at least until the

[95] *Passions* 94, AT XI 399–400, italics mine: "la représente à l'âme comme un bien qui lui appartient, en tant qu'elle est unie avec le corps" and "il lui représente l'un et l'autre comme des maux qui lui sont toujours désagréables."

[96] Unlike Chamberlain (2016, p. 225), I am not inclined to think that the affective character of these sensations *results* in the mind's appropriating the body, experientially, as its own or as itself. As discussed above, I think it's the interoceptive character that accounts for the experiential appropriation of the body, and it's because this body feels like mine or like me that I experience its benefits and harms as my own.

[97] *Discourse* 5, AT VI 59: "des sentiments et des appétits semblables aux nôtres, et ainsi composer un vrai homme."

[98] *Treatise,* AT XI 131: "[l'âme] sera là [dans le cerveau] comme le fontainier, qui doit être dans les regards où se vont rendre tous les tuyaux de ces machines, quand il veut exciter, ou empêcher changer en quelque façon leurs mouvements." Patients with peripheral deafferentation are, perhaps, the closest real-world examples, and there is some evidence that this is roughly the relation they have to their bodies in producing movements in it. See Cole, 1995.

[99] Descartes to Arnauld, 29 July 1648, AT V 222: "ac etaim propter mentis cum corpore unionem, cuius sane mens conscia est; alioquin enim ad membra movenda voluntatem suam non inclinaret."

behavior is automated and not driven by the mind at all). As anyone who has had to re-habilitate a part of their body after an injury or learned a sport or learned to play an instrument knows, the feedback we get about our bodily positions and motions through interoceptive touch helps us to adjust them in order to do what we want with them. (Even if we're not always successful.) Interoception is thus an important guide to voluntary movement of our body.[100]

7.5.5 Interoception and Exteroception

Interoception, of course, does not occur in a vacuum. It interacts with the traditional exteroceptive sense modalities. In fact, it is interestingly intertwined with those traditional sense modalities. In some cases, the exteroceptive senses *depend on* interoception in one way or another. And so interoception looks to be critical even to our exteroceptive access to the world outside of us. Haptically or visually (and so exteroceptively) perceiving the position of an external object relative to me, Descartes proposes, depends on my sensing interoceptively the direction in which my grasping hand is reaching or my looking head and eyes are turned. In these cases, Descartes suggests that we start with interoception: "[W]hen our eye or our head is turned toward some direction *our soul is informed of this* by the change caused in our brain by the nerves embedded in the muscles that make these movement."[101] By shifting its attention (*transférer de là son attention*), Descartes continues, the soul can perceive the position of the external object we are reaching for or looking at. Thus in the case of the blind man with his stick, "the nerves embedded in this hand cause a certain change in his brain, which enables his soul to know *not only the place of his hand but also all the other places that lie in the straight line that extends out form it*; in this way the soul can turn its attention to the objects and thereby determine where they are."[102] In this case, Descartes seems to suggest that the relation between interoception and exteroception is inferential, at least in the psychological sense that we transition from interoceptive feeling of the hand's position to exteroceptive perception of the distal body's position with a shift of attention.[103]

---

[100] There has been a good deal of discussion in the philosophical literature about the relationship between interoceptive bodily awareness and agency. See, e.g., O'Shaughnessy (1995), Grush and Springer (2019), and de Vignemont (2021).

[101] *Dioptrics* 6, AT VI 135, italics mine: "lorsque notre oeil ou notre tête se tournent vers quelque côté, notre âme en est avertie par le changement que les nerfs insérés dans les muscles, qui servent à ces mouvements, causent en notre cerveau en sorte qu'elle peut porter son attention jusqu'aux objets B et D, et déterminer les lieux où ils sont."

[102] *Dioptrics* 6, AT VI 135, italics mine: "les nerfs insérés en cette main causent un certain changement en son cerveau, qui donne moyen à son âme de connaître, non seulement le lieu A ou C [les mains], mais aussi tous les autres qui sont en la ligne droite AE ou CE."

[103] For details and some challenges see Simmons (2003, pp. 405–407).

Interoception and exteroception are particularly intertwined in touch itself. That is because exteroceptive touch occurs through contact between our own body and external objects. At a minimum, the two are intertwined causally: we have already seen Descartes suggest that sometimes one and the same causal chain underwrites both an interoceptive and exteroceptive sensory experience, viz., exteroceptively feeling the warmth of a fire and interoceptively feeling the coolness of the hand that exteroceptively feels the warmth of the fire. They are also causally intertwined to the extent that exteroceptively perceiving the tactile properties of an external object typically (though not always) depends on exploratory haptic behavior that is felt interoceptively: if I want to perceive the tactile properties of the water bottle over yonder I have to reach out and grab it, press it, stroke it, support it, move it, try to bend it, and so on; along the way I will typically feel the actions of my arm and hand.[104]

When I reach for my friend's hand in a dark movie theater, interoceptive and exteroceptive touch are also intertwined phenomenologically as two sides of a single experience: when I find my friend's hand, the exteroceptive feeling of its shape and the interoceptive feeling of my own hand's configuration seem to be two aspects of a single experience. It's hard to say that perceiving the shape of my friend's hand is inferred in any way from perceiving the configuration of my own hand (even in the minimal psychological sense of it following the perception of my hand). They seem rather to be both present at once, though I can shift my attention to focus now on one, now on the other.[105]

Interoception, then, is lurking around and playing a role even in the more familiar exteroceptive sensory perceptions that guide the embodied mind's quest for self-preservation.

7.5.6 Interoception and the Passions

The passions are an important part of Descartes' story of the embodied human mind's pursuit of self-preservation. By now, you will not be surprised to learn that interoception has an important role to play in the workings of the passions. According to the tripartite perceptual taxonomy of the *Passions*, passions proper, like love and fear, are perceptions we refer to our soul, rather than to external or our own bodies. They assist with the sensory mission of self-preservation as follows:

[104] See Fulkerson (2014, p. 55) for discussion of the core exploratory procedures we use in exteroceptive touch.

[105] See Fulkerson (2014, ch. 4) for an extended treatment of the various ways in which we might think of the dependence between interoceptive and exteroceptive touch. See Batty (2016) for some pushback. Note also that the connection between interoception and exteroception in exploratory behavior is not unique to touch: we move our head and eyes around to see things, we sniff things to better smell them, we swish wine around in the mouth to discern its flavor profile; all of these behaviors involve an interoceptive experience of our bodies as well as an exteroceptive experience of an external object.

"their natural function is to incite the soul to consent and contribute to the actions that can serve to preserve the body or make it more perfect."[106] More specifically, they incite the will to set in motion the pursuit of goods and avoidance of evils that the senses represent to the mind.[107] The passions proper are simply modes of mind, and in this case modes we actually manage to "refer", correctly, to the mind; that is, we experience them as modes of mind. But they are always accompanied by bodily sensations, which Descartes takes some care to outline, as I mentioned above. This is an important element to the passions. Physiologically, when a passion is induced in the mind, the body is simultaneously being readied for appropriate action. In the face of bear in the woods, my body is preparing itself to run at the same time that it produces in the mind the sensation of fear. I feel both. The passion of fear inclines the will to give the green light to what we simultaneously feel the body is prepared to do to protect itself. Here too, then, the passions do not do their job alone. They work hand in hand with interoception.

On reflection, then, there seems to be no part of Descartes' depiction of the embodied mind's endeavor to preserve its bodily half that does not involve interception in one way or another. From singling out that body as mine (or as part of me) and so as a body I need to protect, to voluntarily moving it around, to perceiving other bodies, to responding appropriately to them, interoception is playing a (usually) quiet but absolutely critical role in the life of the embodied human mind.

7.6 Future Inquiries

What I hope to have done here, is, in part, to overwhelm you a bit with the presence of interoceptive touch in Descartes and to explore its importance in the cognitive economy of the Cartesian human mind. Admittedly, interoception is not a perfectly well-defined category in Descartes' texts. It is not explicitly foregrounded. Its boundaries are a bit messy: thermal perception, e.g., seems to be both an exteroceptive and interoceptive touch experience. And Descartes' scattered and sometimes elliptical discussion of it invites many questions for further philosophical and scholarly research, some of which I've gestured at along the way. These include:

- Given that all sensations (of pain, of color, of shape alike) are modes of mind, why do we experience *any* of them as modes of body at all?
- Why do we experience some of them (interoceptive sensations) as being in one body and others (exteroceptive sensations) as being in all bodies?
- Why do we experience the body of interoceptive sensations as "my" body or "part of me" and all other bodies as "external" bodies? (And is there a difference between experiencing a body as "my" body and experiencing it as "part of me"?)

---

[106] *Passions* 137, AT XI 430: "leur usage naturel est d'inciter l'âme à consentir et contribuer aux actions qui peuvent servir à conserver le corps ou à le rendre en quelque façon plus parfait."

[107] *Passions* 40 and 52, AT XI 359 and 372.

- Is there a salient line to draw between my body and external bodies *prior* to or *independent of* our experiencing interoceptive sensations in one but not the other, or is the distinction the (constitutive?) *result* of this experience?
- If the distinction between my body and external bodies is the (constitutive) result of experiencing interoceptive sensations in one body in particular, is that distinction merely *epistemological* or is it also *metaphysical*, i.e., is there any *other* way of marking the distinction such that interoception is our epistemic access to that? Or is it rather the case that what it *is* for a body to be my body is just for me to experience it, interoceptively, as such?[108]
- Is the line between my body and external bodies a sharp one or is it vague and is it plastic, e.g., can functional and possibly sentient prosthetic devices become part of my body?
- What is the nature of the dependence between exteroceptive and interoceptive touch in particular: is it constitutive? causal? phenomenological?

These questions and challenges have generated a recent explosion of research (empirical and, more recently, philosophical) into interoception. I've begun to address some of them in the Cartesian context here. As this field of study takes off, I hope that we historians will dive into our texts to provide a better understanding of the ways in which this curious concept has developed over time and the ways in which it has been theoretically framed in our attempt to understand our embodied human experience.

Acknowledgements I'm grateful to the other contributors to this volume for their lively discussion of an earlier version of this essay at *The Sense of Touch: Medieval and Modern Debates in Philosophy and Science* conference in Nijmegen in June 2023. The essay benefited hugely from a robust discussion with the *History of Philosophy Roundtable* at UCSD in March 2024 and from a colloquium discussion at Cornell in October 2024. I'm particularly grateful to Colin Chamberlain, Keota Fields, Matt Fulkerson, Rick Grush, Vili Lähteenmäki, and Matthias Somers for their discussions of this material with me.

References

Primary Literature

Descartes, R. (1964-1974). In C. Adam & P. Tannery (Eds.), *Oeuvres. Nouvelle presentation*. Vrin.
Descartes, R. (1984-1991). In J. Cottingham, R. Stoothoff, D. Murdoch, & A. Kenny (Eds.), *The Philosophical Writings of Descartes*. Cambridge University Press.
Malebranche, N. (1980). In T. M. Lennon & P. J. Olscamp (Eds.), *The Search After Truth*. Ohio State University Press.

[108] In his letter to Mesland of 9 February 1645, for example, Descartes suggests that the human body is whatever portion of *res extensa* united to its soul (AT IV 166). Is interoception simply the epistemic guide to this metaphysical phenomenon?

Secondary Literature

Akins, K. (1996). Of Sensory Systems and the "Aboutness" of Mental States. *The Journal of Philosophy, 93*(7), 337–372.

Alsmith, A. J. T., & Longo, M. (Eds.). (2023). *The Routledge Handbook of Bodily Awareness*. Routledge.

Baker, G. P., & Morris, K. J. (1996). *Descartes' Dualism*. Routledge.

Batty, C. (2016). Exploring Touch: A Review of Matthew Fulkerson's *The First Sense*. *Philosophical Psychology, 29*(1), 138–146. https://doi.org/10.1080/09515089.2015.1010197

Bermúdez, J. L. (2017). Ownership and the Space of the Body. In F. De Vignemont & A. J. T. Alsmith (Eds.), *The Subject's Matter* (pp. 117–144). The MIT Press.

Bertnson, G. G., & Khalsa, S. S. (2021). Neural Circuits of Interoception. *Trends in Neurosciences, 44*, 17–28.

Billon, A. (2017). Mineness First: Three Challenges to Contemporary Theories of Bodily Awareness. In F. de Vingemont & A. J. T. Alsmith (Eds.), *The Subject's Matter* (pp. 189–216), The MIT Press.

Brewer, B. (1995). Bodily Awareness and the Self. In J. L. Bermúdez, A. Marcel, & N. Eilan (Eds.), *The Body and the Self* (pp. 291–309). The MIT Press.

Brown, D. (2011). Cartesian Functional Analysis. *Australasian Journal of Philosophy, 90*(1), 75–92.

Carriero, J. (2009). *Between two Worlds*. Princeton University Press.

Casini, L. (2014). Renaissance Theories of Internal Senses. In S. Knuuttila & J. Sihvola (Eds.), *Sourcebook for the History of the Philosophy of Mind* (pp. 147–157). Springer.

Ceunen, E., Vlaeyen, J. W., & Van Diest, I. (2016). On the Origin of Interoception. *Frontiers of Psychology, 7*(743). https://doi.org/10.3389/fpsyg.2016.00743

Chamberlain, C. (2023). Not a Sailor in His Ship: Descartes on Bodily Awareness. In Alsmith & M. R. Longo (Eds.), *The Routledge Handbook of Bodily Awareness* (pp. 83–94). Routledge. https://doi.org/10.4324/9780429321542-9

Chamberlain, C. (2020). What am I? Descartes's Various Ways of Considering the Self. *Journal of Modern Philosophy, 2*(1), 1–30.

Chamberlain, C. (2018). The Body I Call 'Mine': A Sense of Bodily Ownership in Descartes. *European Journal of Philosophy, 27*(1), 1–22.

Chamberlain, C. (2016). A Bodily Sense of Self in Descartes and Malebranche. In J. Kaukua & T. Ekenberg (Eds.), *Subjectivity and Selfhood in Medieval and Early Modern Philosophy* (pp. 219–234). Springer.

Clarke, D. (2003). *Descartes's Theory of Mind*. Oxford University Press.

Cole, J. (1995). *Prides and a Daily Marathon*. The MIT Press.

Craig, A. D. (2015). *How Do You Feel? An Interoceptive Moment with Your Neurobiological Self*. Princeton University Press.

Craig, A. D. (2002). How Do You Feel? Interoception: The Sense of the Physiological Condition of the Body. *Nature Reviews: Neuroscience, 3*(8), 655–666.

Damasio, A. (1994). *Descartes's Error: Emotion, Reason, and the Human Brain*. Grosset/Putnam.

De Vignemont, F. (2007). *Habeas Corpus: the Sense of Ownership of One's Own Body. Mind and Language, 22*(4), 427–449.

De Vignemont, F. (2018). Was Descartes Right After All? An Affective Background for Bodily Awareness. In M. Tsakiris & H. De Preester (Eds.), *The Interoceptive Mind: From Homeostasis to Awareness*, (pp. 259–271), Oxford University Press.

De Vignemont, F., Serino, A., Wong, H. Y., & Farne, A. (Eds.). (2021). *The World at Our Fingertips: A Multidisciplinary Exploration of Peripersonal Space*. Oxford University Press.

De Vignemont, F. (2020). *Mind the Body: An Exploration of Bodily Self-Awareness*. Oxford University Press.

De Vignemont, F., & Alsmith, A. J. T. (Eds.). (2017). *The Subject's Matter*. The MIT Press.

Des Chene, D. (2001). *Spirits and Clocks: Machine and Organism in Descartes*. Cornell University Press.

Fulkerson, M. (2014). *The First Sense: A Philosophy Study of Human Touch*. The MIT Press.

Gallagher, S. (2017). Deflationary Accounts of the Sense of Ownership. In F. De Vignemont & A. J. T. Alsmith (Eds.), *The Subject's Matter* (pp. 145–162). The MIT Press.

Grush, R., & Springle, A. (2019). Agency, Perception, Space and Subjectivity. *Phenomenology and the Cognitive Sciences, 18*(5), 799–818.

Gueroult, M. (1953). *Descartes selon l'ordre des raisons*. Aubier.

Haggard, P., Chen, T., Beck, B., & Fardo, F. (2017). Spatial Perception and the Sense of Touch. In F. De Vignemont & A. J. T. Alsmith (Eds.), *The Subject's Matter* (pp. 97–114). The MIT Press.

Hatfield, G. (2017). Descartes: New Thoughts on the Senses. *British Journal for the History of Philosophy, 25*(3), 443–464. https://doi.org/10.1080/09608788.2016.1214908

Hatfield, G. (2015). On Natural Geometry and Seeing Distance in Descartes. In V. De Risi (Ed.), *Mathematizing Space* (pp. 157–191). Springer.

Hutchins, B. (2016). Does Descartes Have a Principle of Life?: Hierarchy and Interdependence in Descartes's Physiology. *Perspectives on Science, 24*(6), 744–769.

Knuuttila, S., & Kärkkäinen, P. (2013). Medieval Theories of Internal Senses. In S. Knuuttila & J. Sihvola (Eds.), *Sourcebook for the History of the Philosophy of Mind* (pp. 131–146). Springer.

Koivuniemi, M., & Curley, E. (2015). Descartes on the Mind–Body Union: A Different Kind of Dualism. In D. Garber & D. Rutherford (Eds.), *Oxford Studies in Early Modern Philosophy 7*. Oxford University Press. https://doi.org/10.1093/acprof:oso/9780198748717.001.0001

Longo, M. R. (2017). Body Representation and the Sense of Self. In F. De Vignemont & A. J. T. Alsmith (Eds.), *The Subject's Matter* (pp. 75–95). The MIT Press.

Mantovani, M. (2022). The Institution of Nature: Descartes on Human and Animal Perception. In D. Rutherford (Ed.), *Oxford Studies in Early Modern Philosophy* (pp. 1–30). Oxford University Press.

Martin, M. G. F. (1995). Bodily Awareness: A Sense of Ownership. In J. L. Bermúdez, A. Marcel, & N. Eilan (Eds.), *The Body and the Self* (pp. 267–289). The MIT Press.

Marion, J.-L. (2013). *Sur la pensée passive de Descartes*. PUF.

Matthen, M. (2021). The Dual Structure of Touch: The Body vs Peripersonal Space. In F. De Vignemont, A. Serino, H. Y. Fong, & A. Varne (Eds.), *The World at Our Fingertips: A Multidisciplinary Exploration of Peripersonal Space* (pp. 197–214). Oxford University Press.

O'Shaughessy, B. (1995). Proprioception and the Body Image. In J. L. Bermúdez, A. Marcel, & N. Eilan (Eds.), *The Body and the Self* (pp. 175–204). The MIT Press.

Robson, D. (2021, August 15). *Interoception: The Hidden Sense that Shapes Wellbeing*. The Guardian.

Sherrington, C. S. (1906). *The Integrative Action of the Nervous System*. Yale University Press.

Simmons, A. (2001). Sensible Ends. *Journal of the History of Philosophy, 39*(1), 49–75.

Simmons, A. (2003). Spatial Perception from a Cartesian Point of View. *Philosophical Topics, 31*, 395–423.

Simmons, A. (2008). Guarding the Body: A Cartesian Phenomenology of Perception. In P. Hoffman & G. Yaffe (Eds.), *Contemporary Perspectives on Early Modern Philosophy: Essays in Honor of Vere Chappell* (pp. 81–113). Broadview Press.

Simmons, A. (2014). Sensory Perception of Body: *Meditation* 6.5. In D. Cunning (Ed.), *The Cambridge Companion to Descartes'* Meditations (pp. 258–276). Cambridge University Press.

Simmons, A. (2016). Representation. In L. Nolan (Ed.), *The Cambridge Descartes Lexicon* (pp. 645–654). Cambridge University Press.

Simmons, A. (2017). Mind-Body Union and the Limits of Cartesian Metaphysics. *Philosophers' Imprint, 17*(14), 1–36.

Tsakiris, M., & De Preester, H. (2018). *The Interoceptive Mind: From Homeostasis to Awareness*. Oxford University Press.

Chapter 8
From Fingertips to Papillae: Marcello Malpighi on the Sense of Touch

Luca Tonetti

Abstract In his two works on the tongue and the organ of touch (*De lingua* and *De externo tactus organo*), both published in 1665, Malpighi revealed the existence of tiny structures called "papillae" that, like nerve receptors, he argued to be responsible for taste and touch perception. The papillae represented the extremities of nerve branches that, emerging from the skin, were directly accessible to external stimuli. This discovery was the result of a wealth of evidence collected during his stay in Messina as well as Malpighi's ongoing exploration of ancient and early modern sources. By offering a new interpretation of the operations of these sense organs, Malpighi challenged the Aristotelian notion of a "medium" for senses and also questioned the debates surrounding the structure of the tongue and skin that characterised anatomical research in the sixteenth and early seventeenth centuries. This chapter explores how Malpighi approached touch through a series of lectures on skin and membranes in 1663–1664 and how this paved the way for his demonstration of the existence of "tactile papillae" in *De externo tactus organo*. Drawing on Malpighi's personal unpublished notes, lavishly enriched by drawings, this chapter aims to analyse how Malpighi developed his research on touch, highlighting his dependence on previous research on the tongue, and how his conceptualisation of touch perception as "unmediated" and strictly related to the role of papillae was highly debated by his contemporaries.

Keywords Marcello Malpighi · Papillae · Touch · Tongue · Nerve branches

---

L. Tonetti (✉)
Department of Historical and Geographic Sciences and the Ancient World (DiSSGeA), University of Padua, Padua, Italy
e-mail: luca.tonetti@unipd.it

C. Beneduce et al. (eds.), *The Sense of Touch*, Studies in the History of Philosophy of Mind 32, https://doi.org/10.1007/978-3-032-16308-0_8

8.1 Introduction

Before embarking upon his own series of private anatomical lessons in Padua, the chemist Jacob Barner (1641–1686) sought the advice of the celebrated anatomist Marcello Malpighi (1628–1694) on which dissection techniques it would be appropriate to employ in studying the organs of taste and touch.[1] Malpighi kindly provided Barner with the use of his microscope from Bologna together with a full set of instructions regarding the dissection of the tongue and skin of a bovine specimen:

> The organ of taste is visible to the naked eye at the tip of the tongue and near the base of the papilla with a head, like a fungus, so that if the skin and flesh are gradually scraped or cut perpendicularly with a knife, branches of nerves emerge, continuing into the papillae, which terminate in the trunk. However, they are more likely to be found near the base in the beef tongue that has not been cooked. If an uncooked beef tongue is chosen, pulling away the cuticle and sieving the underlying substance will reveal the hidden papillae of touch.[2]

In this passage from his reply, Malpighi briefly describes the technique that had allowed him a few years earlier to discover, first in the tongue and then also in the rest of the skin of the body, the existence of specific new anatomical structures, the "papillae," that he deemed responsible for the senses of taste and touch. The evidence was set forth in two essays: the first, *De lingua*, dealt with the tongue, while the second, *De externo tactus organo*, addressed the external organ of touch. Both were published in Messina in 1665.

During Malpighi's tenure as a professor of anatomy in Messina, supported by Giovanni Alfonso Borelli (1608–1679), he conducted significant research.[3] Corrado Dollo identified two key factors that may have influenced Malpighi's appointment in Messina. Firstly, Malpighi's reputation as an anatomist was established by his groundbreaking dissertation on the lungs (*De pulmonibus*) in 1661, which constituted a significant advancement in the understanding of respiratory anatomy and

---

[1] Adelmann (1975, vol. II, p. 518), letter no. 244, Barner to Malpighi, February 6, 1671: "Sed et velim Clarissime Vir, ut plenius ad me perscribas modum secandi tum pulmones tum gustus et tactus organon secundum inventionem tuam […]." On Barner, see Adelmann (1975, vol. II, p. 519, fn. 1), Virgilio (1994), Clericuzio (2013, p. 199). In Padua, Barner gave private lessons in chemistry from 1670 through 1672. "Dominus Berner Chymicus" is mentioned by Malpighi in an entry of his "Diary" dated November 16, 1667: see Biblioteca Universitaria di Bologna (henceforth, BUB), MS. 2085/II, f. 40r. Barner paid two visits to Malpighi in Bologna, in 1667 and 1670. For details of their discussions on chemical and therapeutic matters, probably in 1670, see Biblioteca Comunale dell'Archiginnasio (henceforth, BCA), MS. B 3237, f. 22rv.

[2] Adelmann (1975, vol. II, p. 521), letter no. 245, Malpighi to Barner, February 1671: "Gustus organum nudis oculis patet in apice linguae et prope basim papillae cum capitulo, fungi instar, unde si cultro sensim proxima cutis et caro abradatur vel perpendiculariter secetur emergunt rami nervorum in papillas continuati qui in trunctum [*sic*] desinunt. Facilius tamen occurrent prope basim in bovina lingua nundum cocta. Haec eadem si elixetur dum evellitur cuticula et subiecta substantia cribrosa reconditas tactus papillas exhibebit." All translations are mine unless otherwise stated.

[3] On Malpighi's stay in Messina, see esp. Adelmann (1966, vol. I, pp. 203–290) and Dollo (1984, 2005).

physiology. Secondly, the *studium* of Messina was in need of a successor to Pietro Castelli (c. 1570–1661) for the chair of medicine, and finding one was a challenging task due to the university's rigorous requirements for medical and legal chair holders. Malpighi's appointment was also facilitated by the support of a local influential figure, Giacomo Ruffo (1628–1674), who was impressed by Malpighi's successful treatment of his wife's tumours and sores.[4] Notwithstanding Messina's peripheral status within the Republic of Letters, which was deeply rooted in the Galenic tradition, Malpighi's four-year tenure from 1662 to 1666 was remarkably productive. He conducted research on the structure of the tongue, the optic nerve, and the brain. His investigations into touch contributed to the establishment of a mechanistic "esthesiology" through an examination of the anatomy and physiology of sensory organs.[5]

In his reply to Barner, Malpighi subtly suggested that a full understanding of the sense of touch required an examination of his earlier research on the tongue and taste, particularly with regard to the dissection techniques used. Unlike his detailed writings on the brain and tongue, however, Malpighi's work on touch lacks visual aids or illustrations. Surprisingly, there are no illustrations of the "tactile papillae." Could this be because his observations on touch were based on his earlier studies of the tongue, which gave an idea of the appearance of the papillae? Indeed, the study and elucidation of the structure and function of the tactile papillae, even the determination of their shape and size, are closely related to those of taste. Perhaps contextual factors prevented Malpighi from including illustrations. Remarkably enough, he did not involve Borelli, with whom he had collaborated on experiments and observations on sensory organs, including sharing and discussing preparatory visual material for the optic nerve.[6] While he may have collaborated with the painter and naturalist Agostino Scilla (1629–1700) on illustrations of the tongue, there was no such collaboration on touch.[7]

Malpighi's note-taking practices are however characterized by a strong emphasis on visual elements, an aspect that has been relatively unexplored. His personal notes, often accompanied by drawings, are thus crucial in understanding the development of his ideas on touch.[8] They also help in reconstructing his descriptions of the anatomical structures involved in the process, including the tactile papillae.

[4] See Dollo (1984), Calabrese (2016, 2018).

[5] See Bertoloni Meli (2011, pp. 88–93). Bertoloni Meli clarified some philosophical aspects of Malpighi's discourse on the sense of touch, in particular, his engagement with ancient and early modern sources such as Aristotle, Descartes, and Gassendi. On touch in early modern culture, see: Harvey (2003), Classen (2012), Maurette (2013), Pogliano (2015), Maurette (2018), von Hoffmann (2022, 2023). On intersensoriality and on the entanglement of touch with other senses, see Hsiung et al. (2024). On taste and touch, see the inspiring paper by Stabile (2008).

[6] For Malpighi's illustration of the optic nerve, see Tonetti (2022) and Conforti & Tonetti (2026). See also BCA, MS. B 3237, f. 4r, which contains a fragment of the drawing of the optic nerve of a swordfish that Malpighi sent to Borelli on 21 July 1663 ("Transmissa Ecc.mo Borellio 21 Julii 1663"). However, this fragment only contains the captions of the illustration.

[7] On Malpighi's collaboration with Scilla, see Bertoloni Meli (2011, p. 89) and Giallombardo (2016).

[8] A description of Malpighi's unpublished manuscripts is provided by Dollo (1986) and Bertoloni Meli (1998).

Notably, explicit representations of the skin's papillae are in fact only found in Malpighi's posthumous work, published around 30 years after *De externo tactus organo*. How did Malpighi continue his investigation of touch after publishing his essay? How did his methods of observation and testing of papillae change over time? Can we gain insight into how he represented and depicted them? My study aims to explore Malpighi's research on touch primarily through the lens of his personal notes.[9]

This chapter first explores how Malpighi approached touch through a series of lectures on skin and membranes in 1663–1664 and how this paved the way for his demonstration of "tactile papillae" in *De externo tactus organo* (1665). Subsequent studies on the skin and papillae from 1660 to 1670 further elucidated Malpighi's quest to pinpoint a distinct organ of touch and unravel the intricate sensory processes at play. Critiques of Malpighi's work in the 1670s and 1680s challenged his findings, prompting a re-evaluation of his research methods and results. In his posthumous work, *Opera posthuma* (1697), Malpighi buttresses his stance on the organ of touch and sensory mechanisms, drawing on a wealth of evidence accumulated throughout his career.

8.2 Unveiling the Complexity of Skin: Malpighi's *Lectiones*

Many of Malpighi's lectures, covering his entire career from Pisa to Bologna, Messina, and back to Bologna over a period of more than 30 years, can be found in MS 2085/III, housed in the University Library of Bologna.

The manuscript begins with the lectures delivered by Malpighi in Pisa between 1656 and 1659. This section, although incomplete, provides insight into the main topics discussed during this period and appears to align with the prevailing academic standards of the time. The lectures from 1656, which were numbered by Malpighi himself, covered Hippocrates' *Prognostics*. Lecture number 20 concentrated on hands but solely from a medical standpoint; that is, Malpighi examined the predictive value of their movements.[10] Hippocrates had posited that in the presence of acute fevers, pneumonia, phrenitis, and headaches, specific delirious gestures of the patient's hands (such as when they are waved in front of the face or in the air) invariably foreshadow an unfavourable outcome of the disease.

---

[9] Malpighi's dissertation on touch received considerable attention in the second half of the twentieth century. The text was translated into Italian by Vincenzo Busacchi (1966). A more accurate and annotated version of the text, however, was published by Luigi Belloni in his abridged edition of Malpighi's works (Malpighi, 1967). Belloni also replicated some of Malpighi's anatomical observations by dissecting a cow's snout in collaboration with the Institute for the History of Medicine at the University of Milan. See also Belloni (1970).

[10] BUB, 2085/III, p. 9. Malpighi wrote: "Manus igitur sunt instrumentum animae operantis in quibus si animales optime edantur(?) functiones sanum et optimus erit animal sin minus egrum et praecipue in morbis acutis pessimum."

The hand, often referred to as the "instrumentum instrumentorum" after Galen's observations, had not yet been the subject of a comprehensive study by Malpighi. In order to understand the tactile sensitivity that enables the hand to perceive the nature of the objects it grasps, it was therefore essential to further investigate the layered structure of its skin. Malpighi was aware that the physical structure of the skin was still a very difficult subject to understand at that time. The use of the microscope and the development of dissection techniques were instrumental in this process.

Nevertheless, the emphasis on the skin as a field of inquiry amongst anatomists was already becoming apparent by the late sixteenth century.[11] From the very outset of his magnum opus, *De humani corporis fabrica*, Vesalius placed considerable emphasis on the importance of the physician's manual work (*manuum munus*), as evidenced by the symbolic portrayal of him dissecting an arm.[12] In the second book, he devoted an entire chapter to the skin, distinguishing between the skin itself (*cutis*), the cuticle or epidermis (*cuticula*), and the membranes that surround it. Vesalius provides detailed instructions on the correct techniques to employ when studying the skin, emphasising the crucial role of the separation of the skin layers. The cuticle is akin to a "small skin" that is extremely thin and devoid of blood. It resembles, as Vesalius pondered, the thin and transparent skin of onions, except in certain places, particularly on the palm of the hand and the sole of the foot, where it appears thicker and harder. It is ultimately comparable to an efflorescence of the skin (*efflorescentia cutis*), a definition that recurs throughout anatomical treatises. The most intriguing aspect pertains to the matter of sensitivity. Indeed, Vesalius challenged the notion put forth by Aristotle that the skin is an inert and insensitive organ due to its lack of nerves, in contrast to the flesh. Vesalius claimed that in fact, the skin is capable of sensation due to the nerves that run through it:

> Aristotle must certainly be disregarded when he deprives the skin of sensation as if its sense of touch came only from the flesh beneath the skin; for we discover from dissection that a definite series of nerves belonging specifically to the skin extends to the skin, particularly that of the face, arm, forearm, hand, legs, and feet. Throughout the rest of the body, although nerves come to the skin in no particular pattern, yet many end in the skin or beneath it,

---

[11] See especially Gadebusch Bondio (2005) on the study of the skin in early modern medicine. This is included in a special issue of *Micrologus* (XIII, 2005) devoted to "human skin." On the relationship between physicians and the skin, especially with regard to the emergence of skin diseases in the sixteenth century, cf. Murphy (2020) and Savoia (2024a). As Murphy points out, the question of skin disease was a central concern in the sixteenth century, largely because it was not included as a distinct category of disease in classical Galenic theory. Mercuriale provided the first published treatise on the skin, *De morbis cutaneis* (1572): his definition of the skin, still rooted in the Platonic tradition, did not imply that the skin could be "diseased." According to Murphy, Jan Jessen's *De cute et cutaneis affectibus* (1601) shows the scholarly consensus on the importance of the skin as a site of medical practice at the time, even within a Galenic framework. See also the 5-year research project *Renaissance Skin*, led by Murphy and Evelyn Welch, at https://renaissanceskin.ac.uk/about/ (last accessed on December 31, 2024). On the practical knowledge about "surfaces" of things and bodies, see Savoia (2022, 2024b).

[12] See e.g. Rippa Bonati (2004), Mandressi & Claire (2015), Maurette (2018). Cooley (2024) shows how the anatomist Hieronymus Fabricius of Aquapendente sought to train students and readers "haptically."

> especially at the sides of the thorax [...] Thus the skin's power of sensation is not due to the muscles (or the flesh, as Aristotle called them) but, as we said, to the fact that it has its own proper nerves.[13]

From this perspective, the cuticle, which is attached to the skin, may serve as a "medium" for the sense of touch.

As the seventeenth century approached, the central role of the skin was no longer in doubt.[14] Anatomists such as André du Laurens (1558–1609), Caspar Bauhin (1560–1624), and Giulio Cesare Casseri (1552–1616) aimed not only to align themselves with or against Aristotle or Galen, but, following in the footsteps of Vesalius, to reconcile old theories with new empirical data derived from direct observation of the body.

In the view of Bauhin, as in that of Vesalius, the cuticle lacks sensation: it develops above the skin to protect the latter from external attacks and moderate its acute sensitivity. In this sense, the cuticle can be considered the medium of the sense of touch.[15] In contrast, the skin is regarded as the organ (*instrumentum immediatum*) that is responsible for the perception of *external* touch. The internal membranes are likewise involved in the perception of *internal* touch.[16] Interestingly, Bauhin also argued that the skin does not possess its own nerves, but rather is traversed by nerve fibres, the quantity of which determines the sensitivity of the skin in different parts of the body.[17]

In the opening section of the *Pentestheseion, hoc est de quinque sensibus liber*, Casseri (1609) provided a detailed examination of the functions of the skin as an

[13] Vesalius (1999, p. 142): "Neque sane audiendus est Aristoteles, cutem sensu destituens, non secus, quam si carnis cuti subditae duntaxat beneficio tactus sensus perficeretur, quum cuti proprios, certaque serie nervos exporrigi ex Anatome condiscamus, illi praecipue, quae in facie est ac brachio, cubito extremaque manu, deinde & cruribus & pedibus. Per reliquum autem corpus, etiam si non ordinata serie nervi ad cutem deducantur, attamen complures in illam desinunt, sub ipsaque & potissimum ad thoracis latera exporriguntur [...] Atque ita musculorum (quos Carnem Aristoteles appellabat) gratia cutis nequaquam sentit, verum uti diximus propriorum nervorum occasione."

[14] Cf. Gadebusch Bondio (2005).

[15] Bauhin (1605, p. 22): "[...] sensus expers cuti supernascitur, quo eam ab externis iniuriis tueatur, & exquisitum eius sensum attemperando fiat sensus tactus medium; (siquidem, ut ait Arist. omnia quae sentimus per media sentiamus, & nulla fiat sensatio per immediatum contactum sensibilis & sensorii) hac enim avulsa dolorem sentimus, illaque parte rerum contactum ferre non possumus, quod in ulcere quotidie experimur, ut cute subiecta sentiatur quidem, sed sine iudicio & cum dolore."

[16] *Ivi*, p. 28: "Usus tactus externi immediatum esse instrumentum, ut externas iniurias, antequam interna laedantur, persentiamus: quemadmodum interni tactus sunt membranae internae."

[17] This aspect has been emphasized by Gadebusch Bondio (2005, p. 560). See Bauhin (1605, p. 28): "Nervos non quidem proprios et definitos, sed ex nervis qui ad subiectas partes mittuntur, fibrae producuntur, quae inaequaliter disseminantur, et in exilitates expansi in cutim absumuntur: duo tamen insignes circa papillas, unde earum exquisitus sensus, conspiciuntur: sic cuti volae plurimae nervorum fibrae porriguntur, cum exquisitissimo sensu sit praedita, quemadmodum quoque ad unguium radices et pudendi extremum: in aliis vero hebetior est sensus."

organ of touch.[18] He challenged the conventional wisdom of the necessity of a medium for both taste and touch and goes on to propose that touch is the fundamental sense from which all others derive. Notably, this section is accompanied by the representation of the palm of the hand and the sole of the foot, in which the real skin (*cutis*) and the outermost layer (*cuticula*) are clearly distinguished.

The influence of all these traditions on Malpighi's teaching in Messina is clearly discernible. In 1663–1664, he delivered a notable course of lectures on the subject of tegumentary, adipose, and muscular tissues.[19] In defining the cuticle, Malpighi drew on classical sources and the work of Bauhin, emphasising the widely shared view that the cuticle is devoid of sensation.[20] Conversely, there was considerable divergence of opinions on its function:

> In regard to its function, a number of hypotheses have been put forth by various authors. One such hypothesis is that the cuticle serves as the medium for touch, as otherwise, sensation would not occur. This view is supported by the opinion of Aristotle. This, however, presents a challenge, as touch is a form of stimulation that does not necessitate any specific mediation. Consequently, the cuticle, in conjunction with the skin, can be considered the primary organ of touch. Some authors posit that the cuticle serves to moderate the exquisite sensitivity of the skin, such that the sensation of pain occurs in the parts where the cuticle is absent.[21]

[18] On Casseri, see De Ferrari (1978), Riva et al. (2001), Cunsolo (2008), Klestinec (2011), von Hoffmann (2022).

[19] BUB, MS. 2085/III, pp. 211–234. A transcription is provided by Dollo (2005). In a letter to Borelli probably dated December 1663 (unfortunately the letter has not been located), Malpighi expresses frustration with the reluctance of his students in Messina to embrace the novelty of his approach. In his reply, Borelli, following Gassendi's example, offers Malpighi advice on how to organise his own lectures: "[…] si vede che per utile anche de scolari non bisogna comunicarli alla bella prima tutte le cose nuove contrarie alle dottrine vecchie, ma bisogna imitar il medico, il qual s'accomoda alla debolezza dell'infermo, e pian piano va insinuando fra la dolcezza del miele qualche poco di medicamento amaro, questo l'osservò squisitamente il Gassendo, quand'era lettor di filosofia, che leggeva in cátedra la comune, e volgar dottrina peripatetica, et in fin di ciascheduna lezione vi aggiugneva per modo di dubbi, e scrupoli tutte le cose, che egli poteva arrecar in contrario […] una cosa simile vorrei che facesse Vostra Signoria, che dicesse tutta la dottrina comune, e poi in fine v'aggiugnesse i suoi scrupoli." See Adelmann (1975, vol. 1, pp. 190–191), letter no. 92, Borelli to Malpighi, December 21, 1663. In the same letter, Borelli provides a report on Fracassati's teaching experience in Pisa. See also *ivi*, letter no. 93, Borelli to Malpighi, January 10, 1664.

[20] BUB, MS. 2085/III, p. 217: "Praemissis universalibus de continentibus totius corporis partibus, breviter est dicendum et primo occurrit cuticula apud Graecos epidermis idest supra cutim et Hippocrates libro 6 de carnibus eam dermam appellat; a Bauhino dicitur cutis ultima et ab eodem describitur quod sit pellicula cuti supereminens ambiens alba et densa et tenuis, sensus omnino expers."

[21] BUB, MS. 2085/III, p. 218: "De usu plura habentur apud auctores primo enim ex Aristotelis sententia censent aliqui medium esse tactus, alias non fieret sensatio; hoc tamen patitur difficultatem quia tactus est species titillationis unde non requiritur medium quare potius una cum cute est verum organum tactus. Addunt aliqui additam fuisse ut exquisitus sensus cutis moderetur unde in partibus carentibus ex casu cuticula sensus doloris emergit."

Malpighi's analysis of the role of the cuticle challenged the prevailing view, namely, that of Aristotle, who posited that any sensory experience necessitates a medium between the object and the sense organ. Malpighi proposed that touch does not inherently necessitate such mediation. The cuticle and the skin both serve as organs of touch, thereby demonstrating a profound and complex interdependence between these anatomical structures and the experience of touch.

On the subject of skin, Malpighi pointed out that its substance varies from one part of the body to another and that this determines different degrees of sensitivity:

> The skin is soft and highly sensitive to touch on the face, penis, and scrotum. However, it is denser in the legs, the neck, back, and soles of the feet. It is of medium thickness to the touch of the fingertips, which may become calloused through manual labour. It is also thin on the sides, which can cause a tingling sensation, and thinner on the palm of the hand and thinnest on the lips.[22]

The skin is a continuous "tissue" throughout the body, yet it is permeable due to the presence of openings.[23] Although touch was acknowledged as a primary function of the skin ("Usus cutis seu actio est ad tactum"),[24] the precise mechanism underlying this sensory process remained elusive. A marginal note in the manuscript calls for further elucidation of the occurrence of touch ("Exponere quando fiat tactus"),[25] but the page that follows is remarkably left blank (!), leaving this aspect definitely unexplored.

8.3 The Journey from Taste to Touch: Exploring the Discovery of Papillae

The essays on the brain, tongue, and organ of touch, all published in 1665, were written at different times and with distinct approaches. *De cerebro* and *De lingua*, both dated October 31, 1664, were dedicated to Carlo Fracassati (c. 1630–1672) and Giovanni Alfonso Borelli, respectively.[26] Borelli played a pivotal role in coordinating the research, facilitated by continuous communication with Malpighi. This collaboration enabled Fracassati (who had joined Borelli and Lorenzo Bellini in Pisa in 1663)

---

[22] BUB, MS. 2085/III, p. 221: "Mollis est et exquisite sentit in facie, pene et scroto; densior tamen in cruribus, cervice, dorso et planta pedis; mediae naturae est in apice digitorum ad tactum, ni laboribus sit reddita calosa, est etiam tenuis in lateribus, unde titilatio, tenuior in vola manus et tenuissima in labijs."

[23] *Ibidem*: "Continua est per universum corpus definitis tamen foraminibus pervia unde ad oculos, nares, aures, os, mamillarum papillas, penis glandem, podicem, uteri sinum, ad umbilicum, et extremos digitos pervia est."

[24] *Ibidem*.

[25] *Ibidem*.

[26] See Malpighi's two essays were included in *Epistolae anatomicae de cerebro, ac lingua... Quibus Anonymi accessit exercitatio de omento, pinguedine, & adiposis ductibus...* (Bononiae, Typis Antonij Pisarrij, 1665), which incorporates also his work on fat. On Malpighi's brain research, see Belloni (1966), Clarke & Bearn (1968), Bertoloni Meli (2011), Tonetti (2022).

and Bellini to concurrently investigate the problem of sensation.[27] In contrast, the study on touch appears to have been conducted somewhat independently, with limited involvement from Borelli.[28] Although it was published in isolation and a false location was given (Naples instead of Messina), the work is intricately linked to the findings of *De lingua*, and it is therefore to this work that we are to turn for guidance (Fig. 8.1).

8.3.1 The Tongue Unveiled: Discovering the "Taste Papillae"

In *De lingua*, Malpighi aimed to clarify the differing opinions of anatomists, including Vesalius, regarding the structure of the tongue. Indeed, Vesalius himself had acknowledged that he had less knowledge of the tongue's anatomy than that of other body parts.[29] Various anatomists had described the tongue as having loose, spongy flesh, glandular characteristics, or a unique substance distinct from other body

[27] In 1665, Fracassati wrote two short treatises. The first, *Exercitatio epistolica de lingua*, was addressed to Borelli, while the second, *Dissertatio epistolica responsoria de cerebro*, was dedicated to Malpighi. Bellini also published *Gustus organum* in the same year. Fracassati's two works, along with Malpighi's *De lingua* and *De cerebro*, were later compiled into a single volume titled *Tetras anatomicarum epistolarum de lingua et cerebro* in 1665 in Bologna. For more information, see Bertoloni Meli (2011, pp. 93–101). Borelli played a crucial role in informing Malpighi about Fracassati's research and interactions with the Tuscan court. He thus significantly influenced Malpighi's research by replicating and circulating his observations and findings. Further insights on Borelli's network can be found in Favino & Giannini (2022). On Borelli's activity in the Pisa years, see esp. Conforti (2009). Initially, Fracassati planned to include Malpighi's observations on the optic nerve in his brain study. However, Malpighi decided to publish them separately. The genesis of the two treatises *De cerebro* by Malpighi and Fracassati is intertwined with a dispute involving anatomists John Finch (1626–1682) and Thomas Baines (1622–1680), who visited Pisa in 1664. During a public demonstration in front of the Grand Duke of Tuscany, Fracassati replicated Malpighi's observations, which initially faced skepticism from Finch and Baines. They claimed that Malpighi's discovery had been anticipated by Bartolomeo Eustachi. In response, Borelli wrote a paper, *Scrittura fatta l'Anno 1664 dal Sig Alfonso Borelli sopra opposizioni delli Sig. Finchio e Fava Inglesi anatomici del Serenissimo Gran Duca Toscana*, that was included in Malpighi's *Opera posthuma*. Cf. Adelmann (1975, vol. I), letter no. 93, Borelli to Malpighi, January 1, 1664, for more details. See Bertoloni Meli (1997) and Campinoti (2006).

[28] After completing his work on touch (*De externo tactus organo*), Malpighi sent a printed copy to Borelli on 14 December 1665 (see Adelmann, 1975, vol. I, letter no. 154). This action may have surprised Borelli, who was completely unaware of Malpighi's work on skin. Adelmann (1966, vol. I, pp. 255–256) suggests that Malpighi may have been concerned about Borelli's openness in sharing his research and results with others. This concern is evident in Malpighi's later correspondence with Bonfigliuoli in the 1670s, where Bonfigliuoli contrasts Borelli's behavior with that of Adrien Auzout (1622–1691), who was seen as more reserved (see Bonfigliuoli to Malpighi, March 2, 1671, in Adelmann, 1975, vol. II, letter no. 255).

[29] Malpighi, *De lingua*, in Malpighi (1686, vol. 2, p. 13, page numbering varies throughout the volume): "De hac enim divinus Vesalius habuit: *Quandoquidem verò universam linguae fabricam cæterarum corporis partium constructione mihi minus, ut verum fatear, cognitam necessario, ac serie quadam complexi sumus*." Italics in the text. Quotations are from the London edition of Malpighi's *Opera omnia* (1686).

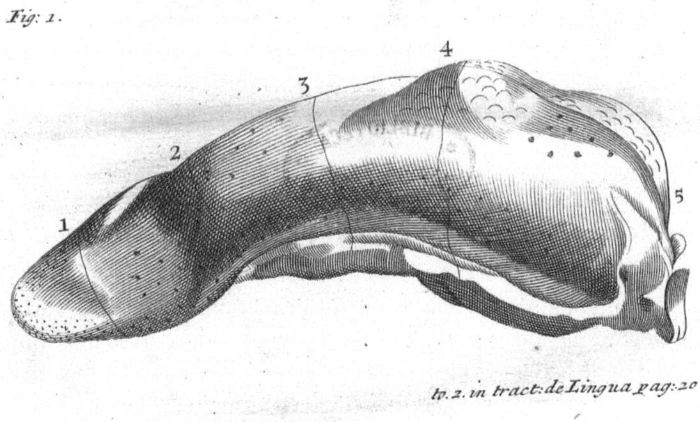

Fig. 8.1 A beef tongue with four lines marking the exact points where it was cut. From Malpighi (1686, vol. 2). Biblioteca del Real Jardín Botánico, RJB-CSIC, https://bibdigital.rjb.csic.es/idurl/1/14214. Public domain

parts.[30] Malpighi, influenced by discussions with Borelli, conducted experiments to determine the nature of the tongue's substance, emphasising its muscular nature over its glandular features. Accordingly, the dissertation's illustrations highlighted the muscular composition of the tongue, specifically, the arrangement and orientation of muscle fibres (Fig. 8.2).[31] Malpighi's primary focus was, however, on determining the specific area of the tongue responsible for taste, which he approached analytically; indeed, this is the sole function of the tongue that he examined in detail. In fact, while anatomists agreed that the tongue is responsible for the sense of taste, the structure through which this is produced was highly disputed.[32]

Malpighi's observations on this matter stemmed from a technique inspired by culinary practices, specifically, the method of boiling veal tongue to facilitate meat stripping and layer isolation with a knife. By peeling off the outermost layer (the cuticle) and an intermediate reticular and mucous layer, Malpighi revealed the "papillary body," a uniform layer consisting of three types of papillae. The first type, resembling the protractile and retractile horns (*cornua emissilia & conductilia*) seen

[30] *Ibidem*: "Unde nihil mirum, si apud Anatomicos varia & opposita in hac pertractanda occurrant placita, dum alii Linguæ carnem, molem laxam, spongiæ instar, statuunt, nonnulli glandulosam naturam ei attribuunt, aliqui musculoso & glanduloso componunt corpore, & tandem cæteri ad peculiarem substantiam, qualem in reliquo corpore nullam reperies, reducere conantur."

[31] These illustrations are likely based on a black-ink drawing, found in BCA, MS. B 3237, f. 53r, that depicts Malpighi's initial attempt to study the organization of fibers in the apical portion of the tongue. See also *ivi*, f. 60r: "Lingua habet fibras transversas, a superiori ad infimam et a lateribus et in media parte a basi angulariter rursus ad basim." A small picture is included at the end of this sentence, depicting (through the use of lines) the primary pattern of fiber arrangement.

[32] *De lingua*, in Malpighi (1686, vol. 2, p. 13): "Non minora etiam deprehenduntur Anatomicorum dissidia in statuenda peculiari gustus particula: plerique enim linguæ parenchymatis munus credidere; nonnulli nerveæ propaginis aliqui extensæ & circumvolventis membranæ: & vulgatum est novissime tonsillas ad linguæ radicem locatas gustus primarium organum sibi vindicare."

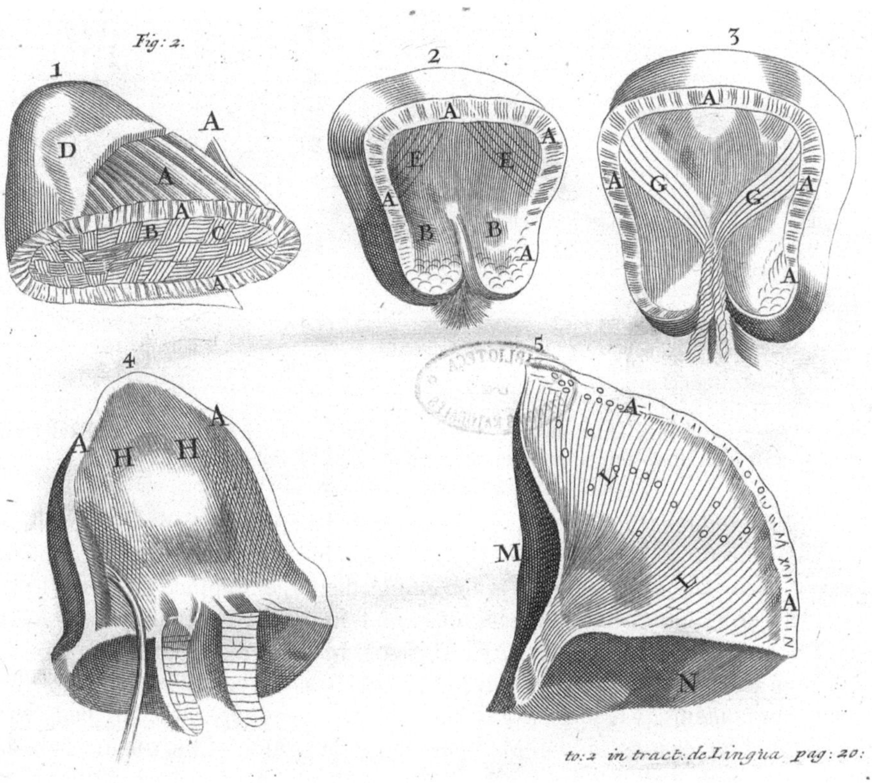

Fig. 8.2 The course of the internal muscle bundles in each section of the tongue. From Malpighi (1686, vol. 2). Biblioteca del Real Jardín Botánico, RJB-CSIC, https://bibdigital.rjb.csic.es/idurl/1/14214. Public domain

in snails or fungi, are few in number but strictly connected to the papillary body and the nerve branches. Then, a second type of papillae, more numerous than the preceding ones but shorter, and a third type, thinner and with a conical shape, follow.[33]

[33] *Ivi*, p. 15. The first mention of Malpighi's work on the tongue can be traced back to a letter from Borelli dated July 12, 1664. Borelli expressed admiration for Malpighi's writing on the subject—probably a draft—and promised to compare his observations on the beef tongue: "La scrittura, che V. S. m'ha mandato sopra la lingua mi è paruta mirabile, e veggo così per ombra, che vi è molto da speculare, io procurerò per ora di confrontar le sue osservazioni in qualche lingua di bue, poi gli soggiungerò particolarmente il mio parere intorno questa sua epistola" (Adelmann, 1975, vol. I, letter no. 106). In subsequent letters dated July 19 and September 11, 1664, Borelli shared his thoughts on the function of papillae and informed Malpighi that he had discussed the topic with Fracassati, who was encouraged to study the cooked tongue of a calf (see Adelmann, 1975, vol. I, letters nos. 107 and 111). In his diary entry on July 12, 1664, Malpighi provided a detailed description of the "fungiform" papillae on sheep and beef tongues, indicating his ongoing research on the topic (see BUB, MS. 2085/II, ff. 11–12). Adelmann (1966, vol. I, pp. 239ff) suggests that Malpighi had been working on this subject for some time before these exchanges. In *Gustus organum*, Bellini described papillae-like structures in the tongue and also identified them as the organ of taste. He acknowledged that his own observations had been inspired by Malpighi's earlier research in this field. On Malpighi's *De lingua*, cf. also Maselli (2023).

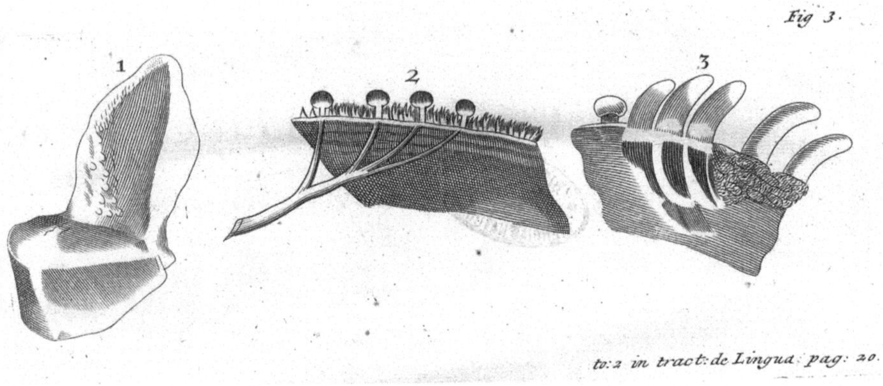

Fig. 8.3 The papillae of the tongue. From Malpighi (1686, vol. 2). Biblioteca del Real Jardín Botánico, RJB-CSIC, https://bibdigital.rjb.csic.es/idurl/1/14214. Public domain

In one of the figures (Fig. 8.3), Malpighi illustrated a cross-section of the papillary body with the three distinct types of papillae, marking the first depiction of the organ of taste. The nerve endings at the end of the larger papillae are also clearly depicted. The origin of this illustration can be traced back to a black ink drawing (Fig. 8.4) found in BCA, MS. B 3237. This confirms that the representation of the papillae in *De lingua*, possibly engraved by Scilla, was conceived by Malpighi. The manuscript contains two sets of drawings. In the first group, labelled "buds with horns" (*papillae cum cornibus*), Malpighi depicts a section of the tongue showing the papillae with their outer surface covered with protuberances or cartilaginous bodies shaped like fangs or horns, which give a rough texture to the touch. The cross-section allows Malpighi to visualise the different layers of the tongue's outer surface and to place the position of the papillae within them. In the second group of drawings, at the bottom of the folio, Malpighi breaks down the image to illustrate first the protuberances that encapsulate the papillae and then the reticular layer, which is characterised by holes of various sizes and shapes, especially around the protuberances. Malpighi indicates this with a dense network of very small circles representing the holes. In the last, more approximate image, the reticular layer is shown unfolded, ready to be viewed under the microscope, as suggested in *De lingua*.

The relationship between the papillary body and the nervous system is a critical point that Malpighi addressed, without yet resolving the issue. He entertained the idea, proposed by anatomist Realdo Colombo (1516–1559), that the membranes of the tongue, thus including the papillary body, may derive from the dura mater, the outermost and thickest membrane enveloping the brain. However, Malpighi also speculated that the papillary body may originate from nerve branches situated at the base of the papillae, drawing parallels to the role of nerves in other sensory organs.[34]

[34] *De lingua*, in Malpighi (1686, vol. 2, p. 16). Cf. also Giorgio Baglivi's "fibre theory," that conceived the body as a bundle of fibers (*fasciculus fibrarum*), with the dura mater serving as the main motor and principle of its functions. See Tonetti (2016).

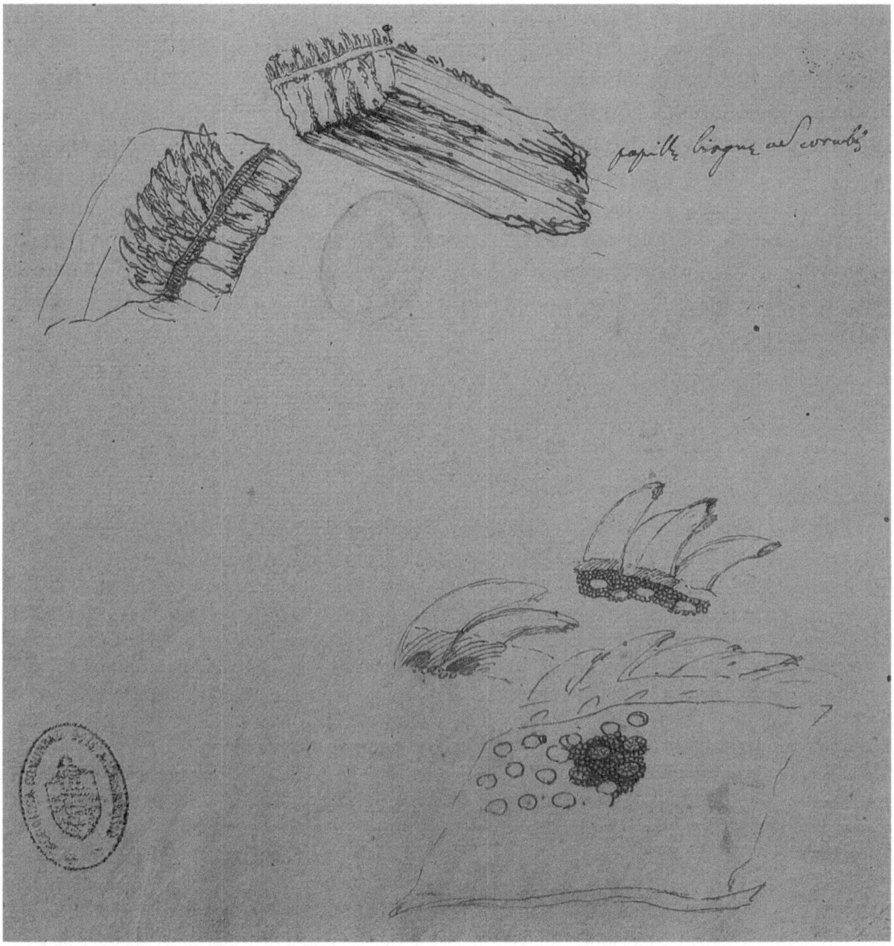

Fig. 8.4 Black ink drawing depicting the papillae of the tongue at various depths, initially (at the top) as a complete structure within the outer layers of the tongue, followed by a close-up view (at the bottom) emphasizing their composition and connection to the reticular layer. The circular perforations typical of the reticular layer are represented as small circles. Biblioteca Comunale dell'Archiginnasio, MS. B 3237, f. 60r. Comune di Bologna.

Malpighi's distinctive understanding of the papillary body's structure, particularly its connection with nerve endings and its relationship with the outer layers of the tongue, led him to propose an alternative theory regarding the sense of taste. He suggested that taste is not generated through the action on the outer membranes of the tongue (in this sense, for example, the organ of taste is not located in the *tenuissima membrana* surrounding the tongue, as previously proposed by anatomist André du Laurens), but rather through direct contact between salt particles mixed with saliva and the papillary body.[35] This theory would explain the purpose of the porosity he observed in the outer layers, allowing salt particles to penetrate deeper and come into contact with the papillae.[36] The theory also purports that the diversity of gustatory sensations is contingent upon the configuration of particles, their trajectory, and their ability to penetrate the structure of the papillary body. Malpighi postulated that all three types of papillae contribute to the sense of taste, yet he did not elucidate how their disparate characteristics might influence perception.[37] Nevertheless, this would render the Aristotelian claim of a "medium" redundant, suggesting that the perception of taste is not, in fact, mediated, given that such receptors are actually in direct contact with particles without the involvement of any intermediaries.[38]

[35] du Laurens (1593, pp. 758–759). See *De lingua*, in Malpighi (1686, vol. 2, pp. 17–18).

[36] *De lingua*, in Malpighi (1686, vol. 2, p. 18). It is notable that Malpighi draws upon Plato's *Timaeus* to elucidate the manner by which salts interact with the nervous receptors. Belloni (1966) highlighted the relevance of a renowned passage from Galileo's *Saggiatore*, which he regarded as a key source of inspiration for Malpighi. Malpighi believed that all matter consisted of tiny and minute particles that governed natural processes depending on their shape, size and motion. However, Gómez López (1997) notes that although Malpighi held a corpuscular view, he did not discuss the atomistic doctrine of sensations or speculate on the structure of matter. In other words, Gómez López argues that the particles mentioned by Malpighi are not the indivisible atoms of the atomists and are not intended to form the basis of a complete "system" of nature. This is not how Borelli and the so-called "Pisan atomists" (such as Lorenzo Bellini and Donato Rossetti) approached the matter. On Bellini, Rossetti and "atomistic anatomy," see Bertoloni Meli (2011, pp. 97–100).

[37] *De lingua*, in Malpighi (1686, vol. 2, p. 18). This is, in fact, a problematic point. Although Malpighi suggests that all three papillae are involved in taste, he expresses doubt as to whether the first ones, due to their size and connection with the nerves, play the most important role. Furthermore, this ambiguity is particularly evident in *De externo tactus organo* (Malpighi, 1686, vol. 2, p. 22): "[…] Papillarum in lingua copiosum deprehendi proventum, ex instituta animantium anatome, elatiorem portionem gustandi ministerio non inepte designaveram, residuum autem vel ad ebibendum tenuiorem cibi succum vel saltem ad juvandum gustum dubitative decreveram." The other papillae seem to have a different function, that of absorbing the softer juices of food and facilitating taste. Malpighi borrowed this idea from Angelo Forte (sixteenth c.), who believed that the finer parts of food reach the brain through the nerves. See Forte (1543, ch. 2): "tenuissimae igitur alimenti particulae, ab ore, per nervorum radiculas, quam primum ad cerebrum perveniunt." See *De lingua* (Malpighi, 1686, vol. 2, p. 18): "Unde non levis oritur conjectura, ex minimis illis papillis, quæ in lingua & palato copiose observantur, exsugi aliquid tenuioris & spirituosioris alimenti, quo fiat tensio quædam in jam delaxatis nervis […]."

[38] *De lingua*, in Malpighi (1686, vol. 2, p. 19): "Ex his postremo elici potest, nequaquam in sensatione gustus medium requiri, ut contendunt Peripatetici; nam si papillari & nervoso corpori debetur gustandi passio, ut hucusque deduximus, cum hoc mediis porulis & sinibus quibus continentur sapidum corpus immediate tangat nulla necessario mediat pars." Interestingly enough, Bertoloni Meli (2011, p. 91) highlights that "the issue of whether Aristotle himself believed taste to require a medium is quite entangled and requires careful handling. Aristotle, however, stated that taste is a form of touch, and in this respect he was closer to Malpighi."

8.3.2 *Do Tactile Papillae Mirror Those of Taste?*

Malpighi's study of the organ of touch was driven by a theoretical hypothesis: he postulated the presence of touch receptors dispersed throughout the body, similar to the taste buds found on the tongue.

The fact that Malpighi initially lacked the option of dissecting animals to test his research hypothesis is a notable, albeit maybe rhetorical, observation and one that provides context for his microscopic examination of his fingertip.[39] While closely examining the ridges of the fingertip, Malpighi noticed the presence of round transparent bodies scattered across the surface. Initially, he postulated that these could be "papillae"; however, upon reflection, he concluded that they were likely to be droplets of sweat released from skin pores—a hypothesis that he had arrived at by the conclusion of his research.[40] This intriguing initial observation prompted Malpighi to pursue a deeper understanding of the diversity of skin, even within the human body.

Besides studying the nature of skin and its physical properties and functions, Malpighi sought evidence from comparative anatomy, consistent with the principle of the unity of nature.[41] He documented a series of observations on hoofed animals (such as the feet of pigs, cattle, and birds) and experimented with different methods of boiling tissues—as he had done to dissect a cow's tongue—to separate the skin layers by heat. From the ungula and feathered leg to the feet and hands of humans, Malpighi extended his research to the entire skin surface of animals, concluding that the same papillae found on the tongue were abundant in tactile areas, particularly those of high sensitivity. He finally recognised that these papillae originate from the nerves and extend to the outer cuticle. He also noted that similar papillae are present throughout the body that, although smaller in size, have the same structure and distribution.[42]

---

[39] *De externo tactus organo*, in Malpighi (1686, vol. 2, p. 22): "Mens de ambiguo usu, pyramidalibus in lingua descriptis papillis assignato, anxia torquebatur; oculorum igitur aciem microscopio munitam, veluti auxiliares convocat copias, & quia brutorum non aderant illico perquirenda membra, extremum digiti lustro apicem, & dum attentive inæquales illas rugas quasi in gyrum, vel in spiras ductas contemplor, en e quibusdam alveolis, & sinibus subrotunda, ac veluti diaphana emergunt corpora, miro ordine per interiorem totius digiti faciem copiose dispersa [...]."

[40] See also BUB, MS. 2085/II, f. 141v. On August 8, 1691, Malpighi reported: "Observavi guttas sudoris erumpentes a foveis rotundis in parte exteriori manus evidentissime a foveis pilorum non vidi sudorem."

[41] Malpighi's methodology posits a constant, uniform operation of nature, whereby observed variations represent only subtle adjustments within this fundamental pattern. However, as Bertoloni Meli (2013) notes, this view faces significant challenges.

[42] *De externo tactus organo*, in Malpighi (1686, vol. 2, p. 26): "Ex his, & similibus videbatur animus abunde certior redditus, earundem papillarum pyramidalium copiam, quas alias in lingua descripsi, in locis præcipue exquisitiori tactui dicatis reperiri, eodem progigni nervoso, & cuticulari corpore, simulque circumvolvi reticulari involucro, & extimam cuticulam, veluti ultimum terminum attingere, rursus in corporis etiam ambitu easdem conspici, minores tamen, sed prorsus eadem servata structura, & situ, quod ultimum in homine ipso comprobavi, dum brachii cutim

Malpighi's repeated dissections led him to conclude that these papillae serve as an effective organ of touch, particularly in areas that are highly responsive to external stimuli. Although he did not provide a comprehensive anatomical account of the manner in which papillae are connected to nerves, he employed a series of analogies, including one comparing nerve branches to the tentacles of an octopus, to illustrate how sensory transmission works. Malpighi purposefully embraced the Stoic idea of a centralized soul that extends to the animal's peripheral sense organs, likening it to the tentacles of an octopus, and extended this analogy to the nature of the brain to emphasise the body's ability to perceive stimuli through nerve branches, regardless of the strength of the stimulus.[43] As a result, even small impulses (*levissimus ictus*) can trigger significant sensory responses.[44] This process is further explained in a note that clarifies the role of the papillae as follows: "when the papillae are stimulated the nerves are moved and from the movement thus communicated to the brain a sensation is produced."[45] As suggested by Bertoloni Meli, Malpighi seems to have revisited the Cartesian notion of the blind man's cane as an extension of the arm. Although Malpighi does not explicitly mention the example of the blind man from Descartes' *Dioptrique*, he employs a similar analogy to elucidate the processes occurring in the papillae of ungulate animals—the same movement that is induced in the palm of the blind man by the stick when he touches objects also occurs in the papillae. This suggests that objects can come into contact with the papillae via the fibres of the stick. Building on this, Malpighi places significant emphasis on the role of the papillae as extensions and appendages of the nerves.[46] This implies that the tactile papillae, as an organ of touch, are more immediate than the skin itself. More significantly, the papillae are nerve endings rather than components of the skin, which serves to reinforce the role of nerves in sense perception. Malpighi referred to Gassendi's example of a woman who experienced pain in her

perquirerem […]." See also the observations recorded in BCA, MS. B 3237, f. 19rv, which are likely related to the dissections conducted before the publication of *De externo tactus organo*.

[43] *De externo tactus organo*, in Malpighi (1686, vol. 2, p. 27): "Et Stoicorum dogma fuit, a principaliori animæ parte septem veluti polypi brachia extendi, a quibus exiens spiritus quinque vulgares sensus, semen insuper, & vocem promeret: satius tamen, relicta animæ divisione, cerebro polypi naturam congruentius adaptare possumus, cum extremis nerveis propaginibus, veluti oblongis exporrectis brachiis, externorum ictus & concussiones, quæ a se longe contingunt, ex inducta titillatione noscere valeat, ita ut vel alteri communicare possit, vel saltem terminata in seipso nova illa motione reagendo sentiat."

[44] *Ibidem*.

[45] BUB, MS. 936/II, A, f. 16v: "De tactus organo. […] Ad M.(?) per vellicatis papillis nervi moveantur et ita comunicato motu cerebro fit sensatio."

[46] *De externo tactus organo*, in Malpighi (1686, vol. 2, p. 29): "[…] probabile enim est in animalibus ungula præditis, & in cæterorum cute, eandem accidere motionem in continuatis mollioribus nervis, seu papillis, quæ in vola cæci manus ex baculo, objectorum extrema tangente, indies contingit; nam manus papillæ continuis baculi fibris, veluti medio quodam, vel appendice elongari videntur, vel mavis, objectum illud his interpositis, absque eo quod approximetur, & loco moveatur, ferire & pulsare easdem tentat." See Bertoloni Meli (2011, pp. 92–93).

finger despite having gangrene in her foot, attributing this to the nerve endings in the finger misleading her perception.[47]

The central question is whether touch requires some form of mediation or can occur through direct contact.[48] Malpighi's perspective offers a possible reconciliation of these two viewpoints: the extremities of the papillae are intimately connected to the particles of the cuticle, causing them to thicken through exposure to air and constant interaction with external objects.[49] Consequently, these extremities also act as a kind of medium that can regulate the sensory impulse. However, Malpighi left unanswered the pivotal question of how animals perceive when their nerve endings are stimulated or affected.[50]

Throughout the 1660s, Malpighi continued to work on tactile papillae. In a note dated 6 August 1666, Malpighi recorded some observations on the lips of oxen. In particular, he noted that by injecting a dye, he was able to highlight the nerve endings branching out into the papillae.[51] Human fingers, nails, and, more generally, the structure of human skin are the subject of another detailed note, probably from the same day. In a black ink drawing (Fig. 8.5), Malpighi focused on the subcutaneous level of the skin, below the papillary body, to illustrate the relationship between the glands and nerve endings distributed in a network in the subcutaneous tissue of the human foot and hand as well as in the lip of an ox.[52] This drawing is quite fascinating, surpassing even the engravings of the papillae of the tongue, where Malpighi attempted to demonstrate, albeit subtly, the connection between the papillae and the nerves. Delving deeper, this illustration reveals the intricate structure and composition of the skin and its appendages. It is worth noting that Malpighi did not provide any information on the visualization methods employed in the accompanying note, rendering replication nearly impossible. Could it be that Malpighi utilized coloring techniques to depict the reticular pattern created by the nerves?

---

[47] *De externo tactus organo*, in Malpighi (1686, vol. 2, p. 28): "Extat pulcherrima illa ex Gassendo mulieris historia […]" Malpighi's mention is specifically from the chapter *De Tactu, & Tactione* of the *Syntagma philosophicum* (Gassendi, 1658, p. 355). Bertoloni Meli (2011) has emphasized Malpighi's engagement with Gassendi's work, especially regarding discussions on the sense of sight and touch. A list of *notatu digna* from Gassendi's *De plantis* (which I found in BUB, MS. 2085/II) further underscores the significance of the *Syntagma* as a reference source, even in natural history subjects such as botany.

[48] *De externo tactus organo*, in Malpighi (1686, vol. 2, p. 29): "An in tactu exigatur medium, disputatum est, antiquitus plures enim admisere diversa media, propter vulgatum illud, sensibile supra sensum positum non facere sensationem. Non desuere tamen, qui in tactu mutuum, & immediatum contactum requiri crediderunt […]."

[49] *Ibidem*: "Utrumque conciliari posse videtur; nam papillæ in brutis ungula pollentibus evidenter protrahuntur usque ad extimam superficiem, cum ea tamen conditione, ut mutata quasi natura non amplius nervi speciem exhibeant, sed quid medium, ad reddendum forte moderatiorem, & magis consonum impulsum."

[50] *Ivi*, p. 30: "Sed anxium te video in perquisitione miri illius artificii quo pulsato vel vellicato extremo nervorum fine sentiat animal quod quidem a teipso facilius attinges quam a me speres."

[51] BUB, MS. 2085/II, f. 11v: "affundatur atramentum emergunt nervorum ramificationes desinentes in papillas."

[52] BUB, MS. 2085/II, f. 12r.

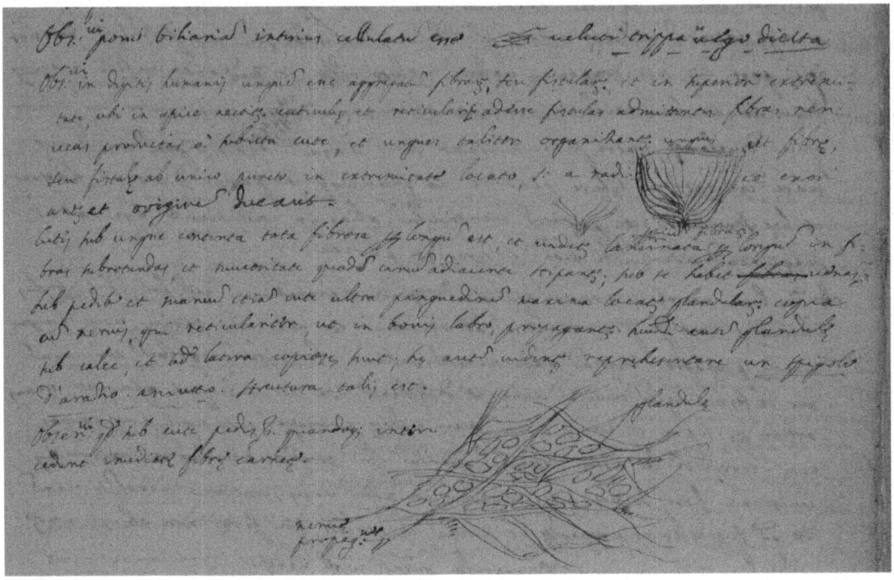

Fig. 8.5 Black ink drawing showing the relationship between the glands (*glandulae*) and nerve endings (*nervorum propagationes*) at the subcutaneous level of the skin. Bologna, August 6, 1666. Biblioteca Universitaria di Bologna, MS. 2085/II, f. 12r. ©Alma mater studiorum Università di Bologna. All rights reserved

More interestingly, on 25 February 1668, Malpighi recorded a fascinating observation on the structure of a dog's paw, stating that he had identified papillae similar to those in the human and ox tongue. He noted the perfect correspondence between the cuticle and the papillary body, emphasising, in particular, the role of the perforated capsules that make up the cuticle, which serve as a kind of involucre for the papillae as they emerge from the skin. The note is accompanied by a small picture (Fig. 8.6), arguably the first to show the tactile papillae, albeit in a stylised form.[53] A few months later, on 10 June, he began an extended observation on human hands.[54] An entire page (Fig. 8.7), dated February 1671, is then dedicated to the structure of the human skin, with a particular emphasis on hair follicles and the function of hair.[55] It is notable that among all the skin adnexa, Malpighi specifically examines hairs (and feathers), highlighting how deeply they penetrate the skin—as shown by the details of these drawings. Once again, Malpighi puts forward the hypothesis, previously mentioned in *De externo tactus organo*, that nerve branches (*portio nervorum*) could reach the base of the hair follicle. However, his focus on hairs is also justified by their physiological role. As he later elaborated in his autobiography,

[53] BUB, MS. 2085/II, f. 29r.

[54] BUB, MS. 2085/II, f. 30r.

[55] BUB, MS. 2085/II, f. 32v.

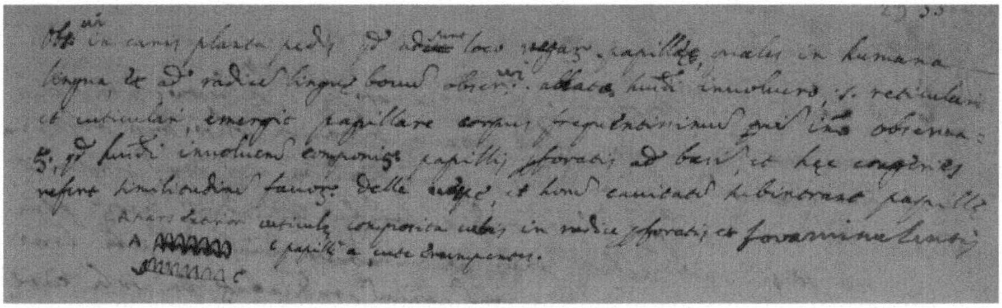

Fig. 8.6 Black ink drawing showing tactile papillae (*papillae a cute erumpentes*), albeit in schematic form. Bologna, February 25, 1668. Biblioteca Universitaria di Bologna, MS. 2085/II, f. 29r. ©Alma mater studiorum Università di Bologna. All rights reserved

since hairs grow near mucous-secreting glands under the skin, their primary function is to aid in the discharge of fluids.[56]

8.4 Tactile Papillae under Fire: Do they Really Exist?

In the 1670s, Malpighi faced some opposition from local physicians in Bologna, notably Giovanni Girolamo Sbaraglia (1641–1710) and Paolo Mini (1642–1693).[57] Not only did they question the role of anatomical research in "clinical" medicine, but they also challenged Malpighi's interpretation of some of his anatomical findings.[58] Malpighi regularly attended their public demonstrations in the anatomical theatre of Bologna and actively participated in the debates.[59] These public dissections, especially in Bologna, consisted not only of lectures and demonstrations, but also of heated

[56] Malpighi (1697, p. 28): "Circa pilos certum est id, quod indicavi, penetrare scilicet cutem, & corium, & peciolo sub cute condito firmari. Et quoniam in glandularum loculis, ut in facie, & alibi, a quibus mucosa eructatur materia, pili implantantur, ideo probabile est his humoris exitum facilitari, cum per pilos transducatur, ut in lucernis observatur."

[57] On Sbaraglia, see: Cavazza (1990, 1997), Bertoloni Meli (2007, 2011), Bresadola (2018a), Cassoli (2023), Tonetti (2024). See also: http://badigit.comune.bologna.it/stemmi/dettaglio.asp?lettera=2088 (last accessed on December 31, 2024). On Mini, see Fantuzzi (1788, pp. 27–30), and http://badigit.comune.bologna.it/stemmi/dettaglio.asp?lettera=358 (last accessed on December 31, 2024).

[58] On Malpighi's medical practice, see Bertoloni Meli (1999, 2011), Bresadola (2006, 2011).

[59] See BUB, MS. 2085/II, f. 52v (January 14, 1679: *Ex Anatome Domini ter Maximi Mini* [sic]), f. 53rv (February 16, 1676: *Ex anatomia Domini Manzij Bononiae*), f. 54rv (January 15, 1673: *Compendium dictorum Sbaragli publici anatomici*), f. 54v (*Compendium dictorum Domini Pauli Mini publici anatomici 1674*), f. 121r (April 9, 1689: *D. Paulus Minus finem imposuit anatomiae publicae…*). Mini's last public anatomy was conducted in 1693: see letter of Antonio Maria Valsalva to Malpighi, January 24, 1693, in Bertoloni Meli (1997b).

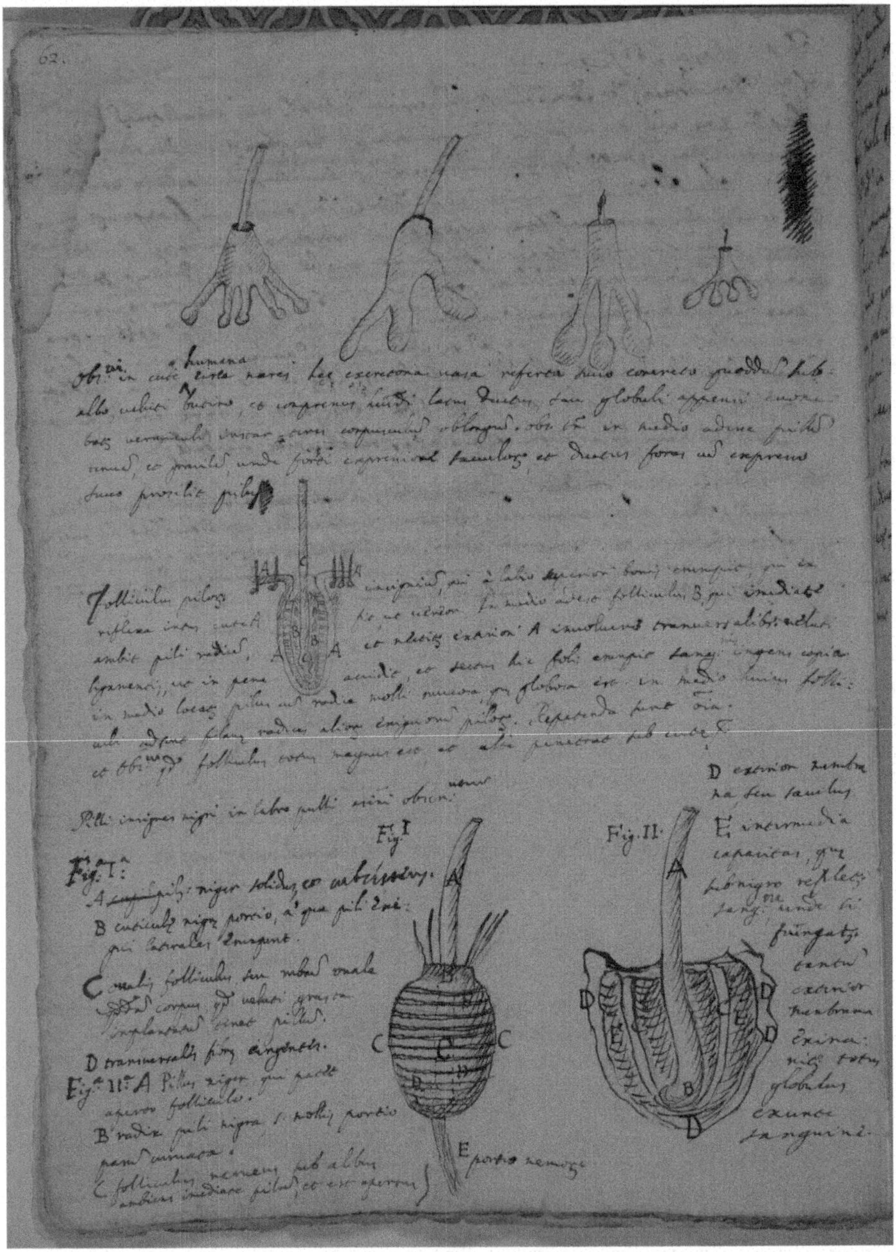

Fig. 8.7 The manuscript page is entirely devoted to the subject of the human skin, with a particular focus on hair follicles. Bologna, February, 1671. Biblioteca Universitaria di Bologna, MS. 2085/II, f. 32v. ©Alma mater studiorum Università di Bologna. All rights reserved

debates in which objections to the lecturer's statements could be raised.[60] One of the most controversial aspects of Malpighi's work was the hypothesis that many physiological processes are regulated by a specific part of the body, notably the "glands," as discussed a few years earlier in his essay *De viscerum structura* (1666–1668). Glands play a crucial role in Malpighi's mechanistic architecture of the body, because their main actions—filtration and secretion—do not imply any faculties or qualities.[61]

8.4.1 Questioning the Organ of Touch in Public Dissections in Bologna and Padua

It is noteworthy that among the earliest notes recorded by Malpighi during Sbaraglia's public dissection on 15 January 1673 were those pertaining to the skin and sense of touch. Sbaraglia had contended that the skin, rather than the papillae, constituted the true organ of touch, given that there are bodily regions that are sensitive despite the absence of papillae.[62] The following year, Mini gave a public lecture on anatomy, and Malpighi immediately noted Mini's attempt to "recapitulate" the alleged discoveries of the moderns within the Hippocratic–Galenic tradition. Mini suggested that the papillae are only used to perceive roughness and heaviness.[63] During Giovanni Galeazzo Manzi's public dissection delivered on February 16, 1676, Malpighi reported that "the skin is composed of nerves, veins, arteries, glands, and papillae, and all this, as it moves in such a way, acts against the pressure of external forces, and thus it feels."[64] A few years later, in 1679, Mini even denied the existence of tactile papillae, proposing instead that touch may depend on the corrugation of the skin, so that ultimately the cuticle plays no mediation in sensation.[65]

Among Malpighi's annotations from the 1670s, there is also a list of arguments on different topics supported by various physicians, including Sbaraglia, Mini, Giovanni Battista Capponi (1620–1675),[66] and Giovanni Andrea Volpari (d.

[60] On public demonstrations in Bologna, see Ferrari (1987a, 1987b). For an overview on teaching and learning in early modern Bologna, see Cavazza (2012) and Lines (2023). On Padua, see Klestinec (2007, 2011).

[61] Cf. Giglioni (1997), Bertoloni Meli (2011, 2019).

[62] BUB, MS. 2085/II, f. 54r: "Tactus organum non sunt papillae, sed cutis nam in uretra non dantur papillae et ipsae etiam non sentiunt."

[63] BUB, MS. 2085/II, f. 54v: "Recentiorum inventa explicanda sunt antiquis theorijs unde Galenus omnia habuit. Papillae cutis inserviunt ad sensum asperi et gravis. Cutis temperie sentit." Galen's theory is briefly summarised by Malpighi as follows: "[…] cum Galeno eam [*scil.* cutem] sentire arbitramur, quatenus sua temperie qualitatum omnium excessus, ut quid medium judicare possit […]." (*De externo tactus organo*, in Malpighi, 1686, vol. 2, p. 27).

[64] BUB, MS. 2085/II, f. 53r: "Cutis componitur nervis, venis, arteriis, glandis, et papillis et hoc totum prout taliter movetur agit contra urgentiam externorum et ita sentit." On Manzi, see Adelmann (1675, vol. II, p. 477, fn. 8).

[65] BUB, MS. 2085/II, f. 52v: "Negavit papillas tactus externi docens esse appendices. Posuit cuticulam non medium tactus. Tactus docuit ex corrugatione cutis humanae."

[66] Mazzetti (1847, p. 83), Betti & Calore (1996), Betti (2006). According to the rotuli of the *studium* of Bologna, Capponi was assigned to anatomy from 1671–1672 through 1675–1676 (cf. Adelmann, 1675, vol. II, p. 575, fn. 5).

1698–9),[67] against Giovanni Carlo Lanzi Paltroni (d. 1687),[68] who had conducted public dissections in 1671–1672.[69] Sbaraglia argued that the brain was the organ that discriminated "on tangible qualities" (*de qualitatibus tangibilibus*), and therefore the skin could not be the organ of touch.[70] Volpari's focus was on the specific case of the human penis, where the glans penis is highly sensitive but does not have the same properties as the skin, and his conclusion, like Sbaraglia's, was that the organ of touch could not be the skin.[71]

More remarkably, in a letter dated 1679, Luigi Ferdinando Marsili (1658–1730) informed Malpighi about a public dissection delivered by Giacomo Pighi (1647–1682) in Padua earlier that year, probably in February.[72] Pighi had assumed the role of Professor of Anatomy at the *studium* of Padua on 18 December 1675, succeeding Antonio Molinetti.[73] Marsili emphasised the differences between the public dissections in Padua and Bologna, noting that Pighi "lectures on each anatomical part in turn and then demonstrates without anyone contradicting him, unlike in Bologna."[74] Pighi, Marsili explained, "is very modern, contrary to Paduan custom, and introduces the teachings and opinions of the moderns, including our Mr Malpighi, the pride of the city of Bologna."[75] Describing integuments, the topic of the first lesson, Marsili wrote:

> [...] And then he went on to the skin [*cute*] and said that this is a nervous structure [*tessitura*] made in such a way that sweat can perspire through it, and then he pointed out, like Mr Malpighi, that he saw certain nervous papillae through which touch is made either pleasant or painful, according to the order in which these papillae are made.[76]

[67] Mazzetti (1847, p. 324). See also: http://badigit.comune.bologna.it/stemmi/dettaglio.asp?lettera=2221 (last accessed on December 31, 2024). On public demonstrations conducted by Volpari, see Adelmann (1975, vol. II, pp. 561–563), letter no. 258, Malpighi to Bonfigliuoli, March 11, 1671. For further details, see, *ibidem*, fn. 15.

[68] Fantuzzi (1786, p. 21), Mazzetti (1847, p. 178). In the academic year 1668–1669, Lanzi Paltroni was for the first time assigned to anatomy. See Adelmann (1975, vol. 1, p. 402).

[69] See BUB, MS. 2085/II, f. 56r (*Argumenta contra Dominum Lanzium Anatomicum 1672 Bononiae*).

[70] *Ibidem*: "Idem [*scil.* Sbaraglia] cerebrum est organum indicans de qualitatibus tangibilibus ergo cutis non est organum tactus."

[71] *Ibidem*: "Vulparius in penis glande non est cutis et sentit ergo haec non est organum externum tactus."

[72] Adelmann (1975, vol. 2, pp. 808–818), letter no. 384, Marsili to Malpighi, [1679]. See also Münster & Gerocarni (1940).

[73] In 1681, Pighi was named prefect of the Botanical Garden at Padua. For further references, see Adelmann (1975, vol. 2, p. 772, fn. 1), esp. Nannini (1964–1965). See also the correspondence with Malpighi in 1678 and 1679: cf. Adelmann (1975, vol. 2), letters nos. 376, 377, 379, 381, 383, 386, 387.

[74] Adelmann (1975, vol. 2, p. 808), letter no. 384: "conforme le parti occorrono espone la lettione, e poi fa l'ostensione senza che vi sij veruno come à Bologna li contradichi."

[75] *Ibidem*: "è moderno assai al contrario dell'uso di Padova portando dottrine, e opinioni de Moderni, e tra gl'altri del nostro Sigr. Malpighi splendore della Città di Bologna."

[76] *Ivi*, p. 809: "[...] e di poi passò alla Cute dicendo, che questa, è una nérvea tessitura fatta in guisa tale da poter per ivi traspirare il sudore, e poi avvertì come il Sigr. Malpighi, che si vedevano certe

Marsili's account documents the impact of Malpighi's theory on the debate of the time, even on the practice of public dissections in Bologna and Padua. One example of its impact is shown by Antonio Giovanni Stancari (1668–1748), who from 1701–1702 onwards served initially as an assistant and then as *anatomicus ordinarius*, participating on multiple occasions in public dissections in the anatomical theatre in Bologna. In his lectures on the skin, Stancari devoted a portion of the material to Malpighi's theory, specifically focusing on the role of the papillae as a "sensorium tactus."[77] Ippolito Francesco Albertini (1662–1738), a student of Volpari and Malpighi, authored the dissertation entitled *De cuticula*, likely one of his lectures delivered in Bologna, on the description of the structure and function of both the skin and cuticle of the human body.[78] He began with the skin, considered the noblest part, thus deviating from the usual order of parts in dissection. The distinction between skin and cuticle was only virtual, as Albertini, drawing on Francis Glisson's work, viewed them as intimately connected. In Glisson's *Tractatus de ventriculo et intestinis*, the skin (*cutis*) was divided into cuticle, "true skin" (*cutis vera*), hair, nails, and moles. While the true skin is deemed nobler, the cuticle is essential for the integrity of the skin. The two are complementary and constitute a whole. Influenced by Glisson, Albertini emphasised the role of fibres in defining the skin as the primary organ of touch and a protection for the underlying parts.[79] The cuticle, as in previous anatomical traditions, was described as an efflorescence and a complement to the skin that, although devoid of sensation and blood, is crucial for enhancing tactile perception and shielding the body from external agents.[80] Albertini's description of these two structures referenced Steno, particularly regarding the role of sweat glands, and once again Malpighi, from whom he adopted the theory of tactile papillae as specific *instrumenta* of touch.[81]

papille nervee medianti le quali si fa il tatto, ò grato, ò doloroso conforme con che ordine vengono queste."

[77] "Quid sit cutis et ex quibus partibus constituatur. An papillae proprium tactus sensorium statuantur. De unguibus et pilis. De glandulis miliaribus et sebaceis, quae cutis ipsius glandulae sunt [...] Malpighii sententia fusius exponitur, quae observationibus [...] confirmantur." Quoted in Vernazza (1961, pp. 159–160). On Stancari, cf. Savoia (2019).

[78] BUB, MS. 2085/XIV, ff. 9r–20r. On Albertini, see Mazzetti (1847, pp. 14–15), Jarcho (1972, 1989), Bresadola (2018b), Tonetti (2024).

[79] BUB, MS. 2085/XIV, f. 9v: "Cutis latinis δέρμα grecis a δέρο, idest ex corio, seu, et fortasse melius δεσμά a δέω, idest in fasciculos ligo, cum vulgatior nostrorum temporum opinio cutim fibrarum et cuiuscumque generis, et vasculorum miro ordine sese intexentium collectam in fasciculos aggeriem depredicet, est commune corporis integumentum, exposita methodo, additis tamen milliaribus glandulis, iisque fortasse, in quibus pili radicantis, a natura fabrefactum, ut sit praecipuum sensus tactus organum, et subiectarum partium tutamen."

[80] *Ivi*, f. 15r: "Graecis ἐπιδερμίς, seu δερματον [*sic*], cuticula Latinis, cutis item efflorescentia, cutis operculum, cutis complementum, est pellicula quaedam, quae cum vita sensuque careat, exanguis est, tenuis, densa, et aliqualiter diaphana imediate supra cutem, ipsique firmo nexu tenaciter adhaerens ad commodiorem sensus tactus perceptionem, et contra externorum ambientium iniurias non inutile tutamen, quare ablata cuticula de partis dolore semper conqueri solemus."

[81] *Ivi*, f. 11v, on the structure and function of *pyramidales papillae*, which Albertini considered to be "nervorum, aut tendinum quaedam propagines" following the definition given by Jean-Baptiste du Hamel's *Philosophia vetus et nova ad usum scholæ accommodata*.

8.4.2 Representing the "Exquisite Sense"

In a brief note, Malpighi documented the objections he had received, likely in preparation for a reply.[82] This reply was later included in his autobiography, which was published posthumously. One objection was that tactile papillae could not function as the organ of touch, as there are areas of the body where touch can occur without papillae. One notable example is the glans penis, whose sensitivity is of critical importance, as it involves the very nature of the sense of touch.[83] Malpighi compared it to other sensitive parts of the body, such as the palm of the hand, due to its papillary structure. However, there is a clear difference in sensitivity between these two areas that cannot be solely attributed to papillae, which is a point of contention.

For Malpighi, the existence of tactile papillae could be further confirmed through an observation made on 16 January 1690 of the corpse of a man who had died in the hospital of Santa Maria della Vita in Bologna. Malpighi was particularly intrigued by the case, as the right foot had developed into a highly distinctive structure.[84] It featured monstrous nails, which he later documented in his notes, along with a thickened cuticle on the sole and papillae emerging from the skin at the heel:

> I observed the foot of a deceased man in the Santa Maria della Vita hospital [...] this had the monstrous oblong nails which I have described elsewhere. The cuticle under the sole of the foot was very thick, especially under the heel, where the reticular body surrounding the tactile papillae, once resected and made friable, split and dissolved into pieces, and there were observed abundant tactile papillae, oblong, round, and sometimes slowly growing, ending in a cone at the apex. Inside [the papillae] were full and not perforated. The thickness of the cuticle of the heel, in which I observed these papillae, was of this thickness —.[85]

A detailed description of the accompanying red chalk drawing is also included:

> So I have outlined the thickness of the cuticle, at the base of which are the holes or sheaths of the papillae, which are revealed when the skin is peeled away. The papillae emerge, often curving towards the apex and planted in a mucous reticular body, which, once dried from

[82] BUB, MS. 936/II, A2, f. 21: "Papillae non sunt organum tactus." See also MS. 936/II, A, f. 10: "Objectiones ab Anglo. Papillae cutaneae non cooperiunt omnes partes ita ut adsit portio quae sine fibris sit sensitiva. Rursus cornea est sensitiva sive papillis."

[83] Malpighi (1697, p. 28): "Papillarum a cute erumpentium a me assignatum usum, quidam in Theatro nostro anatomico infirmare conatus est, asserens in labiis & in glande penis exquisitum persentiri sensum tactus, & tamen in his non observari papillas, ut in vola manus, &c."

[84] BUB, MS. 936/I, K, f. 49r. See also MS. 2085/II, f. 127r. Cf. Malpighi (1697, p. 28): "Papillarum existentiam confirmat observatio a me anno 1690 facta in cadavere cujusdam defuncti in Xenodochio Sanctæ Mariæ de Vita Bononiæ in cujus dextero pede monstruosi ungues observabantur [...]."

[85] BUB, MS. 936/I, K, f. 49r: "Observavi pedem viri defuncti in Hospitali Vitae [...] hic habebat ungues monstruosos oblongos a me alibi descriptos. Cuticula sub planta pedis valde crassa erat praecipue sub calce, ubi resicatum reticulare corpus ambiens papillas tactus et friabile redditum dividebat et solvebatur in frustula et observabantur copiosae papillae tactus oblongae rotundae et interdum sensim gravilescentes et in apicem in conum desinebant. Interius plenae nec fistulosae erant. Crassities cuticulae in calce, in qua observavi huiusmodi papillas erat huius crassitiei —." It sounds as if a measurement is given at the end of this sentence.

> the outside, is removed leaving the ends of the papillae exposed. The colour of the papillae is whitish and almost transparent.[86]

The separation of the skin layers is consistent with the description provided for the tongue. The illustration (Fig. 8.8), later printed in the *Opera posthuma*, is the first cross-sectional view of the skin layers with their papillae. It can therefore be usefully compared with the representation of the taste buds provided 30 years earlier in *De lingua* (see Fig. 8.3). It is noteworthy that, in contrast to his practice in *De lingua*, Malpighi did no longer differentiate between three distinct orders of papillae, nor did he emphasise the connection with the nerve endings. In fact, even the layers of the skin, apart from the reticular layer, are poorly depicted.

The issue of the sensitivity of the glans penis is closely connected to Malpighi's broader interest in male genitalia, particularly in the structure of both testicles and spermatic vessels. This interest originated during his initial research period in Pisa under the supervision of Borelli and subsequently evolved over time.[87] From correspondence with Silvestro Bonfigliuoli (?1637–?1696), we know that in March 1671, Malpighi was engaged in the dissection of the penis of various animals in order to study its anatomical structure and the mechanism of erection.[88] An undated observation of the penis of a mule, with a figure of a cross-section, may date from this period.[89] Although Malpighi referred to the skin and cuticle, which he claimed form the sheath from which the foreskin is formed, his description does not go into detail about the characteristics of the glans penis. An observation on the man's penis ("de pene virili") dated 11 March 1685 seems to be the most important. On this occasion, a public dissection was conducted in the anatomical theatre in Bologna with the participation of Domenico Guglielmini:

> In the year 1685, on the 11th of March, in the public Anatomical Theatre, together with Guglielmini and other friends, I observed in a man's penis, namely in the glans, the continuous skin, the cuticle, and beneath it notable, numerous, pyramidal papillae of touch, arranged almost like silky velvet in series running lengthwise. I observed the same in human lips.[90]

---

[86] *Ibidem*: "Delineavi itaque crassitiem cuticulae, in basi apparent foramina, seu vaginae papillarum, quae avulsa cute occurrunt. Ascendunt papillae, quae versus apicem frequenter curvantur et conduntur in mucoso reticularique corpore, hoc vero exterius resiccatum deciderat, unde fines papillarum patebant. Papillarum color subalbus et quasi transparens."

[87] During his first stay in Pisa, Malpighi took part in the experiments and dissections carried out in Borelli's house, together with the anatomist Claude Aubry. In his brief paper entitled *Testis Examinatus*, published in 1658 under the pseudonym Vauclius Dathirius Bonglarus, Aubry demonstrated that the testicle of a boar possesses a vascular structure. See Belloni (1964), Bertoloni Meli (2011). On Borelli's interpretation of these results, see Conforti (2001, 2009). On Malpighi's study of spermatic vessels, see Tonetti (2023).

[88] See Adelmann (1975, vol. II), letters nos. 260 and 262.

[89] BUB, MS. 2085/II, f. 41r.

[90] BUB, MS. 2085/II, f. 100r: "Anno 1685 die 11 Martij in publico Theatro Anatomico cum D. Guglielmino, et alijs amicis observavi in pene virilis sexus s. in glande cutim continuatam, cuticulam, et subiectas papillas tactus piramidales insignes, copiosas, et quasi sericum velutum, dispositas seriebus per longum locatis. Idem observavi in labijs humanis."

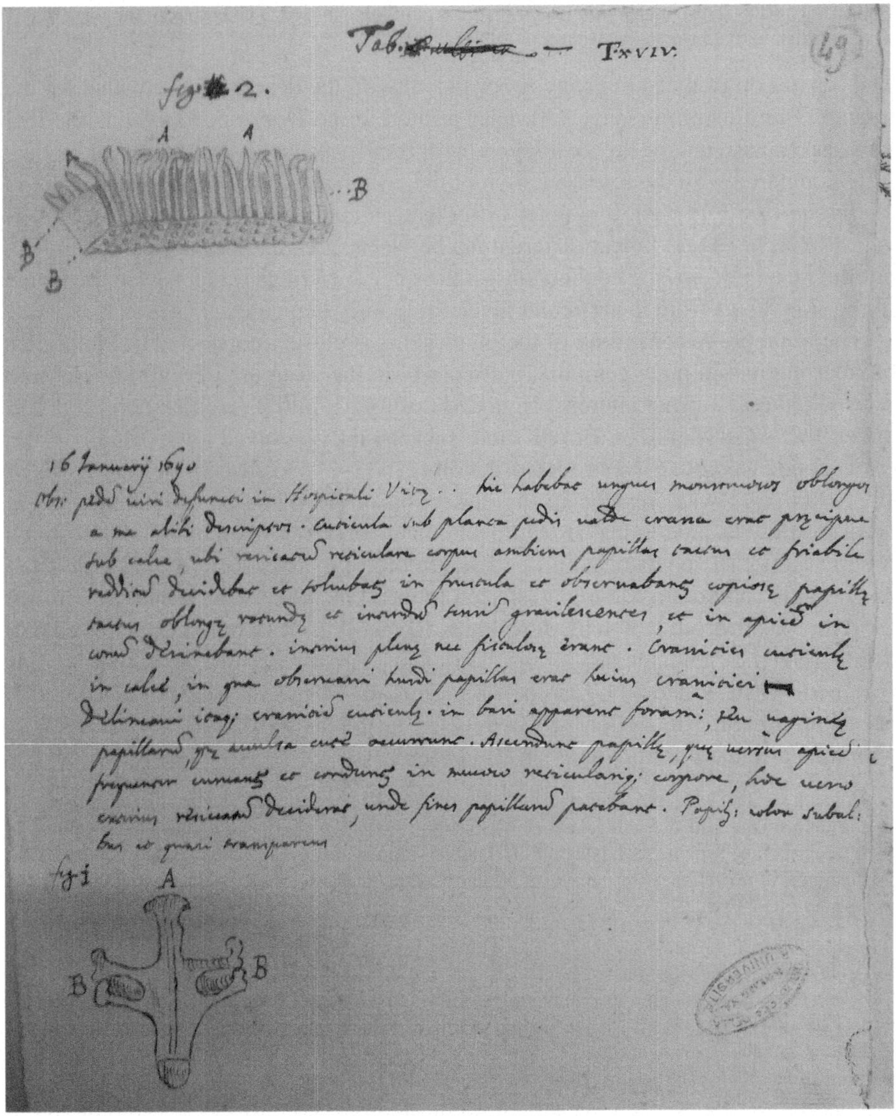

Fig. 8.8 Red chalk drawing showing the tactile papillae, as observed in the monstrous foot of a man in the Santa Maria della Vita hospital in Bologna, on January 16, 1690. Biblioteca Universitaria di Bologna, BUB, MS. 936/I, K, f. 49r. ©Alma mater studiorum Università di Bologna.

Very interestingly, again, the note is accompanied by a small drawing (Fig. 8.9) in black ink in which the "almost innumerable" pyramidal papillae on the surface of the glans penis are clearly visible, "whose brushing or touching elicited acute

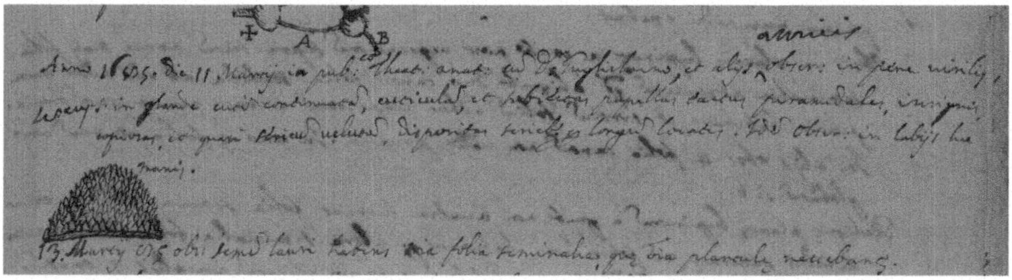

Fig. 8.9 Black ink drawing in which papillae are apparently depicted on the surface of the glans penis. Bologna, March 11, 1685. Biblioteca Universitaria di Bologna, MS. 2085/II, f. 100r. ©Alma mater studiorum Università di Bologna. All rights reserved

sensations."[91] The number and organisation of these structures, which can be found in both the lips and glans penis, would explain their "exquisite sensitivity" (*exquisitus sensus*) in comparison to other parts of the body. Consequently, based on the evidence provided by comparative anatomy, Malpighi reached the conclusion that "if nature were to multiply and accentuate these papillae, which are more evident in areas with heightened sensitivity, this would indicate that these papillae are in fact the organ of touch."[92]

It is, however, the specific nature of this "exquisite sensitivity" that formed the basis of one of Sbaraglia's most striking objections to Malpighi. This is evident in his treatise *Oculorum, et mentis vigiliae* (1704), which was designed to dismantle Malpighi's entire research programme. Sbaraglia provides a detailed analytical review of Malpighi's entire *oeuvre*, including the *Opera posthuma*, identifying critical issues and contradictions as well as ambiguities and inaccuracies within the text. The analysis of the *De externo tactus organo* aimed to evaluate the arguments provided by Malpighi in support of the thesis that the tactile papillae are the "appropriate" organ for touch. This evaluation was conducted on three different levels: the position of the papillae, the number and size of these structures, and the range of "tactile objects" with which the papillae could interact.

Sbaraglia argued that sexual pleasure, which was believed to depend on the stimulation of the glans penis, is fundamentally distinct from the sensation of touch experienced in other parts of the body. It would be difficult to understand how a common structure such as the papilla could also be responsible for sexual pleasure.[93] Despite repeated observations, he eventually denied the existence of papillae,

---

[91] Malpighi (1697, p. 28): "Si tamen Dominus Contradictor [*scil.* Sbaraglia] dignatus fuisset hujusmodi partes perscrutari, talia non objicisset; constat enim ablata cuticula cutim subjectam locari, a qua innumera fere papillæ pyramidales attolluntur, quibus contrectatis, & tactis, sensus excitatur exquisitus."

[92] *Ibidem*: "[…] si Natura in his partibus multiplicat, & magis manifestat papillas, ubi exquisitior percipitur sensus tactus, signum est hujusmodi esse organum tactus."

[93] Sbaraglia (1704, p. 97): "In glande ponuntur pyramidales papillae pro sensu tactus, sed difficultas proposita in nostro theatro non fuit pro sensu tactus, sed pro sensu venereo. Dicamus itaque si papillae pyramidales aequales sunt in labiis, & glande penis, immo & in universo corporis

particularly in relation to certain parts of the body (such as the glans penis) that were known to be extremely sensitive, which, in his view, made the papillae highly implausible as the external organ of touch.

However, Sbaraglia is only one of a number of figures engaged in this debate about the senses and one whose role is relatively marginal, in particular, in relation to the identification of specific sense receptors. The subject is, in fact, considerably more intricate and encompasses a broader array of issues, including the structure of the skin and the functions of its fibres. Malpighi was among the first to emphasise this aspect, building upon the work of Johann Theodor Schenck (1619–1671).[94] However, this area of research underwent a notable expansion during the eighteenth century, affecting not only the broader field of medical research but also anatomical iconography.[95]

The existence of the papillae remained a topic of debate in the eighteenth century, as evidenced by the conflicting opinions on their role in taste and touch perception as well as their purported connection with nerves.[96] To scholars of the time, it seemed implausible that the vast number of papillae that emerge from the entire surface of the body were, in fact, portions of the nerve branches that originated in the spinal cord, brain, and cerebellum and were directed to various parts of the skin. If they were to be aggregated, their collective mass would far exceed that of the relatively modest bundle constituted by the marrow and the brain. In order to facilitate comprehension of the ratio between the spinal cord, the nervous branches, and the papillae, Malpighi employed an analogy. He likened the situation to that of the visual light rays, which, despite their multitude, intersect and converge at the pupil of the eye. However, the example of plants was of greater interest:

> From a thin, narrow caulis a profusion of fibres emerges, giving rise to very large leaves, flowers, and even fruit. And yet it is certain that the number of fibres does not multiply, but they simply extend from the trunk.[97]

ambitu, aequalis debet esse sensatio in omnibus praedictis partibus, & aequalis voluptas, quae omnia sunt commenta inania. [...] Quomodo igitur concedere poterimus (admissa hypothesi Antagonistae) papillam pyramidalem utrobique tam differenter affici? [...]."

[94] *De externo tactus organo*, in Malpighi (1686, vol. 2, p. 28): "Et licet adhuc dubia videatur cutis accurata compositio, ut videre poteris ex Doctissimo Jo. Theodoro Schenckio [...]" Malpighi had also mentioned Schenck in his 1663–1664 lectures in Messina.

[95] On multiple views of the papillae, see e.g. Margócsy (2011) and Fend (2019).

[96] For an overview, see the entry entitled *De lingua* in the Albrecht von Haller's edition of *Methodus studii medici* (Boerhaave, 1751, pp. 478ff). In the original version, Boerhaave dedicated a few lines to the topic, focusing exclusively on the works of Malpighi, Bellini, and Fracassati: "Duo Auctores notissimi sunt nimirum Laurentius Bellini *de gustus organo*, Lugduni Batavorum 1711 editus, qui tamen nihil habet, si conferatur cum Malpighio *de Lingua*, ubi ea. pars adcurate descripta habetur. Quæ Fracassatus addidit, multum molis sed parum nervorum habent." Haller significantly expanded the bibliographical references in this revised edition.

[97] *De externo tactus organo*, in Malpighi (1686, vol. 2, p. 29): "[...] ab exiguo etenim, & arcto caule fibrarum ingens proventus erumpit, ut prægrandia folia, flores, & fructus etiam emergant; & tamen certum est, non multiplicari fibrarum numerum, sed easdem a trunco tantum elongati." On the use of analogy in Malpighi, see Baldassarri (2023).

This plant analogy was also used by other anatomists, including the aforementioned Albertini. In the context of his view of the role of nerves, which was crucial to his hypothesis of the articulation of sensory processes, Malpighi's use of analogies between the visible and the invisible could provide valuable insights.

8.5 Conclusions

Malpighi's belief that the tactile papillae, as the organ of touch, were not part of the skin, but rather represented the ends of nerves *penetrating* the skin, moved the debate on touch entirely into the realm of anatomy, particularly the structure of the skin and its appendages. It was only through the meticulous removal of the layers of the skin that Malpighi was (apparently) able to detect such papillae. Consequently, his study required the implementation of specific dissection techniques, which Malpighi adapted from his earlier research on tongue.

His essay on touch presents a compilation of anatomical evidence, primarily drawn from animal models, that aims to definitively establish the presence of tactile papillae and their function in transmitting sensory stimuli. However, Malpighi seems to suggest that identifying an anatomical structure is sufficient to fully understand the complexities of perception. Is the issue resolved? In fact, his approach raises more questions than it answers. As we have seen, Malpighi's argument relies largely on a comparison between the organ of touch and his observations of the tongue. This strategy, however, blurs the distinction between tactile and gustatory stimuli and leaves unaddressed key questions: how touch differs from taste, whether the two types of papillae can be clearly distinguished, or whether some degree of overlap exists. Moreover, interaction with external objects operates differently in the case of touch; unlike taste, it does not involve salt particles mixed with saliva entering the cavities of the reticular body to reach the papillae. Yet, Malpighi did not provide a comprehensive explanation of how papillae function in relation to touch. At the time of *De externo tactus organo*, he had not yet developed a full theory of sense perception, nor had he described the mechanism of "neurosensory" transmission. He also left unexplored the question of how animals perceive. It is clear that Malpighi's main goal was to establish the existence of papillae rather than to delve into the intricacies of sensory processing.

The controversy surrounding the existence and function of tactile papillae, especially after the publication of *De externo tactus organo*, highlights the uncertainty surrounding this structure. It is evident that a clear and accurate understanding of the tactile papillae had not yet been achieved. Equally significant was the failure to develop accurate and effective visual representations of them. All the images found in the manuscript notes are only visual aids; they neither describe the structure of the papillae in detail nor situate them within a precise anatomical context. Interestingly, in the descriptions accompanying these notes, Malpighi made comparisons not only between the papillae of particularly sensitive areas (hand, foot, lips, glans), but also between the tactile papillae and those of the tongue. This comparative approach reinforces the conclusion that Malpighi consistently interpreted his findings by analogy with the tongue. Yet this raises a fundamental question: were these truly the same papillae?

Although Malpighi's "visual program" was refined in the 1670s, when his focus shifted to natural history (namely, animal generation, insects, and botany) due to his collaboration with the Royal Society, his account of sense perception remained incomplete. The only illustration of tactile papillae, found in his posthumous work, was based on a 1690 observation of a man with a foot deformity. This observation was considered definitive by Malpighi for the existence of papillae. It is noteworthy that this observation was of a morbid condition, with a thickened cuticle and a monstrous nail on the foot. However, the accuracy and "ecological validity" of this illustration are open to question. Unlike his detailed depictions of tongue papillae, Malpighi's representation of these tactile papillae in the foot lacks context and does not show how the skin should be dissected, as he had done for the tongue. This singular piece of evidence raises broader questions about the reliability and the degree of "theory-ladenness" of Malpighi's illustrations.

It is also intriguing that in this section of his autobiography, Malpighi references printed sources, specifically the *Anatome animalium* by Dutch physician and anatomist Gerard Blaes (1627–1682), to address Sbaraglia's objections regarding the papillae of the glans penis. Malpighi had studied Blaes' work extensively, as shown by the reading notes found in his records, and he cited this source to support his argument that the pyramidal shape of the papillae on the glans penis gives them a rough texture when touched, similar to the papillae on the tongue.

An undeniably interesting aspect of Malpighi's analysis is the correlation between sensitivity and nerves—the body parts most sensitive to touch were found to be those with the highest concentration of papillae. This finding played a crucial role in the discourse on sensitivity in eighteenth century medicine.

Acknowledgements Financial support for this work has been provided by the University of Padua, Department of Historical and Geographic Sciences and the Ancient World (DiSSGeA), DOR 2023 "La medicina a Padova nella prima età moderna" and DOR 2024 "Marcello Malpighi e la ricerca anatomica nella prima età moderna."

Preliminary research on Malpighi's manuscripts at the Bologna University Library has been conducted under the project PRIN 2017 "Material and Visual Culture of Science: A longue durée Perspective" (201727TRJX).

References

Primary Sources

Manuscripts

Bologna, Biblioteca Universitaria (BUB):
MS. 936/I, K.
MS. 936/II, A.

MS. 936/II, A2.
MS. 2085/II.
MS. 2085/III.
MS. 2085/XIV.

Bologna, Biblioteca Comunale dell'Archiginnasio (BCA):
MS. B 3237.

Printed Editions

Bauhin, C. (1605). *Theatrum anatomicum. Novis figuris æneis illustratum et in lucem emissum opera & sumptibus Theodori de Bry, p.m. relicta viduæ & filiorum Ioannis Theodori & Ioannis Isrælis de Bry*. Francofurti at Moenum: Typis M. Beckeri.
Boerhaave, H. (1751). *Methodus studii medici. Emaculata & accessionibus locupletata ab Alberto ab Haller*. Sumptibus Jacobi a Wetstein.
Casseri, G. C. (1609). *Pentaestheseion, hoc Est, De Quinque Sensibus liber, Organorum Fabricam variis iconibus... Illustratam*. Apud Nicolaum Misserinum.
Forte, A. (1543). *De mirabilibus vitae humanae naturalia fundamenta*. Impressum per Bernardinum de Vianis de Lexona Vercellensem.
Gassendi, P. (1658). *Opera omnia in sex tomos divisa... Tomus secundus*. Sumptibus Laurentii Anisson & Ioan. Bapt. Devenet.
du Laurens, A. (1593). *Opera anatomica in quinque libros divisa...* Sumptibus Ioannis Baptistae Buyssoni.
Malpighi, M. (1686). *Opera omnia: Figuris elegantissimis in æs incisis illustrata. Tomis duobus comprehensa*. Apud Robertus Scott & Georgius Wells.
Malpighi, M. (1697). *Opera posthuma, figuris aeneis illustrata. Quibus praefixa est ejusdem vita a seipso scripta*. A. & J. Churchill.
Malpighi, M. (1967). Sull'organo esterno del tatto. In L. Belloni (Ed.), *Opere scelte* (pp. 129–152). UTET.
Sbaraglia, G. D. (1704). *Oculorum et Mentis Vigiliæ ad distinguendum studium anatomicum, et ad praxin medicam dirigendam. Accedit mantissa subsidiaria de vi indicationis a parte, et de usu microscopii*. Typis Petri-mariae Montii.
Vesalius, A. (1999). *On the fabric of the human body. Book II, the ligaments and muscles: A translation of De humani corporis fabrica libri septem*. (W. F. Richardson trans.). Norman Pub.

Secondary Sources

Adelmann, H. (1966). *Marcello Malpighi and the Evolution of Embryology, 5 vols*. Cornell University Press.
Adelmann, H. (1975). *The Correspondence of Marcello Malpighi, 5 vols*. Cornell University Press.
Baldassarri, F. (2023). From the Analogy with Animals to the Anatomy of Plants in Medicine: The Physiology of Living Processes from Harvey to Malpighi. In F. Baldassarri (Ed.), *Plants in Sixteenth and Seventeenth Century: Botany between Medicine and Science* (pp. 121–144). de Gruyter.
Belloni, L. (1964). Dall'occhiale di Galileo all'anatomia microscopica di Marcello Malpighi: il *Testis examinatus* di Claudius Auberius. *Rendiconti dell'Istituto Lombardo, Classe di Scienze Matematiche e Naturali, 98*, 205–226.
Belloni, L. (1966). La neuroanatomia di Marcello Malpighi. *Physis, 8*, 253–266.
Belloni, L. (1970). The Repetition of Experiments and Observations: Its Value in Studying the History of Medicine (and Science). *Journal of the History of Medicine and Allied Sciences, 25*(2), 158–167. https://doi.org/10.1093/jhmas/XXV.2.158

Bertoloni Meli, D. (1997). The Posthumous Dispute between Borelli and Malpighi. In D. Bertoloni Meli (Ed.), *Marcello Malpighi, Anatomist and Physician* (pp. 245–273). L. S. Olschki.
Bertoloni Meli, D. (1997b). Additions to the Correspondence of Marcello Malpighi. In D. Bertoloni Meli (Ed.), *Marcello Malpighi, Anatomist and Physician* (pp. 275–308). L. S. Olschki.
Bertoloni Meli, D. (1998). The Archive and Consulti of Marcello Malpighi. In M. Hunter (Ed.), *Archives of the scientific revolution* (pp. 109–120). Boydell.
Bertoloni Meli, D. (1999). Francesco Redi e Marcello Malpighi: Ricerca anatomica e pratica medica. In W. Bernardi & L. Guerrini (Eds.), *Francesco Redi. Un protagonista della scienza moderna. Documenti, esperimenti, immagini* (pp. 73–86). L. S. Olschki.
Bertoloni Meli, D. (2007). Mechanistic Pathology and Therapy in the Medical Assayer of Marcello Malpighi. *Medical History, 51*(2), 165–180. https://doi.org/10.1017/S0025727300001174
Bertoloni Meli, D. (2011). *Mechanism, Experiment, Disease. Marcello Malpighi and Seventeenth-Century Anatomy*. The Johns Hopkins University Press.
Bertoloni Meli, D. (2013). Of Snails and Horsetails: Anatomical Empiricism in the Early Modern Period. *Early Science and Medicine, 18*(4–5), 435–452. https://doi.org/10.1163/15733823-1845P0006
Bertoloni Meli, D. (2019). *Mechanism: A Visual, Lexical, and Conceptual History*. University of Pittsburgh Press.
Betti, G. L. (2006). Giovan Battista Capponi: la "carriera della gloria" di un mago e scienziato nella Bologna del Seicento. Una biografia ufficiale. *L'Archiginnasio, 101*, 91–118.
Betti, G. L., & Calore, M. (1996). L'eredità di Giovanni Battista Capponi, letterato, collezionista, scienziato e bibliofilo. Annotazione intorno al testamento. *L'Archiginnasio, 91*, 31–78.
Bresadola, M. (2006). Pazienti e curatori nella pratica medica di Marcello Malpighi. In P. Govoni (Ed.), *Storia, scienza, società. Ricerche sulla scienza in Italia nell'età moderna e contemporanea* (pp. 16–45). CIS.
Bresadola, M. (2011). A Physician and a Man of Science: Patients, Physicians, and Diseases in Marcello Malpighi's Medical Practice. *Bulletin of the History of Medicine, 85*(2), 193–221. http://www.jstor.org/stable/44451983
Bresadola, M. (2018a). Sbaraglia, Giovanni Girolamo. In *Dizionario biografico degli italiani* (Vol. 91). Istituto della Enciclopedia italiana. Last Accessed Dec 31, 2024, from https://www.treccani.it/enciclopedia/giovanni-girolamo-sbaraglia_(Dizionario-Biografico)/
Bresadola, M. (2018b). The Problems of *Anatomia Practica* and How to Solve Them: Pathological Dissection around 1700. In S. De Renzi, M. Bresadola, & M. Conforti (Eds.), *Pathology in Practice. Diseases and Dissections in Early Modern Europe* (pp. 56–75). Routledge.
Busacchi, V. (1966). Il trattato di Marcello Malpighi sull'organo esterno del tatto. In *Celebrazioni Malpighiane. Discorsi e Scritti* (pp. 56–76). Azzoguidi.
Calabrese, M. C. (2016). Tra scienza e astronomia. Giacomo Ruffo, visconte di Francavilla. In L. Fregonese & I. Gambario (Eds.), *Atti del XXXIII Convegno annuale / Proceedings of the 33rd Annual Conference della Società Italiana degli Storici della Fisica e dell'Astronomia* (pp. 107–119). Pavia University Press.
Calabrese, M. C. (2018). *L'epopea dei Ruffo di Sicilia*. Laterza.
Campinoti, V. (2006). Tra Studio e Corte: John Finch, Thomas Baines e la Toscana. In F. Abbri & M. Bucciantini (Eds.), *Toscana e Europa: nuova scienza e filosofia tra '600 e '700* (pp. 149–175). FrancoAngeli.
Cassoli, P. (2023). *Marcello Malpighi sospettato di omicidio. Il grande scienziato alle prese con la giustizia*. Edizioni Pendragon.
Cavazza, M. (1990). *Settecento inquieto. Alle origini dell'Istituto delle Scienze di Bologna*. Il Mulino.
Cavazza, M. (1997). The Uselessness of Anatomy: Mini and Sbaraglia versus Malpighi. In D. Bertoloni Meli (Ed.), *Marcello Malpighi: Anatomist and Physician* (pp. 21–62). L. S. Olschki.
Cavazza, M. (2012). Aspetti dell'insegnamento dell'anatomia a Bologna nel Seicento e nel Settecento. In G. Olmi & C. Pancino (Eds.), *Anatome. Sezione, scomposizione, raffigurazione del corpo nell'età moderna* (pp. 59–78). Bononia University Press.

Clarke, E., & Bearn, J. G. (1968). The Brain "Glands" of Malpighi Elucidated by Practical History. *Journal of the History of Medicine and Allied Sciences, 23*(4), 309–330. https://doi.org/10.1093/jhmas/XXIII.4.309

Classen, C. (2012). *The Deepest Sense: A Cultural History of Touch*. University of Illinois Press.

Clericuzio, A. (2013). *Elements, Principles and Particles: A Study of Atomism and Chemistry in the Seventeenth Century*. Kluwer.

Conforti, M. (2001). *Testes alterum cerebrum*. Succo nerveo e succo seminale nella macchina del vivente di Giovanni Alfonso Borelli. *Medicina nei Secoli, 13*(3), 577–595.

Conforti, M. (2009). The Experimenters' Anatomy. In M. Beretta, A. Clericuzio, & L. M. Principe (Eds.), Accademia del Cimento *and its European Context* (pp. 31–44). Science History Publications.

Conforti, M., & Tonetti, L. (2026). Through the Eyes of a Fish: Imagery of Nerves after Descartes. In M. Mantovani & D. Cellamare (Eds.), *Cartesian Imagery. Picturing Philosophy in the Early Modern Age* (pp. 515–538). Brill.

Cooley, M. (2024). Likeness across Nature: The Anatomical Eye of Hieronymus Fabricius of Aquapendente (1533–1619). *Nuncius, 39*(2), 305–337. https://doi.org/10.1163/18253911-bja10101

Cunsolo, E. (2008). Giulio Casserio e la pubblicazione del *De Vocis Auditusque organis* tra Padova e Ferrara all'inizio del '600. *Mélanges de l'école française de Rome, 120*(2), 385–405.

De Ferrari, A. (1978). Casseri, Giulio Cesare. In *Dizionario biografico degli italiani* (Vol. 21). Istituto della Enciclopedia italiana. Last Accessed Dec 31, 2024, from https://www.treccani.it/enciclopedia/giulio-cesare-casseri_(Dizionario-Biografico)/

Dollo, C. (1984). *Modelli filosofici e scientifici nella Sicilia spagnola*. Guida.

Dollo, C. (1986). Gli inediti malpighiani. In G. Canziani & G. Paganini (Eds.), *Le edizioni dei testi filosofici e scientifici del '500 e del '600. Problemi di metodo e prospettive di ricerca. Atti del seminario di studio di Gargnano, 1-3 aprile 1985* (pp. 209–220). Franco Angeli. Available also in Dollo (2005, pp. 49–60).

Dollo, C. (2005). *Filosofia e medicina in Sicilia*. Rubbettino.

Fantuzzi, G. (1786). *Notizie degli scrittori bolognesi...* (Vol. 5). Nella Stamperia di San Tommaso d'Aquino.

Fantuzzi, G. (1788). *Notizie degli scrittori bolognesi...* (Vol. 6). Nella Stamperia di San Tommaso d'Aquino.

Favino, F., & Giannini, G. (2022). Borelli Reloaded: Contexts and Networks in Seventeenth-Century Italy. *Introduction to the Special Issue of* Physis, *57*(2), 289–297.

Fend, M. (2019). Drawing the Cadaver *ad vivum*: Gérard de Lairesse's Illustrations for Govard Bidloo's *Anatomia Humani Corporis*. In T. Balfe, J. Woodall, & C. Zittel (Eds.), Ad vivum*? Visual Materials and the Vocabulary of Life-Likeness in Europe before 1800* (pp. 294–327). Brill. https://doi.org/10.1163/9789004393998_011

Ferrari, G. (1987a). Public Anatomy Lessons and the Carnival: The Anatomy Theatre of Bologna. *Past & Present, 117*, 50–106. https://doi.org/10.1093/past/117.1.50

Ferrari, G. (1987b). La pubblica funzione di anatomia. Origini, significato e fine di una cerimonia. In G. Roversi (Ed.), *L'Archiginnasio. Il Palazzo, l'Università, la Biblioteca* (Vol. 1, pp. 301–318). Credito Romagnolo.

Gadebusch Bondio, M. (2005). La Carne di fuori. Discorsi medici sulla natura e l'estetica della pelle nel '500. *Micrologus. Nature, Sciences and Medieval Societies, 13*, 537–570.

Giallombardo, F. (2016). *Agostino Scilla (1629–1700) e la cultura visuale della historia, fra antiquaria e storia naturale*. Università di Palermo.

Giglioni, G. (1997). The Machines of the Body and the Operations of the Soul in Marcello Malpighi's Anatomy. In D. Bertoloni Meli (Ed.), *Marcello Malpighi, Anatomist and Physician* (pp. 149–174). L. S. Olschki.

Gómez López, S. (1997). Marcello Malpighi and Atomism. In D. Bertoloni Meli (Ed.), *Marcello Malpighi, Anatomist and Physician* (pp. 175–189). L. S. Olschki.

Harvey, E. D. (2003). *Sensible Flesh: On Touch in Early Modern Culture*. University of Pennsylvania Press.

Hsiung, H., Paulino, E., & Serrano, E. (2024). Touching Visions: Intersensoriality and Gender in the History of Science. *Nuncius (special issue), 39*(2), 259–450. https://brill.com/view/journals/nun/39/2/nun.39.issue-2.xml

Jarcho, S. (1972). The "Animadversiones" of Ippolito Francesco Albertini (1748) on Cardiac and Precordial Diseases Which Cause Dyspnea. *Proceedings of the American Philosophical Society, 116*(2), 163–181.

Jarcho, S. (1989). *Clinical Consultations and Letters by Ippolito Francesco Albertini, Francesco Torti and Other Physicians: University of Bologna MS 2089–1*. Francis A. Countway Library of Medicine.

Klestinec, C. (2007). Civility, Comportment, and the Anatomy Theater: Girolamo Fabrici and his Medical Students in Renaissance Padua. *Renaissance Quarterly, 60*, 434–463.

Klestinec, C. (2011). *Theaters of Anatomy: Students, Teachers and the Traditions of Dissection in Renaissance Venice*. Johns Hopkins University Press.

Lines, D. A. (2023). *The Dynamics of Learning in Early Modern Italy: Arts and Medicine at the University of Bologna*. Harvard University Press. https://doi.org/10.2307/j.ctv32nxzh8

Mandressi, R., & Claire, E. (2015). Of the Eye and of the Hand: Performance in Early Modern Anatomy. *TDR/The Drama Review, 59*(3), 60–76. https://doi.org/10.1162/DRAM_a_00491

Margócsy, D. (2011). A Museum of Wonders or a Cemetery of Corpses? The Commercial Exchange of Anatomical Collections in the Early Modern Netherlands. In S. Dupré & C. Lüthy (Eds.), *Silent Messengers. The Circulation of Material Objects of Knowledge in the Early Modern Low Countries* (pp. 185–215). LIT Verlag.

Maselli, M. (2023). *Linguam in obscura compositione latitare*. Imaginer et représenter le goût dans la médecine de la première Modernité. *Arts et Savoirs, 20*. http://journals.openedition.org/aes/6498

Maurette, P. (2013). *Touch, Hands, Kiss, Skin: Tactility in Early Modern Europe*. University of North Carolina at Chapel Hill.

Maurette, P. (2018). The Organ of Organs: Vesalius and the Wonders of the Human Hand. *Journal of Medieval and Early Modern Studies, 48*(1), 105–124. https://doi.org/10.1215/10829636-4280873

Mazzetti, S. (1847). *Repertorio di tutti i professori antichi, e moderni della famosa Università, e del celebre Istituto delle Scienze di Bologna*. Tipografia Tommaso d'Aquino.

Münster, L., & Gerocarni, B. (1940). Una relazione inedita di L. F. Marsili fatta a M. Malpighi su una "notomia" eseguita dal Prof. G. Pighi a Padova. *Atti e Memorie dell'Accademia di Storia dell'Arte Sanitaria, 6*(2), 76–92.

Murphy, H. (2020). Skin and Disease in Early Modern Medicine: Jan Jessen's *De cute, et cutaneis affectibus* (1601). *Bulletin of the History of Medicine, 94*(2), 179–214. https://doi.org/10.1353/bhm.2020.0034

Nannini, M. C. (1964–1965). Giacomo Pighi, il suo maestro e una data da correggere. *Acta Medicae Historiae Patavina*, 11, 109–118.

Pogliano, C. (2015). *Senso lato: il tatto e la cultura occidentale*. Carocci.

Rippa Bonati, M. (2004). *Manuum munus*. Per un'iconografia del "toccare con mano". In A. Olivieri, M. Rinaldi, & M. Rippa Bonati (Eds.), *All'incrocio dei saperi: la mano. Atti del Convegno di studi (Padova, 29–30 settembre 2000)* (pp. 325–336). Cleup.

Riva, A., Orrù, B., Pirino, A., & Riva, F. T. (2001). Iulius Casserius (1552–1616): The Self-Made Anatomist of Padua's Golden Age. *The Anatomical Record, 265*, 168–175. https://doi.org/10.1002/ar.1151

Savoia, P. (2019). Stancari, Giovanni Antonio. In *Dizionario biografico degli italiani* (Vol. 94, pp. 22–24). Istituto della Enciclopedia italiana. Last Accessed Dec 31, 2024, from https://www.treccani.it/enciclopedia/giovanni-antonio-stancari_(Dizionario-Biografico)/

Savoia, P. (2022). Knowing Nature by its Surface: Butchers, Barbers, Surgeons, Gardeners, and Physicians in Early Modern Italy. *Centaurus, 64*(2), 399–420. https://doi.org/10.1484/J.CNT.5.129636

Savoia, P. (2024a). Skin on Trial: Experts and Expertise on 'Skin Diseases' in Early Modern Italy. *Social History of Medicine, 37*(2), 429–446. https://doi.org/10.1093/shm/hkad045

Savoia, P. (2024b). *Superfici. Corpi, pratiche e modelli cognitivi nella chirurgia di età moderna.* Officina Libraria.

Stabile, G. (2008). Sapor-sapientia: Tatto e gusto tra cultura agraria, medicina e mistica. In C. Leonardi & F. Santi (Eds.), *Natura, scienze e società medievali: Studi in onore di Agostino Paravicini Bagliani* (pp. 287–344). SISMEL–Edizioni del Galluzzo.

Tonetti, L. (2016). *Corpus fasciculus fibrarum*: Teoria della fibra e pratica medica nel *De praxi medica* di Giorgio Baglivi. *Physis: Rivista Internazionale di Storia della Scienza.*, n.s., *51*(1–2), 379–392.

Tonetti, L. (2022). Malpighi's Early Views on Brain Anatomy: A Look at His Unpublished Drawings and Notes. *Physis: Rivista Internazionale di Storia della Scienza, 57*(2), 357–375.

Tonetti, L. (2023). "… figuram ipsam mentem concepisse videtur": i limiti della rappresentazione microscopica nel dibattito Swammerdam-Malpighi sull'anatomia del baco da seta. In C. Addabbo, E. Canadelli, L. Ingaliso, D. Musumeci, L. Tonetti, V. Vignieri, & M. Vilardo (Eds.), Ad limina. *Frontiere e contaminazioni transdisciplinari nella storia delle scienze* (pp. 548–560). Editrice Bibliografica.

Tonetti, L. (2024). "Fingeranno casi e favole": dicerie e reputazione del medico nella Bologna di età moderna. Il caso di Marcello Malpighi. *Micrologus. Nature, Sciences and Medieval Societies, 32*, 593–614.

Vernazza, G. (1961). La crisi barocca nei programmi didattici dello studio bolognese. In *Studi e memorie per la storia dell'Università di Bologna.* (nuova serie, vol. II, pp. 95–177). Istituto per la storia dell'Università.

Virgilio, G. (1994). Jacob Barner e le sue lezioni di chimica e di chimica farmaceutica a Padova (1670–1672). *Atti dell'Istituto Veneto di Scienze, Lettere ed Arti. Classe di Scienze Fisiche, Matematiche e Naturali, 152*(pt. II), 145–154.

Von Hoffmann, V. (2022). Epistemologies of Touch in Early Modern Holy Autopsies. *Renaissance Quarterly, 75*(2), 542–582. https://doi.org/10.1017/rqx.2022.107

Von Hoffmann, V. (2023). Can Mixtures be Identified by Touch? The Reception of Galen's *De complexionibus* in Italian Renaissance Medicine. *Early Science and Medicine, 28*(3–5), 498–525. https://doi.org/10.1163/15733823-20230084

Chapter 9
Mary Shepherd on Proprioception and Knowledge of the External World

Antonia LoLordo

Abstract Mary Shepherd says that motion is a sixth sense and that the power of motion is a sixth organ of sense. She argues that this sixth sense plays a crucial role in acquiring the concepts required for knowledge of the external world, including the concepts of extension, externality, and the self—the thing that external objects are external *to*. I argue that Shepherd's sense of motion is a form of what later comes to be called proprioception; that Shepherd goes well beyond her predecessors in developing an complex account of the way proprioception gives us notions of extension, externality, and the self; and that she goes well beyond her predecessors in recognizing that proprioception is distinct from the sense of touch.

Keywords Berkeley · Bodily awareness · Condillac · Hume · Kinesthesia · Knowledge of the external world · Proprioception · Reid · Touch · Shepherd

9.1 Introduction

Eighteenth-century philosophers from Berkeley on argued that, in order to acquire knowledge of the external world, we first need to acquire the relevant ideas through sense-experience. There was some disagreement about which ideas are required for knowledge of the external world, but the three main candidates were the idea of extension, the idea of externality or "outwardness", and the idea of a mind or self. (External objects must be external *to* something, and it was typically assumed that the thing they are external to is the mind or self.) There was a broad eighteenth century consensus that knowledge of the external world was possible and that the ideas involved in such knowledge were acquired through sense perception. At the same time, there was significant disagreement over *how* these ideas could be acquired via sense-experience and in what order they had to be acquired.

A. LoLordo (✉)
University of Virginia, Charlottesville, VA, USA
e-mail: lolordo@virginia.edu

C. Beneduce et al. (eds.), *The Sense of Touch*, Studies in the History of Philosophy of Mind 32, https://doi.org/10.1007/978-3-032-16308-0_9

Mary Shepherd (1777–1847) joins this debate in her *Essays on the Perception of an External Universe* (1827). She presents a new theory of knowledge of the external world and lays it out in opposition to a wide range of predecessors: George Berkeley, David Hume, Étienne Bonnot de Condillac, Antoine Destutt de Tracy, Thomas Reid, Dugald Stewart, Thomas Brown, Joseph Priestley, David Hartley, and more.[1] Shepherd's theory assigns a crucial role to what she calls "the power of motion", which she says may be regarded "as a sixth organ of sense",[2] and to motion itself, which she describes as "a sort of sense".[3] I argue that this "power of motion" is a form of what's now called proprioception[4]—perception or awareness of the position and motion of our body in space—although Shepherd does not use the term.[5] Proprioception plays a crucial epistemic role for Shepherd, serving as the source of the idea of the self as well as the ideas of extension and externality.

One can find a hidden appeal to proprioception in many of her eighteenth century predecessors' accounts of knowledge of the external world, although they see themselves as appealing simply to the sense of touch.[6] Shepherd takes a significant step forward by making the epistemic role of proprioception explicit and showing that proprioception is needed to explain not only the idea of objects external to the self but also the very idea of the self. She also takes a significant step forward by seeing that proprioception is a distinct sense modality and not simply part of the sense of touch.

9.2 Berkeley

To make sense of Shepherd's theory, it's helpful to look at it against the background of some of her opponents. Due to space considerations, I'll focus on three such opponents, Berkeley, Condillac, and Reid, starting with Berkeley.

Shepherd's discussion of Berkeley is mainly an attempt to refute his idealism. But for the purpose of this story, the Berkeley of the early *New Theory of Vision* (1710) is what matters. There Berkeley's central claim is that if a man born blind were newly made to see, "Objects […] would all seem to be in his Eye, or rather in his Mind" just as pleasure and pain seem to be in our minds: for, Berkeley claims,

[1] See Berkeley (1998), Hume (2001), Condillac (1930), Destutt de Tracy (1824), Reid (1764, 1859), Stewart (1829), Brown (1835), Priestley (1782) and Hartley (1775).

[2] Shepherd, *Essays on the Perception of an External Universe* (1827, p. 230); henceforth: EPEU, here quoted from the edition by Lolordo (2020).

[3] EPEU, p. 104. See Smith (2023) for an account of the early history of the sense of movement, albeit one that does not mention Shepherd's contribution.

[4] Deborah Boyle (2022, pp. 123–24) makes a similar suggestion, in passing.

[5] The term itself dates to the beginning of the twentieth century, although later nineteenth century thinkers developed complex theories of the different forms of bodily awareness.

[6] In her chapter within this volume, Alison Simmons argues that the distinction between proprioception and touch is present in Descartes.

"our judging Objects perceiv'd by Sight to be at any Distance, or without the Mind, is [...] intirely the Effect of Experience".[7] Vision itself cannot provide us with the ideas of extendedness and externality.

The Berkeley of the *New Theory* assumed that we do have the ideas of extendedness and externality, and that touch provides us with those ideas. (Of course, Berkeley later rejected this assumption.) But if the sense of touch provides us with the ideas of extendedness and externality while vision does not, then the objects of vision and touch must be distinct. The idea of visible shape I acquire from looking at the coffee cup and the idea of tangible shape I acquire from looking at it are two different ideas, and there is no resemblance or necessary connection between them.

Why then do we tend to think that we see and touch the same things? Only, Berkeley argues, because experience has built up associations between tangible ideas and visible ideas:

> Having of a long time experienced certain Ideas, perceivable by Touch [...] to have been connected with certain Ideas of Sight, I do upon perceiving these Ideas of Sight, forthwith conclude what Tangible Ideas are [...] like to follow. Looking at an Object I perceive a certain visible Figure and Colour, with some degree of Faintness and other Circumstances, which from what I have formerly observed, determine me to think, that if I advance forward so many Paces or Miles, I shall be affected with such and such Ideas of Touch [...] I believe whoever will look narrowly into his own Thoughts, and examine what he means by saying, he sees this or that thing at a Distance, will agree with me, that what he sees only suggests to his Understanding, that after having passed a certain Distance, *to be measured by the Motion of his Body, which is perceivable by Touch*, he shall come to perceive such and such tangible ideas.[8]

We learn through experience that certain visual ideas, in conjunction with bodily movements, are regularly followed by certain tangible ideas. As a result, we develop associations between visible figures and tangible figures, associations that are so strong that we fail to notice the difference between the ideas.

For this to work, it's not enough for us to be moving our bodies. We must *know* that we are moving our bodies. But how can we know this? Berkeley's quick answer is that "the Motion of [our] Body [...] is perceivable by Touch." This only makes sense if you think that the sense of touch includes what we would now call proprioception.

Berkeley does not develop this point, and he does not have much to say elsewhere about our awareness of the motion of our bodies. He does not have a term for proprioception, let alone a theory of it. This makes it difficult to say for sure why he lumps proprioception in with touch; at any rate, though, we'll see that he is not alone in doing so.[9]

---

[7] NTV, p. 41.

[8] NTV, p. 45, italics mine.

[9] Here's one possible line of thought. My mind is an immaterial substance, so my body is an external object. Hence knowledge of my body must derive from the external senses. Proprioception seems like a contact sense rather than a distance sense. It can't be a form of smell, since it isn't localized to any particular part of the body and gives knowledge of a wide range of qualities. So it must be a form of touch.

Beyond the work done by proprioception, there is one further role for bodily awareness in the *New Theory of Vision*. Berkeley tells us that

> When we look at a near Object with both eyes, according as it approaches, or recedes from us, we alter the Disposition of our Eyes, by lessening or widening the Interval between the Pupils. This Disposition or Turn of the Eyes is attended with a Sensation, which seems to be that which [...] brings the Idea of greater or lesser Distance into the Mind. Not that there is any natural or necessary Connexion between the Sensation we perceive by the Turn of the Eyes, and greater or lesser Distance, but because the Mind has by constant Experience found the different Sensations corresponding to the different Dispositions of the Eyes, to be attended with a different Degree of Distance in the Object.[10]

Berkeley's claim that "the Sensation arising from the Turn of the Eyes is of itself immediately perceived"[11] has struck the vast majority of his readers as implausible. But we'll see that Shepherd adopts a version of this suggestion, along with corollaries for other sense organs.

One more element of the *New Theory* is worth pointing out here. Recognizing that his readers will be surprised to hear that "a Man no more sees and feels the same thing, than he hears and feels the same thing"[12] Berkeley emphasizes the role of attention in our cognitive lives. He distinguishes the primary and immediate objects of vision, colours, from the secondary and learned objects of vision, external objects. In everyday life, he says, we attend only to the secondary objects. The reason for this is straightforward. We are naturally disposed to attend to those objects that are important for our survival, and colors are not the kind of thing that can hurt us or feed us. To navigate the world successfully, we need to pay attention to tangible ideas:

> The Visive Sense seems to have been bestowed on Animals [...] that by the Perception of Visible Ideas (which in themselves are not capable of affecting, or in any wise altering the Frame of their Bodies) they may be able to foresee (from the Experience they have had, what Tangible Ideas are connected with such, and such Visible Ideas) the Damage or Benefit which is like to ensue, upon the Application of their own Bodies to this or that Body which is at a Distance. Which Foresight, how necessary it is to the Preservation of an Animal, every ones Experience can inform him. Hence it is that, when we look at an Object, *the Tangible Figure and Extension thereof are principally attended to; whilst there is small heed taken of the Visible Figure and Magnitude, which, though more immediately perceived, do less concern us*, and are not fitted to produce any Alteration in our Bodies.[13]

We are aware of all our ideas and sensations; in fact, ideas and sensations are by definition objects of awareness. But we do not *attend* to all of our sensations. For instance, we are aware of "the Sensation arising from the Turn of the Eyes" without attending to it. The distinction between the ideas we are aware of and the smaller group of ideas we attend to plays an important role in Berkeley's theory. We'll see

[10] NTV, pp. 16–17.

[11] NTV, p. 18.

[12] NTV, p. 48.

[13] NTV, p. 59, italics mine.

that it plays an equally important role in Condillac's account of knowledge of the external world.

9.3 Condillac

In the *New Theory*, Berkeley assumed that we get the ideas of exteriority and extendedness from the sense of touch. In his *Treatise on Sensation* (1754)—henceforth: TS—Condillac endorses a very similar claim, writing that "the eye in itself is incapable of seeing space outside itself".[14] Only touch, he continues, is "capable of discovering space and also of instructing the other senses to relate their sensations to bodies extended in space".[15] One of Condillac's main goals is to explain *how* touch discovers space, and he does so using the statue example, a variation on the man born blind who is newly made to see. Consider a statue that is immobile and has only the sense of touch. Its sole sensation is "the feeling [...] of the action of the parts of the body upon one another, especially through the movements of respiration".[16] Condillac calls this type of sensation the *sentiment fondamental,* the fundamental feeling, and, since he is supposing that the statue's only sense modality is touch, it seems that he must think of the fundamental feeling as a tactile sensation. Thus Condillac, like Berkeley, seems to assume that bodily awareness is a form of touch.

The fundamental feeling allows the statue to acquire a rudimentary notion of the self.[17] For, Condillac says, it "can say 'I' as soon as some change in its fundamental feeling occurs".[18] However, the fundamental feeling does not give the statue awareness of the different parts of its body, so it has no idea of extension.[19] (This means that its idea of self is not sufficient to form the idea of something external to the self.

Now suppose that the statue touches its chest with its left hand. It will feel two different sensations, and it will be able to distinguish them due to "the sensation of solidities which they mutually give each other, and which necessarily puts them one

---

[14] Condillac, *Treatise on Sensation* 1.11.1 (Condillac, 1930, p. 58). He notes that this claim "will no doubt appear extraordinary" to his readers, which suggests that although he knew Berkeley's *New Theory*, he did not assume his readers did. "Il paroîtra sans doute extraordinare à bien des Lecteurs de dire, que l'oeil est par lui-même incapable de voir un espace hors de lui". Condillac (1803, p. 140).

[15] TS, 1.11.1; p. 59. "[...] il est capable de découvrir cete espace, & d'apprendre aux autres sens à rapporter leurs Sensations aux corps qui y sont répandus". Condillac (1803, p. 142).

[16] TS, 2.1.1; p. 75. "[...] le sentiment qu'elle a de l'action des parties de son corps les unes sur les autres, & sur-tout des mouvemens de la respiration". Condillac (1803, p. 166).

[17] It can also acquire a rudimentary notion of the self through smell (see TS, 1.6.3; p. 44).

[18] TS, 2.1.3; p. 75. "Enfin, nous remarquerons qu'elle pourroit dire *moi*, aussitôt qu'il est arrive quelque changement à son sentiment fundamental". Condillac (1803, p. 167).

[19] TS, 2.2.2; p. 76.

outside the other".[20] However, it will experience them as distinct entities rather than as parts of the same thing. The statue "will find itself in each part as in so many different bodies and it will not yet know that altogether these parts form one body".[21] This is because the sensations "are not shown to be contiguous and consequently they do not form one single continuum".[22] Thus, the statue will not yet acquire an idea of extension.

The statue acquires the idea of extension when it becomes capable of further motion: it "discovers, by means of movement, that it has a body and that outside it there are other bodies".[23] Suppose that the statue runs its hand along its arm and over its chest and head.[24] Then, Condillac explains, "since a body is a continuum formed by the contiguity of other extended bodies, the sensation which represents it must be a continuum formed by the contiguity of other extended sensations".[25] So when the statue runs its hand over its body, the "hand, which will unite in a single continuum parts formerly separated, will give it evident feeling of extension"[26] followed by an idea of extension.

To understand what's going on here, we need to recognize that for Condillac, our sensations have a great deal of content that we are not immediately aware of. If the sensations that represent the body must be a continuum because the body is a continuum, the sensations that represent the hand (when the statue touches its chest with its hand) must be extended because the hand is extended. But until it runs its hand along its body, the statue is not capable of recognizing that its sensations represent or "enclose" extension. Condillac insists that "It is not true that we have all the ideas which are enclosed in our sensations; we have only those which we learned to notice".[27] Initially, we experience our sensations as modifications of our souls. Over time, we come to see these sensations as modifications of sense organs and thus see them as referring to the external objects that are their occasional causes.[28]

---

[20] TS, 2.5.3; p. 87. "Alors sa main et sa poitrine se distingueront à la sensation des solidités qu'elles se renvoient mutuellement, et qui les met nécessairement l'une hors de l'autre". Condillac (1803, p. 186).

[21] TS, 2.5.3; p. 87. "[…] elle se trouvera dans chacune, comme dans autant de corps différens, et elle ne saura pas encore que, toutes ensemble, elles n'en forment qu'un seul". Condillac (1803, p. 187).

[22] TS, 2.5.3; p. 88. "C'est que les sensations qu'elle a éprouvées, ne les lui représentent pas comme contiguës, ni par consequent, comme formant un seul continu". Condillac (1803, p. 187).

[23] TS, 2.5.2; p. 85. "[…] elle découvrira, d'après ses mouvemens, qu'elle a un corps et qu'au delà it y en a d'autres". Condillac (1803, p. 183).

[24] TS, 2.5.3; p. 88.

[25] TS, 2.4.2; p. 81. "[…] puisqu'un corps est un continu, formé par la contiguité d'autres corps étendus, il faut que la sensation qui le représente, soit un continu formé par la contiguité d'autres sensations étendus". Condillac (1803, p. 177).

[26] TS, 2.5.3; p. 88. "[…] cette même main, qui réunira, dans un seul continu, les parties auparavant séparées, en rendra l'étendu plus sensible". Condillac (1803, p. 188).

[27] TS, 1.11.8; p. 67. "[…] nous n'avons pas toutes les idées que nos sensations renferment; nous n'avons que celles que nous y savons remarquer". Condillac (1803, p. 156).

[28] TS, 2.4.3; p. 82.

When we consider sensations as representations of external objects in this way, they are ideas.

The same process that gives rise to the idea of extension also gives rise to the idea of the body—the idea of itself as a single extended object. The statue learns "to recognize itself in all its component parts, because as soon as it places its hands upon one of them, the same sentient being replies [...] from one to the other: *this is myself*'.[29] However, this process cannot give rise to the idea of an external object. That requires one more step.

In order to acquire the idea of an external object, Condillac thinks, the statue must reach out and touch an external object. When it touched itself, the statue felt itself affected in two different parts of its body. When it touches an external object, it feels itself affected in only one place. The statue can compare those two ideas and perceive that they are different—which comparison, on Condillac's view, amounts to judgment.[30] Thus, the statue will judge that it is touching an external object, and, as a result of this judgment, it will form the idea of an external object.

Notice that Condillac's account appeals to several forms of bodily awareness, including the fundamental feeling (which it has even when incapable of movement) and the sense of touch (which it has even when all it can do is touch its chest with its hand). However, the most important form of bodily awareness in Condillac's account is proprioception. Think about how the statue is supposed to acquire the idea of extension by running its hand over its own body. If the statue did this without being aware of the motion of its hand, it would have only a sequence of temporally ordered sensations, without any spatial ordering. In order to acquire the notion of extension, the statue must be aware of the motion of its hand through space. However, Condillac does not seem to recognize the work that proprioception is doing in his account of the acquisition of ideas. What he emphasizes is that the statue needs to move its body in order to acquire the idea of extension, and not that the statue needs to be *aware* that it is moving its body in order to acquire the idea of extension.

9.4 Reid

Berkeley argued that vision cannot provide the ideas of extension, externality, and the body or self that we need for knowledge of the external world, while assuming, in the *New Theory*, that touch *can* give us those ideas. Condillac tried to explain *how* the sense of touch provides us with the ideas of extension, externality, and the body or self. Reid uses a roughly similar thought experiment to show that sensation cannot provide the ideas of extension, externality, and so on. He calls this the

[29] TS, 2.5.4; p. 88. "La statue apprend donc à connoître son corps, et à se reconnoître dans toutes les parties qui le composent". Condillac (1803, p. 188).

[30] TS, 1.2.15; p. 9.

experimentum crucis because it is supposed to show that the 'way of ideas' is entirely mistaken[31]:

> We are commonly told [...] that we get the idea of extension by feeling along the extremities of a body [...] I have sought [...] to find out how this idea can be got by feeling, but I have sought in vain. Yet it is one of the clearest and most distinct notions we have [...] It is true we have feelings of touch, which every moment present extension to the mind; but how they come to do so, is the question; for those feelings do no more resemble extension, than they resemble justice or courage: nor can the existence of extended things be inferred from those feelings by any rules of reasoning.[32]

In fact, Reid thinks, the *experimentum crucis* shows that we have *several* notions that were not acquired through sensation, including "our notions of external existences, of space, motion, and extension, and all the primary qualities of body".[33] Thus the way of ideas is false, and hence there is no reason to take the external world skepticism of Berkeley and Hume seriously. We can rest easy in our belief in the existence of an external world—a belief that every normal human being has, even the philosophers who question its rational credentials when they are in the study.

Here's how the 'experiment' goes. Suppose that there is a blind man whose conceptions have all been removed by "some strange distemper" that leaves his reasoning ability intact.[34] Suppose also that his body is "fixed immovably in place".[35] Now suppose that different sense modalities and experiences are added back in, one by one.

First, "suppose him [...] to be pricked with a pin":

> This will, no doubt, give a smart sensation: he feels pain; but what can he infer from it? Nothing surely with regard to the existence or figure of a pin. He can infer nothing from this species of pain, which he may not as well infer from the gout or sciatica. Common sense may lead him to think that this pain has a cause; but whether this cause is body or spirit, extended or unextended [...] he cannot possibly, from any principles he is supposed to have, form the least conjecture. Having had formerly no notion of body or of extension, the prick of a pin can give him none.[36]

The man will have a new sensation, and he may form the belief that some unknown thing is the cause of that sensation, but no more.

Reid then considers a series of variations. Suppose that "a body not pointed, but blunt, is applied to his body with a force gradually increased until it bruises him".[37] This will produce a new sensation, distinct from the sensation of being pricked with a pin, but it will not produce any conception of an extended thing. Next, suppose

---

[31] See Buras (2015) for a helpful overview of the *experimentum crucis*.

[32] Reid, *Inquiry Into the Human Mind* (henceforth: *Inquiry*), 5.3 (1764, p. 129).

[33] *Inquiry* 5.6, p. 129. He does not have much to say about the notion of the body/the self, but since it has extendedness as a component it too cannot have been acquired through sensation alone.

[34] *Inquiry* 5.6, p. 125.

[35] *Inquiry* 5.6, p. 126.

[36] *Inquiry* 5.6, p. 126.

[37] *Inquiry* 5.6, p. 126.

"that the body applied to him touches a larger or a lesser part of his body." What does this add?

> Can this give him any notion of its extension or dimensions? To me it seems impossible that it should, unless he had some previous notion of the dimensions and figure of his own body, to serve him as a measure. When my two hands touch the extremities of a body; if I know them to be a foot asunder, I easily collect that the body is a foot long; and if I know them to be five feet asunder, that it is five feet long: but if I know not what the distance of my hands is, I cannot know the length of the object they grasp; and if I have no previous notion of hands at all, or of distance between them, I can never get that notion by their being touched.[38]

Or suppose "that a body is drawn along his hands or face, while they are at rest". Again, what does this add?

> How it should convey a notion of space or motion, to one who had none before, I cannot conceive. The blood moves along the arteries and veins, and this motion, when violent, is felt: but I imagine no man, by this feeling, could get the conception of space or motion, if he had it not before. Such a motion may give a certain succession of feelings, as the colic may do; but no feelings, nor any combination of feelings, can ever resemble space or motion.[39]

The addition of various extended stimuli will produce various new sensations, but, Reid insists, it will not produce the notion of extension. Notice the role of resemblance here: the new sensation cannot provide a conception of extension because "no feelings [...] can ever resemble space or motion." Here Reid is replying to the Lockean claim that our ideas of primary qualities resemble those qualities.

Finally, suppose that the man "moves a limb by instinct, without having had any previous notion of space or motion"[40]:

> He has here a new sensation, which accompanies the flexure of joints, and the swelling of muscles. But how this sensation can convey into his mind the idea of space and motion, is still altogether mysterious and unintelligible. The motion of the heart and lungs are all performed by the contraction of muscles, yet give no conception of space or motion. An embryo in the womb has many such motions, and probably the feelings that accompany them, without any idea of space or motion.[41]

Again, new sensations—in this case, what we would call proprioceptive sensations—are produced without any conception of extension. At this point, Reid concludes that the conception of extension cannot be obtained from sensation alone. Since we obviously do have a conception of extension, it follows that not all our conceptions are acquired from experience. Hence the way of ideas is false.

Why do Condillac and Reid reach such different conclusions on the basis of such similar thought experiments? The underlying disagreement seems to be about the nature of sensations. We saw that for Condillac, sensations are very rich: they contain a great deal of content, even though the thinker may not be aware of all of it.

---

[38] *Inquiry* 5.6, p. 127.

[39] *Inquiry* 5.6, pp. 127–128.

[40] *Inquiry* 5.6, p. 128.

[41] *Inquiry* 5.6, p. 128.

Recall what happens when the statue runs its hand over its body: "since a body is a continuum formed by the contiguity of other extended bodies, the sensation which represents it must be a continuum formed by the contiguity of other extended sensations".[42] From Reid's point of view, this begs the question by assuming that sensations represent their causes. Such representation cannot occur in the absence of any resemblance between sensations and external objects.

For Locke, ideas of primary qualities resemble, and hence represent, intrinsic features of bodies, while ideas of secondary qualities represent their causes without resembling them. Since Reid rejects any resemblance between mental entities and external objects, he offers a new version of the primary quality-secondary quality distinction. He maintains, with Locke, that primary qualities are intrinsic properties of bodies and secondary qualities are mere powers or relational qualities. He adds that primary qualities are clearly conceived, while secondary qualities are only obscurely conceived. The reason that secondary qualities are only obscurely conceived is that we conceive of them as powers, without grasping their categorical basis.

The *experimentum crucis* is not supposed to explain the source of our conceptions of extension, externality, and the body or self. But a quick hint at Reid's theory will be useful for later. The key is the distinction between sensations and perceptions, where to perceive an external object is to have a conception of it and to believe in its present existence.[43] There is no necessary connection between sensation and perception, although—except in the case of extension, whose perception has no corresponding sensation—our perceptions tend to be accompanied or triggered by sensations.

So we do not acquire conceptions from sensations (because sensations do not represent objects) or from perceptions (because we can't perceive something without having a conception of it). How then do we acquire conceptions? Reid answers simply that the ability to perceive is a feature of our "natural constitution." We are in effect hard-wired so that sensations trigger conceptions in us, although they are not the source of the content of those conceptions.

Reid is not explicit about the existence of proprioception and other forms of bodily awareness, but he seems to be assuming that these sensations work the same way as external sensation. When the blind man feels the prick of the pin, he has an interoceptive sensation, namely pain, without any corresponding perception. When he moves his limbs, he has a proprioceptive sensation—the sensation that "accompanies the flexure of joints, and the swelling of muscles"[44]—without any corresponding perception. Reid might also be willing to allow that we have certain proprioceptive *perceptions*, but they could not be the source of our conceptions of extendedness and externality because such perceptions require our already having those conceptions.

---

[42] TS, 2.4.2; p. 81.

[43] Reid, *Essays on the Intellectual Powers of Man*, 2.20 (ed. Reid, (1859), p. 226).

[44] *Inquiry* 5.6, p. 129.

9.5 Shepherd on the Source of the Notion of Extension

Shepherd aims to reinstate reason as the ground of human cognition in the face of what she sees as Hume's dangerous skepticism and Reid's unacceptable concessions to it. As part of this project, she wants to show how we know, using reason and the senses, that there is an external world. For Shepherd as for her predecessors, this requires explaining how we get the notions involved in knowledge of the external world: extension, externality, and the self or body.[45] All three notions derive from a process of what Shepherd calls "latent" reasoning.[46] We'll see that this reasoning involves the premise that all new objects have a cause; in the EPEU Shepherd is taking for granted the anti-Humean conclusion she drew in a previous work, namely that "there is no object which begins to exist, but must owe its existence to some cause".[47]

Shepherd has a very different notion of extension than her predecessors. She accepts a version of Berkeley's resemblance thesis: "nothing can be like a sensation, or idea, or perception, but a sensation, idea, and perception".[48] But while Berkeley inferred from this that no idea represents anything external, Shepherd infers only that no idea *resembles* anything external. In particular, our primary-quality sensations do not resemble their causes. For the unperceived causes of our sensations are "not *like* their sensations, or perceptions; yet incapable of being conceived of, except under the images of sensations, and as named by the names given to those appearances".[49] Failure to recognize this fact has caused confusion among previous philosophers and "is the reason of a false philosophy concerning the nature and operations of the mind, in its perception of an extended, material universe, external to, and independent of it".[50]

If primary qualities do not resemble our ideas of them, how do we conceive of them? Shepherd answers that we conceive of them simply as powers, like secondary qualities. Colors are defined as powers to produce certain sensations. Similarly, Shepherd says,

---

[45] The term 'notion' is Shepherd's. She is not always entirely consistent in her usage, but I follow Boyle (2022) in taking Shepherd to use the term 'notion' for ideas constructed by reason, including the notions of extension and externality.

[46] EPEU, p. 14. She does not explain what it means for a piece of reasoning to be latent, but since she cannot allow genuinely unconscious reasoning, she must mean something like reasoning that we do not attend to.

[47] Shepherd, *Essay on the Relation of Cause and Effect* (1824, p. 36). The argument of that work is beyond the scope of this paper, but a great deal of the secondary literature on Shepherd focuses on her notion of causation: see e.g. Bolton (2011, 2019), Cruz (forthcoming), Fantl (2016), Folescu (2021), Landy (2020), LoLordo (2019), McRobert (unpublished), Paoletti (2011), and Wilson (2016).

[48] EPEU, p. 112.

[49] EPEU, p. 41.

[50] Shepherd, "Lady Mary Shepherd's Metaphysics" (henceforth: LMSM) (1832, p. 697).

> I define [...] the *idea* of extension to be a simple sensation of the mind relative to that external extension which is not an idea, and which is its cause. I define external extension to be an unperceived cause, fitted to create or produce the idea of extension on the mind, and also to be a capacity for the admission of unperceived motion.[51]

We conceive of extension as the unknown external quality that causes certain sensations.[52] Here's how Shepherd explains the relevant sensations:

> [T]he very impression of motion consists in the impression of passing through extended space, and as a corollary with it suggests to the mind, *here*, and *there*; and whilst the *mind* requires *no place*, nor *space*, to *comprehend it*, the sensation of passing through different points of space, suggests the notion, or rather inspires the immediate feeling of the *extension of space*, (or of an unresisting medium).[53]

The impression of passing through different points of space is a proprioceptive sensation, rather than a visual or tactile one. So the notion of extension is a relative notion (although Shepherd does not use the term), constructed out of proprioceptive sensations, the notion of a cause, and the notion of externality.

The fact that our notion of extension is a relative notion explains why Shepherd, unlike her predecessors, thinks that we simultaneously learn that extended things exist and acquire the notion of extension. We reason that there must be some unknown thing that causes our sensations of extension, and we call that thing 'extension' because

> The *names* for the qualities may be indifferently applied to the *causes*, or *external objects*, or to the *effects* the *inward perceptions;* or to *both together*, as *compound beings*. It is in the latter sense they are *always popularly* applied, and on account of which circumstances there has been so much confusion in the minds of philosophers upon the subject.[54]

Thus, a single chain of reasoning simultaneously produces a notion of extension and proves that extended things exist.

9.6 Shepherd on the Source of the Notion of Externality

If we understand extension as the external cause of certain proprioceptive sensations, then we must have the notion of externality before we get the notion of extension. Shepherd's notion of externality or 'outwardness' is quite thin, like her notion

---

[51] LMSM, p. 698.

[52] Shepherd says that we also conceive of extension as the "capacity for the admission of unperceived motion", where "motion is when unperceived a capacity or quality of being, in relation to those various objects which are *proved to be continually existing* by their regular reply to its action". EPEU, p. 75.

[53] EPEU, p. 55; italics mine.

[54] EPEU, p. 55.

of extension. She explains that "having the word 'outward' we must have the ideas the word stands for; and the ideas are negative ones".[55]

She delineates three separate ways we can acquire the notion of outwardness.

The first is by reasoning from the fact that external objects are 'ready to appear.' By moving around in the world, we can make certain sensations appear, disappear, and reappear. For instance, Shepherd says, "by returning on their steps men can again recover the image of the house, the tree, they have just passed".[56] When I leave for work in the morning, I see the magnolias in my yard; when I walk down the street, they disappear; and when I return home in the afternoon, I see them again. The sight of the magnolias in the afternoon is a new sensation, so it requires a cause. And unless the magnolias were "*created purposely, ready to appear,* upon each irregular call of the senses, they must CONTINUE *to exist, ready to appear* to them upon such calls".[57]

This argument is supposed to show that there are external, independent, continually existing objects. But it is not supposed to show us much about the nature of those objects:

> [I]n order to the formation of *all* the effects produced on the mind, through the senses, there must be *efficient causes, not included* in the *general essence* of the mind; and these are "*ever ready to appear*" [...] whatever they may be, however unknown, they may well be termed *objects, outward objects,* which the organs of sense, and their associations reveal, according to their peculiar bearings upon the mind. I repeat it, therefore, that the unknown causes of all our perceptions, are as the unknown quantities in algebra, which yet may be measured, valued, reasoned on by their signs; and the signs of these outward objects are the sensations they can create; and they may always be spoken of, and compared together, as though they did *truly exist*, in *these forms* in which they appear to the mind.[58]

I know that my sensations have causes, and I know that those causes are external to me. I know they are extended because they produce various sensations of extension. But knowing this does not give me knowledge of the intrinsic properties of external objects.

Proprioception plays a role here, although Shepherd does not make this explicit. Her claim is that we know that objects exist externally and continuously because we know that they are 'ready to appear' in response to certain movements of our bodies. This means we need to know that we are moving around in the world, and the way we know this is through proprioception.

The second way we acquire the notion of outwardness is through a "comparison of motion with a state of rest". Shepherd explains,

> I consider another (and that perhaps the chief) method which nature takes to impress the notion of outwardness, to be by means of motion. For the intimate sentiment of our own

---

[55] EPEU, p. 51.

[56] EPEU, p. 37.

[57] EPEU, p. 37. How do I rule out the possibility that the magnolias were created anew each time I see them? Because I cannot cause them to appear at will, and because I know that my mind and my sense organs have remained the same.

[58] EPEU, pp. 47–48.

> existence, separated from the ideas of our bodies [...] has no relation to *space* [...] But the very impression of motion consists in the impression of passing through extended space, and as a corollary with it suggests to the mind, *here*, and *there* [...] the sensation of passing through different points of space, suggests the notion, or rather inspires the immediate feeling of the *extension of space* [...] This space [...] appears continually to exist, and to respond regularly to *motion*, as other objects do to other senses. It is hence the immediate consequence of motion also to suggest [...] the *reality* of distance or outwardness from the *sentient being*, the *self*; which [...] knows of outward existence [...] by a piece of reasoning; viz. that it needs must be in order to justify the possibility of motion when in a state of rest, as well as regularly to respond to its action upon demand.[59]

The "impression of passing through extended space" or "sensation of passing through different points of space" is, again, proprioception. So just as we use proprioception to discover that external objects are "ready to appear" to us in response to our bodily movements, we use proprioception to discover that space itself is "ready to appear."

The third and final way we get the notion of outwardness is by reasoning from the fact that certain sensations "require the use of the organs of sense in order to let their specific impressions enter the mind".[60] We reason that these sensations have causes, and that their causes are "*out of* (i.e. *not included in*,) the definitions and limitations of our own bodies".[61] This reasoning starts from knowledge that certain sensations require the use of the sense organs, knowledge that Shepherd thinks is not itself based on reasoning. Rather, she thinks, we are directly aware of using our sense organs:

> *In the general conscious perception of sensible qualities, are included the knowledge that the organs of sense are used [...]* And this conscious use of the mechanical action of the five senses in relation to other beings than the mind, *is a very different part, or kind of the universal idea of perception,* from the *mental consciousness of* particular sensible qualities only [...] The consciousness whether the organs of sense be used or not, in perceiving objects, is the great criterion of a sane, or insane state of mind, of its waking or sleeping condition; the consciousness that the organs of sense are used, makes all the difference between objects of sense, or objects of memory, reason, or imagination.[62]

Shepherd claims repeatedly that we are conscious of using our sense organs, speaking of the "conscious living feeling" of their use[63]; the "conscious use of the organs" of sense[64]; and so on. So in addition to proprioception, we have interoceptive awareness of the use of our sense organs. (Recall Berkeley's claim that our awareness of the rotation of our eyeballs in their sockets allows us to compute distance.). The claim that we have interoceptive awareness of the use of our sense organs may seem odd. Surely you can't just refute the skepticism of Descartes' First Meditation by claiming that we're directly aware of whether we're using our sense organs! But I

---

[59] EPEU, p. 55.

[60] EPEU, p. 56.

[61] EPEU, p. 56.

[62] EPEU, pp. 113–114.

[63] EPEU, p. 42.

[64] EPEU, p. 72.

don't think this is really what Shepherd is doing. She is not claiming to refute the skeptic. In a dream, we might falsely believe that we are using our sense organs because we have a sensation that is internally indistinguishable from ones produced by the sense organs. However, this would be "a mere lunatic illusory call of the organs of sense".[65] It would not actually be the same sensation, just as the sensation of looking at an actual table and the sensation of looking at a table in a dream might be internally indistinguishable despite being different sensations.[66]

9.7 Shepherd on the Source of the Notion of the Self

We've seen that Shepherd understands extension as the unknown external cause of primary-quality sensations. Thus, on her view, the notion of extension depends on the notion of an external object. We've also seen that the notion of an external object depends on the notion of the body or self. So now we need to know how Shepherd thinks we acquire the notion of the body or self. She explains,

> We consider *that as our own body*, which is within a bound, or certain limit, and is the source of *conscious* pleasure and pain, and this *limit* we call the skin, within which, is contained *all we call ourselves*, and being summed up, is the notion of the conscious sensation of the extension of the body, and of a sufficient cause for life and sensation in general. Because without any impression from what are called external things, or the use of the organs of sense, the general sensation of life can go on.[67]

In the continuation of this passage, Shepherd contrasts the "general sensation of life" with particular kinds of sensations elicited by the organs of sense. Shepherd does not really say much about what the general sensation of life consists in, but it may be something like Condillac's "fundamental feeling", which includes awareness of the movements of the body in respiration. At any rate, it is a kind of interoceptive bodily awareness.

This passage requires some unpacking. Shepherd says that we consider everything within the limit of the skin "*as our own body*" and that within this limit "is contained *all we call ourselves.*" Now, the notion of a limit seems like a spatial limit, so you might think this implies that we need to have a notion of extension before we get the notion of the body or self. However, Shepherd can't say this, on pain of circularity, and fortunately she doesn't *need* to say this. When she speaks of "the conscious sensation of the extension of the body," she is referring to our felt awareness of the different points of our body in space, which we may use to construct a notion of external objects but certainly does not presuppose it.

---

[65] EPEU, p. 71.

[66] Shepherd seems to hold not just that like causes have like effects but also that like effects have like causes. So, if the causes are different, the effects cannot be the same: the experience of seeing a table in a dream cannot be the same as the experience of seeing a table in reality. This sort of disjunctivist theory is unsurprising for someone writing in the wake of Reid.

[67] EPEU, p. 63.

Our proprioceptive awareness of our body as extended in space is something like Condillac's experience of sensations in different parts of the body. We also have interoceptive awareness of events within the body as part of the "general sensation of life." We thus conceive of the self or body as the extending continuing cause of such sensations. Like the notions of extension and outwardness, the notion of the self or body is a relative notion.

9.8 The Role of Proprioception in Shepherd's Theory of Knowledge of the External World

We started by looking at Shepherd's claims that the power of motion is a "sixth organ of sense",[68] that "motion is [...] a sort of sense",[69] and that "motion [...] yields to our minds its own peculiar sensation".[70] We've seen that these are really claims about proprioception. We've also seen that proprioception plays a number of roles in Shepherd's account of knowledge of the external world:

1. We use proprioception to construct a relative notion of extension. For we conceive of extension as the unknown, external cause of the bodily sensations associated with moving through space.
2. We use proprioception to acquire knowledge of external objects. For one way we acquire such knowledge is by discovering that objects are "ready to appear" in answer to the movement of our bodies—and we know the movements of our bodies through proprioception.
3. In similar fashion, proprioception helps us discover that space itself is "ready to appear."
4. We use proprioception (as well as other forms of bodily awareness) to construct the notion of the body or self. On Shepherd's view, we are aware of our body occupying space; as in Condillac, we experience different sensations in different parts of the body. We also have awareness of events within the body as part of the "general sensation of life." We thus conceive of the self or body as the extended continuing cause of such sensations. Like the notions of extension and outwardness, the notion of the self or body is a relative notion.

   One more use of proprioception deserves a quick mention:

5. Our proprioceptive awareness of our bodies moving through space helps us to correlate the sense of vision and the sense of touch.

For

> It is motion, as first in order, and first in proof, which is empowered to detect the outwardness of objects: because those things which return upon the application of motion to the

---

[68] EPEU, p. 230.

[69] EPEU, p. 104.

[70] LMSM, p. 701.

> sense of touch, are by that *necessity* of motion in order to apprehend their tangibility *just defined as distant* from the mind which apprehends them.[71]

Extending Berkeley's claim that the eye alone cannot see distance, Shepherd explains that "all sensible qualities whatever [...] are carried *out* by an act of the mind, and considered as proportionally distant from the mind, as is the quantity of motion required to attain them in their *tangible* form".[72] Unlike Condillac, Shepherd insists that the process relies not on mere association but requires "a mixture of reasoning" with the "sense of the resistance of solidity to the sense of voluntary motion.[73] The power of motion, regarded as a sixth organ of sense, is "the method of overcoming distance, and of becoming acquainted with tangible extension",[74] but it does not function alone. Correlating visual and tactile perceptions requires both proprioception and reasoning.

9.9 Conclusion

Many eighteenth century philosophers appeal to a form of proprioception in their account of knowledge of the external world. (I've sketched how this works in Berkeley, Condillac, and Reid, but there are many other examples to be found as well.) But although such philosophers use proprioception to explain other phenomena, they do not seem to see a need to explain proprioception itself. What is more, they do not seem to recognize that proprioception is a distinct sense modality, instead lumping it in with the sense of touch. Writing at the beginning of the nineteenth century, Shepherd develops a rudimentary theory of proprioception and distinguishes proprioception from the sense of touch.[75] She then uses proprioception to do a great deal of work in explaining the source of concepts like extension, externality, and the self.

Scholars have not really recognized the role of proprioception in eighteenth century philosophers' attempts to explain knowledge of the external world. Nor have they noticed that Shepherd makes clear that proprioception is a distinct sense modality and expands its epistemic role. My main goal in this chapter has been simply to demonstrate these two hitherto unrecognized facts. However, I want to close with a brief, and tentative, suggestion about *why* Shepherd sees proprioception as a distinct sense modality when her predecessors assumed it was part of the sense of touch. I suggest that this is because she has a new account of the relationship between mind

---

[71] EPEU, p. 102–177.

[72] EPEU, p. 179.

[73] EPEU, p. 102, n. 5.

[74] EPEU, p. 230.

[75] Writing at around the same time, Thomas Brown develops a notion of "muscular feelings" in his *Lectures on the Philosophy of the Human Mind* (1824). I do not know whether Shepherd read this work, although she discusses another of Brown's works—Brown (1806)—extensively in the ERCE.

and body—and that this led to a new way of thinking about what it is that external objects are supposed to be external *to*.

Shepherd conceives of mind and body as distinct powers with different natures. In particular, she conceives of the mind as a capacity for sensation that cannot be actualized unless it is mixed with the body. Thus, sensations must always belong to a mixture of mind and body, and it is this mixture of mind and body that constitutes the self.[76] One implication of this is striking: my body is not an external object for me.

This suggests a motivation for seeing proprioception as a distinct modality instead of subsuming it under the sense of touch the way Shepherd's predecessors did. For Shepherd, the sense of touch is the sense that tells us about external objects, while proprioception and other forms of bodily awareness tell me about myself. This is a very different picture than the one found in Berkeley and other early moderns, who typically think of external objects as objects external to the *mind*, so that my body is an external object among many others. The similarity to Merleau-Ponty, writing a century later, is striking.[77]

References

Primary Literature

Berkeley, G. (1710). *An Essay Towards a New Theory of Vision*. Printed by Aaron Rhames, at the Back of Dick's Coffee-House, for Jeremy Pepyat, Bookseller in Skinner-Row.
Berkeley, G. (1998). *A Treatise Concerning the Principles of Human Knowledge (J. Dancy, Ed.). Oxford University Press.*
Brown, T. (1806). *Observations on the Nature and Tendency of the Doctrine of Mr. Hume, Concerning the Relation of Cause and Effect. Mundell & Son.*
Brown, T. (1824). *Lectures on the Philosophy of the Human Mind*. John Grigg.
Brown, T. (1835). *Inquiry into the Relation of Cause and Effect*. Constable.
de Condillac, E. B. (1803). *Traité des sensations. Oeuvres Complètes de Condillac (Vol. 4). Chez Dufart, Imprimeur-Libraire.*
de Condillac, E. B. (1930). *Treatise on the Sensations (G. S. Carr, Trans.), University of Southern California Press.*
Destutt de Tracy, A.-L.-C. (1824). *Élémens D'idéologie. Mme Lévi..*
Destutt de Tracy, A.-L.-C. (1972). In F. Rastier (Ed.), *Idéologie et théorie des signes: Analyse structurale des éléments d'idéologie*. Mouton.
Hartley, D. (1775). *Theory of the Human Mind, on the Principle of the Association of Ideas: With Essays Relating to the Subject of it*. Johnson.
Hume, D. (2001). *A Treatise of Human Nature (D. F. Norton & M. J. Norton, Eds.), Oxford University Press.*
Priestley, J. (1782). *Disquisitions Relating to Matter and Spirit*. Johnson.
Reid, T. (1764). *An Inquiry into the Human Mind: On the Principles of Common Sense*. Millar.

[76] For more on Shepherd's account of the self and the mind-body relationship, see Boyle (2020), Fasko (2022, 2023, forthcoming), and LoLordo (2022).

[77] For an account of the nineteenth century antecedents of Merleau-Ponty's claim that the self is not an external object, see Vincenti's chapter.

Reid, T. (1859). *Essays on the Intellectual Powers of Man. Hamilton, W. S & Walker, James (Eds.) Phillips, Sampson and company.*

Shepherd, M. (1824). *Essays on the Relation of Cause and Effect*. Hookham.

Shepherd, M. (1827). *Essays on the Perception of an External Universe*. Botson and Palmer.

Shepherd, M. (1832). Lady Mary Shepherd's Metaphysics. *Fraser's Magazine for Town and Country, 5*(30), 697–708.

Stewart, D. (1829). *The Works of Dugald Stewart*. Hilliard and Brown.

Secondary Literature

Bolton, M. (2011). Causality and Causal Induction: The Necessitarian Theory of Lady Mary Shepherd. In K. Allen & T. Stoneham (Eds.), *Causation and Modern Philosophy* (pp. 242–261). Routledge.

Bolton, M. (2019). Lady Mary Shepherd and David Hume on Cause and Effect. In E. O'Neill & M. Lascano (Eds.), *Feminist History of Philosophy: The Recovery and Evaluation of Women's Philosophical Thought* (pp. 129–152). Springer.

Boyle, D. (2020). Mary Shepherd on Mind, Soul, or Self. *Journal of the History of Philosophy, 58*(1), 93–112.

Boyle, D. (2022). *Mary Shepherd: A Guide*. Oxford University Press.

Buras, T. (2015). *Thomas Reid's Experimentum Crucis. In R. Copenhaver & T. Buras (Eds.), Thomas Reid on Mind, Knowledge, and Value. Oxford University Press.*

Cruz, M. (forthcoming). Shepherd's Case for the Demonstrability of Causal Principles. *Ergo: An Open Access Journal of Philosophy*.

Fantl, J. (2016). Mary Shepherd on Causal Necessity. *Metaphysica, 17*(1), 87–108.

Fasko, M. (2022). 'God Said "Let us Make Man in our Image after our Likeness"'—Mary Shepherd, the *Imago-Dei*-Thesis, and the Human Mind. *British Journal for the History of Philosophy, 31*(1), 1–22.

Fasko, M. (2023). The Compound Mass we Term SELF'—Mary Shepherd on Selfhood and the Difference between Mind and Self. *European Journal of Philosophy, 32*(3), 1–15.

Fasko, M. (forthcoming). Mary Shepherd's Dispositional Notion of God, Matter, and Finite Minds. *Philosophers' Imprint*.

Folescu, M. (2021). Mary Shepherd on the Role of Proofs in our Knowledge of First Principles. *Noûs, 56*(2), 1–21.

Landy, D. (2020). A Defence of Shepherd's Account of Cause and Effect as Synchronous. *Journal of Modern Philosophy, 2*(2), 1–15.

LoLordo, A. (2019). Mary Shepherd on Causation, Induction, and Natural Kinds. *Philosophers' Imprint, 19*(52).

Lolordo, A. (2020). *Mary Shepherd's Essays on the Perception of an External Universe. Oxford University Press.*

LoLordo, A. (2022). *Mary Shepherd*. Cambridge University Press.

McRobert, J. (unpublished). *Mary Shepherd and the Causal Relation—Part One*.

Paoletti, C. (2011). Restoring Necessary Connections: Lady Mary Shepherd on Hume and the Early Nineteenth-Century Debate on Causality. In C. Paoletti (Ed.), *Hume, Nuovi Saggi*. Il Poligrafo.

Smith, R. (2023). *Kinaesthesia in the Psychology, Philosophy and Culture of Human Experience. Routledge.*

Wilson, J. (2016). On Mary Shepherd's Essay upon the Relation of Cause and Effect. In E. Schliesser (Ed.), *Neglected Classics of Philosophy, II. Oxford University Press.*

Chapter 10
Touch *A Priori*? Tactile Perception of Space in Nineteenth-Century Physiology and Psychology

Denise Vincenti

Abstract In 1867 Hermann von Helmholtz described the coeval controversy regarding empiricism and nativism as an opposition between those who believe that space perception results mostly from experience, and those who instead appeal to additional innate elements. This description is more nuanced than it seems. In nineteenth-century Germany, both nativist and empiricist thinkers often resorted to the notion of *a priori* to characterize perception of space—but they defined apriorism in quite different terms. Although scholars have already retraced the history of this controversy, existing literature focuses almost exclusively on the role of vision and sight in space perception, paying little attention to the role of touch. This chapter aims to offer an overview of some German nativist and empiricist accounts of space (by Johannes Müller, Ernst Weber, Hermann Lotze, Wilhelm Wundt, and Carl Stumpf), in an attempt to understand how they conceived the relationship between touch and *a priori* elements and to what extent they regarded it as a leading sense in space perception.

Keywords Sense of touch · Space perception · Nativism *vs* empiricism · German physiology and psychology · *A priori*

10.1 Introduction

The mental experiment on spatial perception that William Molyneux (1656–1698) presented to the "father" of British empiricism, John Locke (1632–1704), is perhaps too well-known to need any introduction.[1] What can be usefully recalled however is

[1] See Locke (1786, vol. 1, pp. 107–8); Locke (1978); Molyneux (1692). The literature on the topic is vast, but a selection would include Morgan (1977); Degenaar (1996); Chottin (2014); Ferretti & Glenney (2020).

D. Vincenti (✉)
Dipartimento di Filosofia, Scienze Sociali, Umane e della Formazione, Università degli Studi di Perugia, Perugia, Italy
e-mail: denise.vincenti@unipg.it

C. Beneduce et al. (eds.), *The Sense of Touch*, Studies in the History of Philosophy of Mind 32, https://doi.org/10.1007/978-3-032-16308-0_10

that the question regarding whether a man-born-blind could, if his sight were suddenly restored, immediately discern a sphere from a cube, catalyzed the later debate regarding the innate or acquired nature of space and constituted a constant point of reference for a number of (often competing) perspectives. Indeed, to affirm that the newly sighted person would hypothetically be able to distinguish, without the use of their sense of touch, a sphere from a cube amounted to acknowledging the existence of some innate ideas at play in the perception of these geometrical forms; to deny this hypothetical possibility implied stating that any notion of spatiality resulted from experience and learning. As Denis Diderot (1713–1784) pointed out, for a long time "[…] training and questioning a man-born-blind would have been a task not unfit for the combined talents of Newton, Descartes, Locke and Leibniz".[2]

As a matter of fact, Molyneux's mental experiment opened up several interesting questions. First and foremost, the interchangeability of visual and tactile data, i.e., the possibility to automatically apply the concepts acquired through the sense of touch to visual images, and vice versa. It was, in other words, a question of understanding whether the figure seen and the figure touched were the same figure; whether there was a certain similarity between them, or an intermediate term which allowed reason to conclude that the two figures were the same. From the position taken in this regard followed a more general implication: the innate or acquired nature of spatial perception. Indeed, the possibility for a man-born-blind to immediately distinguish between a cube and a sphere can be explained by the idea that the same *abstract notion* of roundness or squareness is awakened by the sensory impressions of both sight and touch. From this viewpoint, the concept of space is immediately given in the form of an abstract idea; it is an innate property of the mind.[3] On the other hand, the impossibility for a man-born-blind to distinguish between the two figures can be explained by the idea that the visual cubic figure and the tactile cubic figure are heterogeneous and did not belong to the same idea of roundness. The association of these two sets of impressions is therefore conceived of as derived from experience, learning and judgement.[4] Although the modern rationalism *vs* empiricism controversy involved a much ampler set of philosophical questions, it is the problem of space perception that influenced the epistemological divide between those advocating for the existence of innate elements in the process of knowing and those contending that knowledge was solely derived from sensory experience.

With the advent of Immanuel Kant's (1724–1804) reflection and related reappraisal of the problem of space in light of transcendental philosophy, the physiognomy of the debate changed significantly, with the traditional epistemological opposition rationalism *vs* empiricism undergoing a sort of "rebranding". Hermann von Helmholtz (1821–1894) was arguably the first to signal this change of pace in

[2] "[…] préparer et interroger un aveugle-né n'eût point été une occupation indigne des talents réunis de Newton, de Descartes, de Locke et de Leibniz", Diderot (2011, p. 205).

[3] Examples of this line of reasoning are: Leibniz (1921, pp. 90–6); Hamilton (1861, p. 179).

[4] See, for instance: Locke (1786, p. 108); Berkeley (1709); Mill (1930, pp. 555–62).

contemporary German discussions of space perception,[5] by reframing them in terms of an opposition between *nativism* and *empiricism*:

> Under these circumstances, it can often be quite difficult to determine what, in the representations we gain through vision, is derived directly from sensation, and what, on the contrary, is conditioned by experience and practice. This difficulty is also tied to the most fundamental theoretical opposition that exists between different researchers in this field. Some are inclined to grant experience the widest possible scope of influence, in particular to derive all spatial perception from it; we may call this view the empiricist theory (*empiristische Theorie*). Others, however, though they must acknowledge the influence of experience in a certain range of perceptions, nonetheless believe that, for certain elementary representations that occur uniformly in all observers, it is necessary to presuppose a system of innate representations not based on experience, especially with regard to spatial relations. We may, in contrast to the first, designate this latter view as the nativist theory (*nativistische Theorie*) of sensory perception.[6]

Helmholtz's description is only apparently disregarding of the impact Kant's philosophy had on coeval accounts on space. When he states that "*empiristische*" theory attributes "the widest possible scope of influence" to experience, he implies that experience is not the sole ingredient in the knowing process. Indeed, several nineteenth-century empiricist positions borrow from Kant the idea that perception of space is not acquired exclusively through experience but from some not yet determined *a priori* elements. On the other hand, Helmholtz's reference to "*nativistische*" theory—rather than "rationalist" or "innatist"—suggests that the very terms of the controversy over spatial perception had changed.[7] As James Mark Baldwin's *Dictionary of Philosophy and Psychology* recalls, Kant's philosophy made clear that the "[…] innate cause of necessary knowledge does not *resemble* that knowledge itself, and the great influence of this [Kantian] theory in this respect finally puts an end to the doctrine of 'innate ideas' as such".[8] The influence of Kantian criticism on earlier innatist perspective leads to a range of diverse and nuanced proposals. One of the most significant consequences, however, is that the notions of the innate and the *a priori* in knowledge ultimately came to be linked with the problem

[5] This is for instance the opinion of Ribot (1878, p. 130), and Hoffmeister (1955, p. 421).

[6] "Es kann unter diesen Umständen oft recht schwer werden, zu beurteilen, was in unseren durch den Gesichtssinn gewonnenen Anschauungen unmittelbar durch die Empfindung, und was im Gegentheil durch Erfahrung und Einübung bedingt ist. An diese Schwierigkeit knüpft sich auch der hauptsächlichste principielle Gegensatz, welcher zwischen verschiedenen Forschern in diesem Gebiete besteht. Die einen sind geneigt, dem Einfluss der Erfahrung einen möglichst breiten Spielraum einzuräumen, namentlich alle Raumanschauung daraus herzuleiten; wir können diese Ansicht als die empiristische Theorie bezeichnen. Die andern müssen allerdings den Einfluss der Erfahrung für eine gewisse Reihe von Wahrnehmungen zugeben, glauben aber für gewisse bei allen Beobachtern gleichförmig eintretende elementare Anschauungen ein System von angeborenen und nicht auf Erfahrung begründeten Anschauungen, namentlich der Raumverhältnisse, voraussetzen zu müssen. Wir dürfen diese letztere Ansicht im Gegensatz zur ersteren wohl als die nativistische Theorie der Sinneswahrnehmungen bezeichnen", Helmholtz (1867, p. 435). From now on, if not specified as an English edition, all translations are mine.

[7] For an in-depth analysis of this terminology, see Hatfield (1990, pp. 271–280).

[8] Moore (1957, p. 131).

of biological inheritance, or rather, with some congenital, inborn aspects proper to the knowing subject.[9] The *a priori* thus becomes embodied and physiologically qualified.

A certain form of apriorism in the investigation of space perception brings together empiricism and nativism. Of course, these competing theories interpret the notion of *a priori* in quite different ways, providing definitions of space that hardly overlap. Yet, a common thread woven through the intricate tapestry of nineteenth-century German debates is the reappraisal of *a priori* elements in spatial representations. The history of this controversy is well-known and has captured the attention of many scholars over the years. Their focus, however, has been almost exclusively on the role of vision and the sense of sight, paying little interest to the role of touch.[10] This is not surprising. The opposition between empiricism and nativism is often analyzed by taking the controversy between Helmholtz and Ewald Hering (1834–1918) on *visual* spatial perception as a reference point.[11] The sense of sight is indeed in the spotlight in this debate and becomes a privileged key to determining the innate or acquired nature of space. There is room though to wonder whether the Helmholtz-Hering debate should be held as so paradigmatic of German analyses of space perception. As we will see, analysis of the different positions reveals the sense of touch to be no less important. Already in 1878, Théodule-Armand Ribot (1839–1916) lamented the absence of accounts highlighting the role that *tactile apprehension of space* played in German physiology and psychology.[12] After almost a-hundred-and-fifty years, the situation has changed very little. Studies of how touch affects perception of space continue to be few, and the theoretical supremacy of sight remains unchallenged.

This chapter aims to offer some brief insights on this question, by drawing upon various German analyses of space (from both nativist and empiricist discourses), such as Johannes Müller's, Ernst Weber's, Hermann Lotze's, Wilhelm Wundt's, and Carl Stumpf's. I decided not to delve into Hermann von Helmholtz's reflection (nor into his famous controversy with Hering), because, albeit crucial in the development of a genetic and empiricist theory of space perception, it centers almost entirely on optic perception. Helmholtz occasionally seems to admit that touch educates vision in space perception, but he does not elaborate on this idea further.[13] Through the discussion of these various physiological and psychological positions, the chapter intends to explore how touch interplays with the above-mentioned *a priori* elements and to what extent it can be regarded as having the upper hand in space perception.

---

[9] *Ibidem.*

[10] Fundamental studies, such as Sully (1878); Boring (1929); Boring (1942); Pastore (1971), and Hatfield (1990), focus in particular on visual space and provide only a few insights on touch perception.

[11] On the subject, see Hatfield (1990, pp. 181–193); Turner (1994); Gutiérrez (2021).

[12] Ribot (1878, p. 130).

[13] See Helmholtz (1893); Hatfield (1990, p. 327).

10.2 The "Age of Touch": Johannes Müller and Ernst Weber

The rise of a nativistic position on space in nineteenth-century Germany is often attributed to the theoretical and experimental work of physiologist Johannes Peter Müller (1801–1858). Regarded by his peers as a Kantian thinker—one that endowed Kant's doctrine with an experimental framework[14]– Müller is one of the first to redefine the problem of spatial representation through reference to some innate, congenital, and inborn elements. Without delving into Kant's actual influence on Müller,[15] it can nonetheless be said that his association with Kant's philosophy should be taken cautiously. If it is true that for Müller space is somehow innate, i.e., not acquired via experience, he at the same time opposes the idea of space as a pure form of intuition, attributing the innateness of space to some properties of certain sense organs rather than a subjective disposition of the perceiver. It would thus be better to consider Müller's position as a middle ground "between Herbart's realist interpretation of Kant and the 'operationalized' Kantianism of Helmholtz", and to stress his debts to British empiricism as well.[16] Even more, Müller's position can be interpreted as connected to some aspects of Aristotle's thought, as it will be discussed in the followings.[17]

Although Müller's first account of space—specifically visual space—appears in his 1826 work *Zur vergleichenden Physiologie des Gesichtssinnes des Menschen und der Thiere* (*On the Comparative Physiology of the Human and Animal Sense of Sight*),[18] scholars tend to draw attention to the two-volumes *Handbuch der Physiologie des Menschen* (*Handbook of Human Physiology*) (1835–1840), which offer a more detailed analysis of the problem.[19] The *Handbuch* is also the work in which Müller articulates his word-renowned doctrine of the specific nerve energies, which can be summed up as follows: the nerves of each sense, though affected by different types of stimulations, are capable of only one kind of sensation.[20] This doctrine is not of a little interest for the question of the apriority of space. In fact, what Müller states is that the nerves of each sense organ display a specific energy, or rather, that the *quality* of the sensation produced following the stimulation of a nerve is intrinsic to the nerve itself and not dependent on the type of external

[14] Especially Helmholtz (1867, p. 208).

[15] See for instance: Montgomery (1889); Lenoir (2006); Isaac (2019). For an overview on the influence of Kant's philosophy on the development of experimental physiology and psychology in Germany, see Leary (1978); Leary (1982); Pradelle (2013).

[16] Isaac (2019, p. 83).

[17] Even though Aristotle's name hardly ever appears in Müller's writings, some aspects of his thought seem to closely trace certain positions of the Greek philosopher.

[18] Müller (1826).

[19] Müller (1835); Müller (1840).

[20] Müller (1840, pp. 249–75; Eng. tr. pp. 1059–87). For a detailed analysis of this doctrine, see Isaac (2019).

stimulus causing it.[21] The same vibratory stimulus can give rise to different sensations depending on the sense organ that it affects: a sensation of sound if it affects the sense of hearing, or a sensation of trembling in the skin if it affects the sense of touch. Thus the same stimulus can cause different sensations depending on the specific "energy" of the nerves affected. This suggests an *a priori*, innate power of the nerves (of a biological and embodied kind), which was subsequently considered by Helmholtz as the core of Müller's Kant-inspired nativism,[22] but that should be more fruitfully connected to other philosophical positions.

It is worth pointing out that from this perspective nerves do not only provide the *quality* of a sensation, but can also, in some cases, identify their *position in space*. Not all nerves can deliver this kind of information, however. Resorting to a traditional view of the senses—especially Aristotle's theory of *common sensibles* –,[23] Müller ascribes the ability to spatially localize sensations only to the senses of touch, sight, and taste.[24] In what follows, I will leave aside the role of taste, given that, with regard to the formation of spatial perception, Müller tends to consider it a form of touch (as Aristotle did),[25] and his views on sight and touch are certainly more interesting. Why these senses and not others? As Müller remarks,

> all the senses are not equally adapted to impart the idea of "extension" (*Ausdehnung*) to the sensorium. The nerve of vision and the nerve of touch, being capable of an exact perception of this property in themselves, make us acquainted with it in external bodies.[26]

Noting that, by *Ausdehnung*, he means the spatial disposition of sensations in a *bidimensional* field, we can see that according to Müller, in order to attribute a specific position to each sensation, the sense organ must first perceive the extension of its own nerves. In other words, the organ must possess the property of extension and

---

[21] Müller (1840, pp. 254–8; Eng. tr. pp. 1069–71).

[22] Helmholtz (1867, p. 208).

[23] That is, the theory that there are certain types of sensations that can be perceived by more than one sense. Unlike colors, for example, which are the specific sensations of sight, odors, which are the specific sensations of smell, etc., some others, such as motion, number, figure, and size are proper to more than one sense (Aristotle, *De Anima*, II.6, 418a-b 11–21 [ed. Barnes, 1984, vol. 1, p. 665]). Müller's notion of space can be seen as a common sensible.

[24] Müller (1840, p. 262; Eng. tr. pp. 1073–4). It should be noted that Müller gives a rather broad and fluctuating meaning to the term "sense". Indeed, when he speaks of touch and sight, he sometimes qualifies them in biological terms (i.e., in terms of their sense organs: the retina, the skin) and sometimes in "psychological" terms, i.e., as true sensory modalities capable of conveying certain types of sensation (visual, tactile). The boundary between the biological and the psychological is often lost. This ambiguity stems from the basic approach of his theory. It must be remembered that Müller is a physiologist and as such approaches the subject of space. The psychological element always tends to be reduced to its biological basis, sometimes leading to a certain confusion between the two levels. The same applies to Weber, as we shall see below.

[25] Aristotle, *De Anima*, II.11, 422a 10 (ed. Barnes, 1984, vol. 1, p. 672).

[26] "Zur Mittheilung der Ausdehnung im Raume an das Sensorium sind nicht alle Sinne gleich gut geschickt. Der Gesichtsnerve, der Gefühlswerte zeigen die Ausdehnung im Raume an, weil sie einer genauen Empfindung ihrer eigenen Ausbreitung fähig sind", Müller (1840, p. 262; Eng. tr. p. 1073).

the ability to perceive the relative positions of its extended parts. Touch, sight, and taste possess this property and ability; smell and hearing do not.

In the case of sight, experimental research directed in the 1830s by German naturalist Gottfried Reinhold Treviranus (1776–1837) had revealed the existence of photoreceptor cells in the retina,[27] which work as points of reception for different luminous stimuli coming from different points in the visual field. With respect to touch, on the other hand, Müller's main source was undoubtedly the work of physiologist Ernst Heinrich Weber (1795–1878), whose *De subtilitate tactu*,[28] published in 1834, produced a major turn in the experimental study of touch by demonstrating, among other results, that the surface of the skin was divided into different sensory circles (*Empfindungskreise*) that map the tactile sensitivity of this organ.[29] The size of each circle was experimentally determined by the use of the Weber compass: the experimenter would lean the legs of the compass—spread to a certain angle—on a portion of the skin; when the double stimulation was perceived by the subject as a single impression, it meant that the compass legs were both inside the borders of the same sensory circle. In order for the subject to perceive two different impressions stemming from the compass, the legs would need to be positioned across two different circles, i.e., the two-point threshold.[30]

This rendered the angle of the compass a measuring device to determine the size of a specific circle.[31] Weber also noticed that each circle's dimensions varied significantly based on the portion of the skin tested and that the smaller the circle, the more acute the sensitivity of that tactile region. For instance, sensory circles on the tip of the tongue are 1½ in Paris line (1 Paris line is approximately 0,225 cm or 0,1 inches), whereas those on the back are 12 in Paris line, making the tongue a much more sensitive area.[32] The absence of a one-to-one relation between a single point of the skin and a single impression (i.e., the presence of variably sized circles) was explained by the fact that an impression arises only when a nerve is affected. Since, from a physiological standpoint, each nerve divides at the end into several nerve-fibers running beneath a portion of the skin, the sensitized area was also deemed larger than a punctiform spot.[33] Weber's research hence provided an exhaustive

[27] Treviranus (1837). Müller (1840, p. 262; Eng. tr. pp. 1073–4).

[28] It is the last section of Weber's work *De pulsu, resorptione, auditu et tactu* (Weber [1834, pp. 44–175; Eng. tr. pp. 23–136]). See Müller (1840, p. 263; Eng. tr. p. 1074). Another text on touch was issued by Weber in 1846: *Der Tastsinn und das Gemeingefühl* (Weber [1905]).

[29] On Weber's reflection, see Boring (1929); Bück-Rich (1970); Gundlach (1992); Meischner (1993); Murray & Farahmand (1995); Ross (1995); Nicolas (2002); Ross & Murray (2017).

[30] Weber (1834, pp. 44–59; Eng. tr. pp. 29–45). Unfortunately, in Weber's texts there are no illustrations depicting the tools he used. To have an insight into the various types of devices for measuring tactile acuity, see Nicolas & Vobořil (2019).

[31] Weber introduces here the well-known notion of "differential threshold"—to be distinguished from the notion, later coined by Fechner's psychophysics, of "absolute threshold". The idea is that beneath a certain threshold, our sensitivity is numb; beyond it, we feel a change in our sensory sphere. For touch, such a threshold is determined by the perimeter of each sensory circle.

[32] Weber (1834, pp. 50–57; Eng. tr. pp. 34–42).

[33] Weber (1905, p. 68; Eng. tr. p. 180).

account of the different tactile regions of the body, alongside a (still coarse and sketchy) *mapping* of their relative spatial positions.[34]

Drawing upon Treviranus' and Weber's results, Müller conceived tactile and visual organs as extended surfaces, equipped with sensory nerves that possess an ability to discriminate between different stimulations. The fact that these points display a spatial configuration made them eligible to endow sensations with spatial positions. At this point, it is worth emphasizing the epistemological issue that seems to emerge in Müller's doctrine: the type of correspondence that exists between the sensory field and the external world.[35] For Müller contends that we are aware only of the states of our nerves, which means that, properly speaking, our sensory contents does not include anything more than the stimulation of the nerves[36]; they contain no properties of the external world and no information about the presence or spatial configuration of outer objects. This position might at first be read as either Kantian, in the sense that it denies the knowability of things in themselves, or as a revised form of George Berkeley's (1685–1753) idealism, since it limits perceptual activities to a strictly phenomenological ground. Upon closer inspection, though, Müller proves more in line with John Locke's empiricism: without denying the knowability of the external world or surrendering to skepticism or immaterialism, he opts for a dim form of realism, stating that sensory information actually tells us something about the structural properties of objects.[37] Like Locke, Müller thinks that secondary qualities—such as light, color, the bitter and sweet tastes, pleasant and unpleasant odors, cold and warmth—"are to be regarded [...] as sensations or modes of reaction of the nerves of sense", whereas primary properties, such as extension, progressive and tremulous motion, and chemical change, "belong wholly to external nature".[38]

Extension is thus a structural property of the object that is telegraphed to the brain by the activity and position of the extended nerves located on the retina or a portion of the skin.[39] More than a copy, the sensory nerves are a sort of a *mirror* of the spatially disposed parts of an object. Müller considers, to this end, the case of three luminous points. The luminous points (A, B, C) will stimulate three distinct photoreceptors (*a*, *b*, *c*) on the retina, each with a position on the retinal surface cor-

[34] It can be considered as in a way an antecedent of the *body map theory*, later introduced by Pierre Bonnier (1905). On this point, see De Vignemont & Massin (2015).

[35] This topic is well covered by Isaac (2019).

[36] Müller (1840, p. 254; Eng. tr. p. 1065). As Isaacs (2019, p. 84) shows, several criticisms have been moved to this idea (e.g., Cassirer [1912]; Bridges [1912]; Gibson [1966]).

[37] On Müller's structural realism, see Isaac (2019, pp. 84–6). Lenoir (1993) tends instead to conceive of Müller's position as genuinely Kantian.

[38] "Qualitäten welche den Sinnesnerven mehr durch den Conflict mit dem Sinnesorgan als Empfindungen entstehen, sind die Empfindung des Lichts, der Farbe, des Tons, des Bittern, Süssen, des Gestanks, Wohlgeruchs, des Schmerzes, der Wollust, des Kalten, Warmen; Eigenschaften welche ganz von aussen bestimmt werden können, sind die Ausdehnung, die fortschreitende, die zitternde Bewegung, die chemische Veränderung", Müller (1840, p. 262; Eng. tr. p. 1073).

[39] Müller (1840, p. 261; Eng. tr. p. 1072).

responding to the actual position of A, B, and C. The stimulation of *a*, *b*, and *c* is immediately transmitted to the brain,[40] where three mental sensations (α, β, γ) are produced. What is notable is that, according to Müller, neither the quality nor the spatiality of these sensations is lost in this transmission, since the spatial configuration of *a*, *b*, *c* is what enables the communication to the brain of the position assigned to α, β, γ. The same goes for the sense of touch, with the sole difference that the localization is subject to a certain degree of variation, as shown by Weber's sensory circles.[41] Despite their different degrees of acuity and accuracy, both sight and touch are able to immediately perceive extension, namely a surface or bidimensional figure.

In this respect, the parallelism between the two senses is so tight that Müller is also able to give a positive answer to Molyneux's problem of the man-born-blind without resorting to pure innatism: the blind man whose sight is suddenly restored is capable of distinguishing a sphere from a cube not because he innately possesses abstract ideas of roundness or squareness, but because the sense of touch and sense of sight are operatively akin.[42] It is as if the spatially qualified sensations experienced many times by *touching* a cube now become signs for the new spatially qualified sensations produced by the stimulation of different points of the retina. Yet how is it possible for tactile sensations to immediately *signify* the new visual sensations? Here, I think, Müller is implicitly postulating an innate ability of these two senses to signify one another when it comes to extension and surface perception—an innate ability rooted in the relative position of the parts composing the extension of these sense organs. In other words, he suggests a sort of biunivocal relation connecting a specific point of the tactile field to a specific point of the retina (a sort of isomorphism of tactile and visual sensations). Müller disagrees in this respect with Locke, who attributed the sign-relation between visual and tactile sensations to the faculty of judgment.[43]

The spatial appreciation of a bidimensional surface is thus immediate and innate for touch and sight. Different is the case of the third spatial dimension, depth. Since there is no differentiation for Müller between contact senses and distance senses (since all senses are contact senses),[44] no sense organ can account for immediate depth perception. The idea of a tridimensional object is always acquired through experience and is the result of a mental operation combining a series of tactile or visual sensations with the movements required to produce those sensations.[45] The same can be said for the notion of "distance", which seems to be more immediate

---

[40] Here Müller solves the problem of binocular vision, resorting to an innate ability of the visual organ to unify the images stemming from the two retinas. See Helmholtz (1893, pp. 253–4).

[41] Müller (1840, pp. 262–3; Eng. tr. p. 1074).

[42] Müller (1840, p. 362; Eng. tr. p. 1176).

[43] Locke (1786, vol. 1, p. 108).

[44] Since the vibration of a fluid in space, the ether, is what produces in the retina the sensation of color, Müller included sight (Müller [1840, p. 349; Eng. tr. p. 1162]).

[45] Müller (1840, p. 354; Eng. tr. p. 1167). This idea seems to resort to a Berkeleian conception of space.

than the sense of sight but actually is obtained by outward projection of sensations stemming from a simple contact.[46]

Until this point, the role of touch and sight in spatial perception appears to be almost identical for Müller. Both senses have the ability to provide sensations through a bidimensional configuration (innate power) and to combine with movement to produce sensations of depth and distance (via experience). There are nevertheless, in Müller's views, other aspects affecting the perception of space which apply only to touch: *exteriority* and *solidity*.[47] For an object to be in a space, it must occupy a portion of this space in such a way that it is impossible for another object to occupy the same position. This connection between the idea of solidity (or impenetrability) and the notion of "space" can already be found in Locke's *Essay Concerning Human Understanding*[48] and is also reiterated by Thomas Reid (1710–1796), who considers these notions responsible for the belief in the existence of external objects.[49] The property of solidity is then a mark for something that is exterior to the perceiver.[50] As Locke affirmed, it is firstly given by touch in the form of *resistance*.[51] The very same idea can be found in Müller, who reclaims the primacy of touch in the perception of resistance (*Widerstand*) from an ontogenetic standpoint. If we consider the fetus in utero, which has never had any visual impressions, it is clear that it can obtain only obscure ideas of an external object (the utero) and only via touch; at the same time, alongside the idea of external resistance, the fetus begins forming the notion of "internal resistance", from the activity of its own body.[52] This activity is referred to, by Müller, as *coenesthesia*, or common sensibility (*Gemeingefühl*), which falls in the spectrum of the sense of touch.[53] This leads Müller to a crucial conclusion:

> This consciousness of our own corporeal existence is the standard by which we estimate in our sense of touch the extension of all resisting bodies.[54]

The primary standard is thus touch, not sight. It is also worth considering that the distinction between the self and the external world (through the distinction of the ideas of internal and external resistances), while innate and immediate for touch, is

---

[46] Müller (1840, p. 268; Eng. tr. p. 1079).

[47] Müller (1840, pp. 268–9; Eng. tr. pp. 1080–1).

[48] It is true that Locke states that space should not be equated to the notion of impenetrability (or matter), since he advocates for the existence of empty space (vacuum). Yet solidity and impenetrability are concepts deeply connected to space perception (Locke [1786, vol. 1, pp. 86–90]).

[49] Reid (1994, 5.3, 61/122b).

[50] For a more in-depth discussion of this issue, also in relation to Mary Shepherd's thought, see Antonia LoLordo's chapter in this volume.

[51] Locke (1786, vol. 1, p. 86).

[52] Müller (1840, pp. 269–70; Eng. tr. pp. 1080–1).

[53] Müller (1840, p. 494; Eng. tr. p. 1324).

[54] "Die Leiblichkeit unserer selbst ist das Maass, nach welchem wir sofort im Gefühlssinn, die Ausdehnung aller widerstandleistenden Körper beurtheilen", Müller (1840, p. 270; Eng. tr. p. 1081).

always acquired for sight. Supposing that touch did not form a notion of internal and external resistance in utero, sight could distinguish the self from the world only by resorting to movement: by moving the eyes, the perceiver acknowledges that certain objects in space remain almost immobile (own body) while others change (external objects).[55] Yet, the differentiation of the self from the world here is acquired, not congenital.

When it comes to spatial perception, Müller emphasizes several innate elements. The spatial disposition of the nerves sensitizing both the sense organs of touch and sight are innate, congenital, or inborn, which enables them to provide their sensations with a spatial (bidimensional) quality. In touch specifically, coenesthesia, or *Gemeingefühl*, also plays a fundamental role in giving an immediate idea of external and internal resistance. It is clear, from these remarks, that for Müller the notion of "innate" acquires a specific meaning: it is equated with immediacy, instinctiveness, and embodiment. It is not, as Helmholtz claimed, an outright form of Kantian apriorism. This is further demonstrated by Müller's criticism of Kant's concepts. According to Müller, what enters into play in the formation of a concept is not an *a priori* activity of the perceiver's thinking, but rather a form of abstraction (*Abstraktionsvermögen*)—as Locke and, in a sense, Johann Friedrich Herbart (1776–1841)[56] had already claimed. For Müller, abstraction is the faculty of discovering *generalities* among sensations, and hence of producing *general abstract ideas*: a faculty proper only to human beings, since it is connected to language, and not to animals.[57] Interestingly enough, Müller also ranks the idea of space among these general abstract ideas. Indeed, he identifies an additional notion of space: the intellectual and conceptual idea of space obtained by abstracting similarities among tactile and visual spatial sensations and combining them into a new, more comprehensive, idea of space. In this respect, abstraction is an innate faculty only if considered as a synonym for *generality* or *universality*.

It should be noted, however, that these two conceptions of space are not superimposable. In this respect, I tend to disagree both with Isaac (2019), who contends that Müller's space is solely a *general abstract idea*, and Lenoir (1993), who interprets Müller's space in continuity with Kant's intuitionism. I instead hold that Müller postulates two kinds of space: one that is immediately given in sensation, and hence is also possessed by animals (sensory notion of space), and one that is obtained through the faculty of abstraction and represents an intellectual acquisition proper only to human beings (conceptual notion of space).

A rather similar conception is presented by Ernst Weber some years later, in *Tastsinn und Gemeingefühl* (1846). Here, Weber illustrates the process by which spatial representations are formed:

[55] Müller (1840, p. 271; Eng. tr. pp. 1082–3).

[56] Müller (1840, pp. 518–9; Eng. tr. pp. 1347–8). Locke (1786, vol. 1, p. 120); Herbart (1816, pp. 137–46). See also Isaac (2019, p. 85).

[57] This idea is shared by Locke and Müller. See Locke (1786, vol. 2, pp. 1–3).

1. Movements in bodies surrounding us, which are transmitted into the material of our sense-organs;
2. Movements in our nerve-fibers, caused by the above movements, not being of a different kind;
3. Changes in our consciousness excited by the movements in our nerves, and which we call sensations;
4. The ordering of the sensations in categories of time, space, and number;
5. The abstract concepts of these and all other categories, along with concepts derived from combinations of these categories.[58]

As for Müller, conceptualization of space and other categories comes last and only humans can be credited with it. More interesting is the transition from point 3 to point 4, regarding which Weber claims: "[...] the way in which we go about interpreting our sensations is not completely governed by our own free will; we feel obliged, for some unknown reason (*sondern wir sind durch eine unbekannte Ursache genötigt*), to order our sensations with respect to the categories of space, time, and number".[59] This act of ordering sensations in spatial forms is thus "[...] not the result of experience, but rather experience itself".[60] And the "[...] unknown reason" whereby we feel compelled to organize sensations in a spatial form is something that resembles an "instinct" or "intellectual instinct" (*intellektuellen Instinkt*)—not by chance, animals are driven by the same cause to order their sensations in terms of space.[61]

Weber experimentally demonstrates the existence of this unknown reason through his observations on touch. Drawing upon his theory of sensory circles, he formulates the following law: "*the number of elementary nerve-fibers ending within a square inch of a touch-organ endowed with a localization-sense influences the*

[58] "(1) die Bewegungen in den uns umgebenden Körpern, die sich in die Materie unserer Sinnesorgane hinein fortsetzen; (2) die Bewegungen in unseren Nervenfäden, die von jenen Bewegungen verursacht werden, aber anderer Art sind; (3) die Veränderungen in unserem Bewußtsein, welche durch die Nervenbewegungen angeregt werden, und die wir Empfindungen nennen; (4) die Vorstellung der Empfindungen in den Kategorien der Zeit, des Raumes und der Zahl; (5) die abstrakten Begriffe der genannten und aller anderen Kategorien, sowie die durch ihre Zusammensetzung entstehenden Begriffe", Weber (1905, p. 12; Eng. tr. pp. 146–7).

[59] "Die Art und Weise, wie wir bei der Auslegung unserer Empfindungen zu Werke gehen, hängt nicht ganz von unserer freien Selbstbestimmung [487] ab, sondern wir sind durch eine unbekannte Ursache genötigt, die Empfindungen nach den Kategorien des Raumes, der Zeit und der Zahl uns vorzustellen und in einen Zusammenhang zu bringen", Weber (1905, p. 11; Eng. tr. p. 146).

[60] "Diese Vorstellungen sind also nicht das Resultat der Erfahrung, sondern Erfahrung wird erst dadurch möglich", *ibidem*.

[61] "Die Bildung von Vorstellungen oder auf die Hervorbringung von Bewegungen beziehen, so kann man jene Seelenanlage auch als einen intellektuellen Instinkt bezeichnen", Weber (1905, p. 12; Eng. tr. p. 146). It is worth noting that animals do not possess an intellectual instinct, but only a certain form of instinct that enables them to form perceptual space. That animals have a notion of space is evident through observation of their manifest behavior. What animals lack is the intellectual notion of space.

criteria by which we estimate filled space (*erfüllten Raum*)".[62] It follows that, since our sensory viewpoint depends on the smallest standards that we have for measuring space, the square inch of a touch organ becomes the standard by which we estimate the spatiality of objects—and not only.[63] Given that we are always somehow conscious of the positions of our sensory circles and of the distance between them, we are equipped with a *mapping* of our body that allows us to provide sensations with a spatial quality. This mapping is not acquired through experience, but rather given congenitally by the body itself.[64] Unlike Kant, who considered space a pure form of intuition, Weber, in synergy with Müller, attributes it to inborn physiological mechanisms.

Weber contends that the sense of sight, compared to that of touch, displays a much acuter localizing power. This is because the sensory circles of its organ (the retina) are smaller and more compact than those of the skin.[65] Yet, since the concept of sensory circles is always connected to the tactile disposition of a surface, the localizing power of sight depends more on its tactile properties than its visual ones. What type of information do sensory circles immediately convey? Firstly, Weber maintains, the idea of the pure contact (resistance) of a solid body with an undefined form.[66] We immediately perceive solidity, but not necessarily form. Form is provided only by portions of the skin where the size of the sensory circles is so small that the localization sense is acute and hence can discern shapes.[67] To have an exhaustive understanding of the spatiality of an object, it is necessary to combine sensations of localization in the skin with the intentional movements of the limbs. This combination, which is no longer immediate but acquired through experience, gives us notions of form, texture, softness and hardness, distance between objects, and concave or convex surface.[68]

As we have seen, Weber seems to follow in the wake of Müller's nativism, by positing some innate, inborn elements located in the physiological constitution of the perceiver at the origin of spatial representation. He considers extension a function of the number of sensory circles between points of excitation, since each circle is a sort of unit of measure for external space (surface, in particular) embedded in our body. Sensory circles are thus a tool for mapping external space and evaluating its extension in an immediate and *a priori* way.

---

[62] "*Die Zahl der elementaren Nervenfäden, welche auf einem Quadratzoll unseres mit Ortsinn begabten Tastorganes endigen, hat einen Einfluß auf den Maßstab, womit wir den erfüllten Raum messe*", Weber (1905, pp. 70–1; Eng. tr. p. 181).

[63] Weber (1905, p. 71; Eng. tr. p. 182).

[64] Note that this mapping does not include the notion of depth but only surface. What is given immediately is bi-dimensionality, not three-dimensionality. The same applies to Müller: coenesthesia offers the idea of resistance, but not of the depth of one's own body.

[65] Weber (1905, pp. 71–2; Eng. tr. p. 182).

[66] Weber (1905, p. 88; Eng. tr. p. 193).

[67] *Ibidem*.

[68] Weber (1905, pp. 89–91; Eng. tr. pp. 194–5). These notions are the result of an inference (or association). We find a similar idea in Helmholtz's theory.

10.3 A New Twist: Lotze's Theory of Local Signs

Under the impulse of Weber's experimental work and Müller's physiology, the first half of the nineteenth-century became the "age of touch", with not little consequence on the debate on spatial representation. Analysis of tactile sensitivity proved key to ultimately demonstrating the apriority of space and showing the existence of some innate physiological mechanisms underpinning spatially qualified perceptions. Hermann Lotze's (1817–1881) reflection has at times been considered a successful attempt at overcoming this very trend in the name of empiricism[69]—an attempt that also paved the way for other accounts of space representation, such as those of Hermann von Helmholtz and Wilhelm Wundt. Yet though it is undeniable that, by stating that sensations do not immediately "cry" spatial configurations[70] and that the notion of spatiality follows from an additional mental reconstruction, Lotze advocated for a reappraisal of the problem of space, his position can hardly be defined as purely empiricist. Some interpreters have rightfully labelled it as a form of mentalism or constructivism,[71] or highlighted the role of some *a priori* elements in his understanding. It is perhaps now worth investigating Lotze's theory of space in relation to the accounts already analyzed and asking how touch interplays with those aforementioned *a priori* elements.

In a later text, *System der Philosophie* (*System of Philosophy*), published in 1879, we find an interesting remark that can be used as a starting point for our investigation. Implicitly referring to positions like Müller's and Weber's, Lotze claims:

> to some people everything seems especially clear when they allow the mind (*Seele*) to be immediately present in the eye and allow it, *like a groping hand* (*eine tastende Hand*), to grasp immediately there with a thousand nerve endings the single points of color precisely in the position which they simply have there.[72]

For Lotze, theories considering points of stimulation as physiological conveyers of spatial determinations tend to conceive these points as "groping hands",[73] and the sense organs (i.e., the eye for sight, and the skin for touch) as a bundle of micro-hands, each capable of acquiring and immediately localizing the corresponding sensation. This is true of Müller's theory, where the position of the stimulated point on

---

[69] On Lotze's theory of space, see (among others) Woodward (1978); Rollinger (2001); Fisette (2006); Fisette (2014); Centi (2011).

[70] See the brilliant definition William James provides on the subject (James [1887, p. 20]).

[71] Fisette (2014, p. 49).

[72] "Manchen erscheint Alles besonders klar, wenn sie unmittelbar im Auge die Seele vorhanden sein und wie eine tastende Hand sie gleich dort mit tausend Nervenspitzen die einzelnen Farbenpunkte genau in der Lage erfassen lassen, die sie dort eben besitzen", Lotze (1879, § 276, p. 545). Engl. tr. in Rollinger (2001, p. 106) (italics are mine). On Lotze's criticisms of Müller's theory, see Woodward (1975).

[73] It is interesting how the sense of touch is metaphorically used to characterize these positions, reiterating the paradigmatic role this sense played in previous accounts.

the retina or skin immediately communicated to the mind the spatial positioning of the corresponding sensation.

What is wrong with such accounts, for Lotze, is that they disregard a fundamental philosophical problem: the relationship between *extensive* and *intensive*. Indeed, to understand Lotze's psychology of space it is necessary to refer to his philosophical position on the mind, which is intensive, whereas spatial elements are intrinsically extensive.[74] When two points of the retina are stimulated by two luminous sources, the mind is immediately capable of differentiating these sensations in terms of qualitative differences (intensive) but not in terms of spatiality (extensive), since spatial elements are incompatible with the intensive nature of the mind.[75] Thus no immediate correspondence is possible between the two, nor is a relation of the *original-copy* type. Rather, a further *re-creation of space* (*Wiedererzeugung der Räumlichkeit*) is required to reinstate the spatial order between qualitative sensations in the mind.[76]

If Müller could describe retinal sensations as already spatially qualified, it was because he regarded the mind as somehow *extensive*. Since each single nerve projects onto a single point of the brain, the mind becomes a collection of different points, mirroring the points of stimulation on the sense organ. The image on the retina, composed of spatially arranged sensations, is seen by a second eye, the mind, which is morphologically akin to the former.[77] Thus a structural isomorphism among perceptual stimuli, nervous currents, and the contents of the mind takes place. This is why geometrical relations are fully conserved in the transmission of sensations from the sense organ to the brain. Moreover, Müller conceives sensations as presenting themselves to the mind synchronically—not diachronically, as Herbart's psychology contended.[78] He understands them to be projected to the brain as a group without losing their original configuration. Without structural isomorphism, Lotze's theory must account for the spatially minded relationship in a different fashion. Thus, space is not immediately given in sensations but has to be re-created somehow in the mind. To clarify that, we could use the example of two stimuli, A and B. These stimuli have a position in external space, but when they affect the sense organ (the eyes in the case of sight; the skin in the case of touch), they lose their spatial elements and maintain only their intensive qualities A' and B'. To reinstate their spatial relation (for instance, the fact that A is on the left and B is on the right), something has to be added to them (A" and B"). At this point, Lotze introduces the notion of *local signs* (*Lokalzeichens*).

Lotze's theory of local signs is formulated across different texts and reworked over the course of several years, making it far from a consistent and well-defined

---

[74] Rollinger (2001, p. 107).

[75] Lotze (1879, § 276).

[76] Lotze (1852, p. 328); Lotze (1877, p. 346).

[77] Lotze (1852, pp. 327–8).

[78] Isaac (2019, p. 86).

position.[79] It is nonetheless possible to clarify some of its aspects. First of all, what is a local sign? As mentioned above, when two stimuli affect a point on the retina or on the skin, they produce two qualitatively distinct sensations but reveal nothing about their relative positions in space. Yet, they call forth a different and accessory element, the local sign, which is the first step toward re-creating their spatial relation. Local signs of this sort are different according to the sense organ we consider. Regarding sight, Lotze resorts to some observations made by German physicist Jakob Friedrich Fries (1773–1842) in his 1839 treatise *Über den optischen Mittelpunkt im menschlichen Auge* (*About the Optical Center in the Human Eye*),[80] which claimed the existence in the eye of a reflex mechanism of ocular-motor movements (or *tendences* of movement) that defines the coordinates of each point stimulated on the retina. Lotze suggests that there is a central point within the retina, the yellow spot or *fovea centralis*, which displays more sensitivity to light than the other peripheral parts.[81] When it sees an object, the eye automatically moves into a position that allows the light to affect the *fovea*, thereby enabling maximum clarity of vision. It should be noted, however, that ocular movement is not necessary for a visual local sign to arise. It is sufficient, for Lotze as for Fries, that there is a tendency towards a specific movement. In tactile perception, local signs are of a different nature. Instead of calling into play movement or tendences of movement, Lotze turns to Weber's research, claiming that the nerves themselves provide the local sign to tactile sensations.[82]

Up to this point, it seems that Lotze's theory is not that far from traditional nativistic stances. An inborn physiological mechanism, which he calls "local signs", is what provides sensations with spatiality. Indeed, in his 1852 *Medizinische Psychologie oder Physiologie der Seele* (*Medical Psychology or Physiology of the Mind*), he states that a local sign is

> [...] a physical nerve process (*physischen Nervenprocess*), which consistently associates at each point of the nervous system with that variable nerve process that underlies the qualitative content (*qualitativen Inhalt*) of the changing sensations at the same point. The two processes either do not interfere with each other at all, or only to a very negligible extent, and while the mind continues, under the influence of the latter, to form its usual qualitative sensations, each of them is simultaneously accompanied by another excitation (*Erregung*), which, depending on the local sign, determines its later placement at a point in the represented space.[83]

[79] As, for instance, Lotze (1846, pp. 172–190); Lotze (1852, pp. 325–352); Lotze (1856, pp. 330–347); Lotze (1879, pp. 543–573); Lotze (1881, pp. 26–38); Lotze (1873, pp. 315–324); Lotze (1877, pp. 345–365). See Fisette (2014, p. 45).

[80] Fries (1839); Lotze (1937). See Woodward (1978, p. 575).

[81] Lotze (1852, p. 327); Lotze (1879, p. 586). Weber had in turn pointed out this special ability of the eye and also experimentally determined that the *fovea* was about 1/3 Paris lines wide (Weber [1905, p. 13; Eng. tr. p. 147]).

[82] Lotze (1852, pp. 401–2).

[83] "[...] einen physischen Nervenprocess überhaupt bezeichnen, der sich constant für jede Stelle des Nervensystems mit jenem veränderlichen Nervenprocess associirt, welcher an derselben Stelle dem qualitativen Inhalt der wechselnden Empfindungen zugrunde liegt. Beide Processe stören

One could rightfully claim that the idea of a physiological "excitation" determining "its subsequent positioning in space" is something that already characterized Müller's or Weber's positions. Yet, although Lotze seems here to imply that the mapping of our own body via the relative position of the nerves reveals something about the spatialization of sensation, he in fact does not. This becomes even clearer in subsequent formulations of his theory, where his physiological viewpoint gives way to a psychological one.[84] Lotze contends that the eyes' movements (or tendencies) for sight, and the nerves of the skin for touch, are not themselves bestowed with spatial determinations but rather represent merely one of the two components interplaying in re-creation of spatiality. They stand for the physiological part of the local sign, to which another part (psychological) has to be added. Each ocular movement is indeed accompanied by a corresponding *sensation of movement* that is psychological in nature, and the same goes for ocular-motor tendencies (*Bewegungstriebe*), which are accompanied by corresponding *sensations of tendency*.[85] For touch, on the other hand, the psychological part is played by the *sensory circles* sensitized by the tactile nerves: cutaneous space (extension of the skin) is perceived thanks to the local signs stemming from stimulation of the sensory circles or from sensations of distance between sensory circles.[86]

Lotze's similarity with Müller and Weber fades away when we recognize that space is not immediately given in sensory circles or sensations of ocular movements. Rather, the leading protagonist in space perception is the mind (or consciousness). Local signs are a special "coloring" (*Färbung*) that accompanies qualitative impressions and supplements them.[87] This coloration should be considered a sensation *sui generis* which directly affects the mind and only there re-acquires its spatial determinations. Nothing in the sensory circles or movements of the eyes can be

einander entweder gar nicht, oder in höchst unbeträchtlichem Masse, und während die Seele fortfährt, unter dem Einflusse des letzteren ihre gewöhnlichen qualitativen Empfindungen zu bilden, wird jede von ihnen zugleich von einer andern Erregung begleitet, welche abhängig von dem Localzeichen ihre spätere Einordnung an eine Stelle des vorgestellten Raumes bedingt", Lotze (1852, p. 331).

[84] Over the years Lotze becomes aware that his theory of local signs risked being interpreted as a form of associationism (qualitative sensation associates with a physiological local sign). This view, although easier to understand, would require a double conduction of two impressions along the same nerve fiber. This is why he subsequently made clear that a local sign is psychological, not physiological (Lotze [1856; Eng. tr. pp. 311–2]; see Woodward [1978, p. 574]). Though we tend to overstate the difference between these "two" formulations of Lotze's theory—following in the wake of Stumpf's (1893) reappraisal of this theory –, Lotze's account is quite inconsistent and "volatile", making it difficult at times to understand if he conceives local signs as physiological or psychological.

[85] Lotze (1852, pp. 333, 370, 444); Lotze (1877, pp. 358–9).

[86] Lotze (1852, pp. 402–5). Woodward (1978, p. 574).

[87] "so muss jede Erregung vermöge des Punktes im Nervensystem, an welchem sie stattfindet, eine eigentümliche Färbung erhalten, die wir mit dem Namen ihres Localzeichens belegen wollen"; "Every excitation must by virtue of the point in the nervous system in which it occurs receive a peculiar coloring to which we want to give the name of its *local sign*" (Lotze [1852, p. 331]; Eng. tr. in Rollinger [2001, p. 108]).

qualified as utterly spatial, since it is only the mind that reinstates the spatial configuration originally displayed by the stimuli coming from an object.

The idea of resorting to consciousness to explain space perception raises various questions. How is it possible that the extensive (spatial) properties of objects are reinstated by the intensive mind? To solve this philosophical conundrum, Lotze appeals to an *innate disposition* of the mind. As he states in *Medizinische Psychologie* (*Medical Psychology*), "For all our physiological considerations, it is sufficient to conceive that spatial perception is a possession originally and a priori belonging to the nature of the mind, which is not produced by external impressions, but is only provoked for certain applications".[88] Similarly, in an 1877 article in *Revue philosophique*, he contends that

> [...] assuming as given, in the nature of mind, the faculty and determination of this disposition of sensations, it is a matter of simply knowing how the mind assigns, in this intuition of space which is necessary to it, to each of its sensations its determined place, in correspondence with the object which is its cause. Our *theory of local signs* is intended to answer this [...] question.[89]

A form of innatism underpins Lotze's account. The general and *a priori* intuition of space that he attributes to the mind goes hand in hand with an "applied" conception of space, aimed at localizing sensations. Lotze underscores that this distinction should not be taken literally—as if it existed in an infinite, empty, and three-dimensional space where we arrange sensations.[90] Still, his theory relies on the innate tendency of the mind to arrange sensations in a geometrical order. Indeed, from an epistemological standpoint, Lotze's psychology tends to attach major importance to the notions of *relation* and *order*, rather than to the traditional distinction between object and subject. For Lotze, the primary function of consciousness is thus to order its contents, since the elements of mental life (sensations, feelings, movements) acquire meaning only if arranged into certain configurations—first and foremost spatial ones.[91]

It should be noted that Lotze does not deny local signs a role in the process of re-creating space—although he assumes that such an operation is a specific disposition of the mind. Local signs explain how the mind, already imbued with an innate intuition of geometrical order, assigns each sensation a determined place. We could say that they work as "occasions" for the mind to reinstate spatial

---

[88] "Für alle unsere physiologischen Betrachtungen reicht die Vorstellung hin, dass die Raumanschauung ein der Natur der Seele ursprünglich und a priori angehöriges Besitzthum sei, das durch äußere Eindrücke nicht erzeugt, sondern nur zu bestimmten Anwendungen provoziert wird", Lotze (1852, p. 335).

[89] "[...] supposant comme données, dans la nature de l'âme, et la faculté et la détermination de cette disposition des sensations, est simplement de savoir comment fait l'âme pour assigner dans cette intuition de l'espace, qui lui est nécessaire, à chacune de ses sensations sa place déterminée, en correspondance avec l'objet qui en est la cause. C'est à cette seconde question seulement que nous prétendons répondre par notre *théorie des signes locaux*", Lotze (1877, p. 352).

[90] Lotze (1877, pp. 364–5).

[91] Lotze (1852, p. 325).

determinations—which is why Lotze's theory has been described as a form of occasionalism (meaning this term in a very broad and wide sense).[92] Sensations of eye movements contain a *direction* which informs the mind of the location of a visual sensation; sensory circles and their distances perform a *mapping* of the relative locations of tactile impressions. Touch and sight are the two senses that make the mind's spatial arrangements of sensations possible. In this respect, Lotze, like many thinkers before him, tends to stress the greater accuracy of sight compared to touch: the localizing power of sight is more acute, since it is not subject to the variability of sensory circles—as Weber had shown.[93] For Lotze, bidimensional space is thus the first information that local signs provide to the mind. As in his description of depth perception, he seems to be in line with both British empiricism and Müller, since he conceives depth as obtained by the association of local signs with movements.[94]

Setting aside their different levels of accuracy, touch and sight appear to have the same weight in Lotze's understanding of spatial perception. By contrast, it is less clear what kind of sensation he thinks that local signs entail. As we have seen, Lotze conceives sensations of movement, sensations of tendency, and sensations of sensory circles as qualifying space. But what are these sensations, specifically? At first glance, local signs seem mostly related to the notion of *movement*. Indeed, it is the movement of the eye to allow light to hit the *fovea* which brings out the local sign in the sense of sight, just as it is sweeping movement on the skin which activates its different sensory circles and provides their relative positions in the sense of touch. Yet, this is not what Lotze claims. He instead argues that local signs are not primarily awakened by movements and that sensations of movement are not necessarily connected to actual motor activity. Since to see an object would require the execution of several movements at once, which is not possible, spatialization should occur even when the eye is stationary. This is why Lotze insists on the notion of *tendencies*. Likewise, sensations related to the relative distance between sensory circles do not necessarily require sweeping movement on the skin; a simple contact (or pressure) is sufficient. Instead of overstating the role of muscular sensations—as would occur in subsequent psychological debates –,[95] Lotze first appeals to some *unconscious elements* as the primary components of local signs,[96] then ultimately rejects this hypothesis.[97] By reading what he writes in 1877, it becomes clear that he considers local signs not linked to innate or inborn notions, or to the unconscious power of the mind, but rather to some properties of the nervous system:

> We do not claim that they [local signs] can always be regarded as sensations of the same nature as the other main sensations […] to which these signs are added; rather, they will

[92] Woodward (1978, p. 577).

[93] Lotze (1877, p. 364).

[94] Lotze (1852, pp. 419–20). See Hatfield (1990, p. 162).

[95] For a history of the muscular sense, see Smith (2019).

[96] Lotze (1852, pp. 336–7).

[97] Lotze (1877, p. 360).

> most often resemble those sensations of fatigue, languor or vigor which accompany the exercise of our activity, and which, whatever they may be, dark or obscure, always constitute the affections we experience, the ways of being which indicate what is, at a given moment, the state of our well-being.[98]

Lotze thus affirms that local signs are a form of *coenesthesia*. With the term coenesthesia, psychological and philosophical reflection had for a long time indicated the fundamental and obscure feeling of the individual's bodily constitution,[99] a feeling different from muscular and voluntary activity, which was referred to as kinesthesia. It is interesting to note that Müller and Weber had already described this feeling of embodiment as a *Gemeingefühl*, a common sensibility, or internal touch. Lotze does not go so far as to define local signs a form of internal touch. Yet, after criticizing positions framing apprehension of space as a form of touch, he ends up qualifying local signs as coenesthetic sensations.

10.4 Touch and Movement: Wundt's Theory of Space

The debate between nativism and empiricism/constructivism on the issue of space can be restated as an opposition between the belief that sensations immediately articulate their spatial disposition to the mind and the belief that it is the mind which superimposes a spatial configuration on these sensations. Indeed, the real point of contention lies in the concept of ordering sensations. Constructivist thinkers, such as Lotze, attach great importance to this concept, and not by chance: they interpret the relationship between the mind and its sensory contents as analogous to the classical opposition between form (*morphé*) and matter (*hyle*), such that the "formal" notion of order can be attributed only to the mind, not to sensory material (which is raw and unorganized). On the other hand, nativist thinkers, though they do not deny space the property of arranging sensations—in Müller's theory, for instance, we find something resembling an organic *morphé*, or an embodiment of spatial configurations—do not stress this concept.

It thus should not come as a surprise that, when Wilhelm Wundt (1832–1920) discussed the theory of spatial perception in his 1896 *Grundriss der Psychologie* (*Outline of Psychology*), his distinction between intensive and extensive ideas pivoted precisely on the notion of *order*. In that text, he claims that "[...] spatial [...] ideas are immediately distinguished from intensive ideas by the fact that their parts

[98] "Nous n'affirmons pas qu'on puisse toujours les regarder comme des sensations de même nature que les autres sensations principales [...] auxquelles ces signes s'ajoutent; ils ressembleront plutôt le plus souvent à ces sensations de fatigue, de langueur ou de vigueur, qui accompagnent l'exercice de notre activité, et qui, quelles qu'elles soient, claires ou obscures, constituent toujours des affections que nous éprouvons, des manières d'être qui indiquent quel est, à un moment donné, l'état de notre santé", Lotze (1877, p. 349).

[99] This notion was crucial in Condillac's theory of sensation, as well as in the theories of many other thinkers. For a historical overview of this concept, see Starobinski (1977).

are united, not in an arbitrarily variable, but in a definitely fixed order (*fest bestimmten Ordnung*) [...]. Ideas with such a fixed arrangement are called in general *extensive ideas* (*extensive Vorstellungen*)".[100] The fixed order or arrangement to which Wundt refers is something that holds only for the relation among parts, not for their relation to the perceiver. Actually, he considers this latter relation rather varied, since the parts of spatial objects change their positions according to the perceiver's perspective. There is, Wundt claims, a "movability and rotability" (*Verschiebbarkeit und Drehbarkeit*) of spatial compounds in relation to the subject.[101] Such variation is not unlimited, however, since the directions in which these movements and rotations can take place are only three (i.e., those of the three-dimensional space).[102] A spatial structure (*Raumgebilde*), by which he means extensive idea, is therefore composed of different parts which display an order that, albeit fixed and immutable *per se*, vary in relation to the perceiver.

Of these two orders (fixed and variable), for Wundt the latter is most important. It is indeed this fluctuating order which tells us that it is

> [...] psychologically impossible that the arrangement of the elements in such an idea should be an original attribute of the elements themselves, analogous to the intensity or quality of sensations: it requires rather that this arrangement (*Zusammensein*) should result from the bringing together of these elements, and should arise from some new psychical conditions (*neu psychischen Bedingungen*) that come with this coexistence.[103]

Here Wundt makes a double theoretical move: on the one hand, by claiming the intervention of "new psychical conditions", he appeals to the theory of local signs[104]; on the other hand, the idea that spatial arrangements result from the *bringing together* of sensations suggests that his adhesion to Lotze's psychological agenda is not unconditional. Indeed, Wundt's stress on the idea of a *process* bringing elements

---

[100] "Von den intensiven unterscheiden sich die räumlichen und zeitlichen Vorstellungen unmittelbar dadurch, dass ihre Teile nicht in beliebig vertauschbarer Weise, sondern in einer fest bestimmten Ordnung miteinander verbunden sind, so dass, wenn diese Ordnung verändert gedacht wird, die Vorstellung selbst sich verändert. Vorstellungen mit solch fester Ordnung der Teile nennen wir allgemein extensive Vorstellungen", Wundt (1896, p. 120; Engl. tr. p. 102). The difference from Lotze's position (which considered sensation intensive, not extensive) is only apparent. What Wundt is saying is that spatial ideas, once formed in the mind from intensive sensations, can only be extensive. This transition from the intensive to the extensive thus requires the intervention of the *mind* (just as Lotze believed it was necessary to postulate a *re-creation* of space).

[101] Wundt (1896, p. 121; Engl. tr. p. 102).

[102] *Ibidem*.

[103] "Dieses in Wirklichkeit in allen räumlichen Vorstellungen zugleich gegebene Verhältniss zu dem vorstellenden Subjecte schließt von vornherein die psychologische Forderung ein, dass die Ordnung der Elemente in einer solchen Vorstellung nicht eine ursprüngliche Eigenschaft der Elemente selbst, analog etwa der Intensität oder Qualität der Empfindungen, sein kann, sondern dass sie erst eine Folge des Zusammenseins der Empfindungen ist, die aus irgendwelchen durch dieses Zusammensein neu entstehenden psychischen Bedingungen entspringen", Wundt (1896, pp. 121–2; Engl. tr. p. 103).

[104] On Wundt's debts to Lotze, see Hatfield (1990, p. 175); Fisette (2014, p. 56). However, scholars have not devoted much attention to Wundt's theory of space, mostly conceiving it as a mere reworking of Lotze's.

together and producing spatial ideas is what most distinguishes his account from Lotze's. As we have seen, Lotze considered local signs *occasions* for the mind to spatially arrange sensations. Wundt similarly bestows the mind with the ability to form spatial ideas, but, in accordance with his psychophysical parallelism, he cannot interpret local signs as occasions.[105] Instead, he focuses on the psychic processes (which he calls *fusions*, *Verschmelzungen*) generating spatial representations from simple mental contents (qualitative sensations).

This theoretical position is reinforced by the definition he gives of his theory of space: "genetic theory" (*genetischen Theorie*).[106] To use a quite effective description of this theory developed some years later by Franz Brentano (1838–1917), genetic psychology is a type of psychology that "[…] informs us about the laws according to which phenomena arise and vanish", and "[…] since the conditions under which phenomena arise, because of the undeniable dependence of psychical functions on processes taking place in the nervous system, are largely physiological, it becomes clear how psychological research here must be intertwined with physiological research".[107] Wundt's theory falls perfectly under this characterization, in that it is aimed to investigate the laws governing the rise and disappearance of psychic phenomena and their physiological counterparts.

Yet Wundt was not the first to have attempted a genetic appraisal of space perception. Before him, thinkers such as Johann Friedrich Herbart and Alexander Bain (1818–1903) had given genetic accounts of the problem, with differing results.[108] In Herbart's theory, space is conceived as the product of an intellectual *fusion* combining a series of sensations (regarded as discrete, punctiform units) into a single representation.[109] Sensations of this sort are provided by two main senses: touch and sight. Indeed, when we move our eyes or hands around an object in space, we experience a series of nonspatial presentations. The strongest among them is the one currently being perceived, since only one representation can be consciously apprehended by the mind at a time. Thus, for Herbart mental life is a battle among representations aspiring to rise above the threshold of consciousness. By reversing the course of eye or finger movements, we obtain an inverse series of past representations—and thus acquire a representation of their spatial simultaneity, and ultimately of space. Bain's position is somewhat similar to Herbart's, but, instead of deriving space from sight and touch alone, he contends they must be combined with another

[105] Woodward (1978, p. 577).

[106] Wundt (1896, pp. 135–6; Engl. tr. p. 115).

[107] "Da die Bedingungen wegen der unleugbaren Abhängigkeit der psychischen Funktionen von den Borgängen im Nervensystem großenteils physiologische sind, so sieht man, wie hier die psychologischen Untersuchungen mit physiologischen sich verflechten muffen", Brentano (1895, p. 34).

[108] Hermann von Helmholtz would in turn develop a genetic and empiricist theory of space perception. As said in the introduction, I will not delve into that reflection here, because it is almost entirely centered on optic perception. Like Wundt's, Helmholtz's position is largely considered a reworking of Lotze's theory of local signs, with the additional idea that sensations are uninterpreted *signs* or *symbols* whose meanings must be discovered through experience.

[109] Herbart (1816).

sense modality: the *muscular sense*.[110] Combining tactile-visual sensations with corresponding sensations of movement is what gives us the ideas of permanence, stability, and coexistence we usually attribute to space.

These genetic theories provide Wundt some interesting insights, such as the idea that space depends on a psychological fusion of discrete elements and the notion of the centrality of the muscular sense to space perception. This does not mean however that his position overlooks or denies the supremacy of sight and touch in spatial perception. On the contrary, he accepts the idea, stated by Müller, that touch and sight are privileged sense modalities because of the "extended surface of [their] peripheral sense-organs"—to which he adds "their equipment with organs of movement".[111] The focal point of his position is hence the idea that space perception is a *fusion* of the data coming from these two modalities, which seems to suggest the necessity of a multi-sensorial experience. Does this mean a blind person is deprived of the idea of space? Certainly not. Wundt's psychology is no doubt adult-centric, paradigmatically focusing on mental operations as they occur in unimpaired adult subjects—and thus shows little interest in pathological, evolutive or comparative psychology. Nonetheless, Wundt does also account for space perception in blind individuals, but since his characterization of space is mainly drawn from observations of individuals with sight, he tends to prioritize a multi-sensorial definition of space.

Although the sense of touch appears earlier in the development of organisms and shows the structural relations between sensations in much coarser, and hence plainer, form,[112] Wundt affirms that

> the localization of tactual impressions (*Localisation der Tasteindrücke*) and the spatial arrangement (*räumliche Ordnung*) of a number of such impressions [...] presupposes spatial ideas of sight, which can be made use of, however, only because the various parts of the tactual organ have certain qualitative attributes, local signs, which arouse for the visual image of the part touched. [...] We may describe the formation of spatial ideas in this case as the arrangement of tactual stimuli in visual images already present.[113]

---

[110] Bain (1855, pp. 192–5). The history of the muscular sense is particularly articulate and interesting. See for instance: Forest (2004); Smith (2011); Smith (2019).

[111] "Bei dem Tast- und Gesichtssinn aber sind offenbar schon durch die flächenförmige Ausbreitung der peripheren Sinnesorgane und durch die Ausstattung dieser mit Bewegungsapparaten, die eine wechselnde Orientierung der Eindrücke zum vorstellenden Subjekte möglich machen, günstige Bedingungen zu einer extensiven räumlichen Ordnung der Empfindungen gegeben", Wundt (1896, pp. 122–3; Engl. tr. p. 104).

[112] *Ibidem*.

[113] "Die Lokalisation der Tasteindrücke und mit ihr die räumliche Ordnung einer Mehrheit solcher beruht, wie die oben geschilderte Mitwirkung der Gesichtsvorstellungen der betasteten Körperteile lehrt, beim sehenden Menschen weder auf einer ursprünglichen Raumqualität der Hautpunkte noch auch auf einer primären raumbildenden Funktion des Tastorgans, sondern sie setzt die räumlichen Vorstellungen des Gesichtssinns voraus, die aber nur dadurch wirksam werden können, dass den Teilen des Tastorgans selbst gewisse qualitative Eigenschaften, die Localzeichen, zukommen, welche die Gesichtsvorstellung des berührten Teils erwecken", Wundt (1896, pp. 124–5); Engl. tr. pp. 105–6).

In sighted persons, the localization of tactile impressions and their spatial ordering is not based on an original spatial quality of the skin or on a primary spatial function of the tactile organ, but rather on spatial representations provided by the sense of sight. In this sense, it is as if the visual field worked as a grid on which tactile sensations are arranged and acquire spatial configurations. Yet, Wundt claims, spatial representations of sight *become active* only because parts of the tactile organ possess certain qualitative properties, namely local signs, which awaken the visual representation of the part touched. More than granting priority to sight—as Wundt initially did—it seems that here there is rather a sort of *operative circularity* between sight and touch, as far as spatial perception in sighted, adult, and unimpaired persons is concerned. In fact, local tactile signs act as qualitative signals of visual images, since tactile and visual sensations are firmly associated with experience.[114] Neither local tactile signs nor visual images are endowed with spatiality; rather, spatial representations are produced by the ordering of tactile sensations within visual images. Thus, for Wundt, qualitative sensations always need an accessory element in order to be spatially ordered; they themselves lack such an arrangement *per se*. In the tactile localization of the sighted person, the accessory element that joins the local tactile signs is the visual image. In the blind person, the role of this accessory element must be fulfilled by something else.

It is at this point that the muscular sense is called into play. In order for a blind person to perceive space, their tactile data must combine, in a psychic fusion, with sensations of movement.[115] In the blind the magnitude and direction of these movements are given by an inner tactile sensation (*inneren Tastempfindung*),[116] which is not immediately linked to a representation of space. The point for Wundt is that two different tactile sensations, *a* and *b*, can be qualitatively differentiated (as Lotze maintained) but not spatially arranged. The spatial arrangement of *a* and *b* is not even immediately given by local signs, since it would require the possession of an innate, *a priori* intuition of body space—something Wundt does not recognize.[117] To determine their position, the transition from *a* to *b* must be accompanied by an internal tactile sensation α.[118] In order to understand this passage of Wundt's reflection, we could imagine that we press two different cutaneous points on a blind person's back (see Fig. 10.1). Nothing in these tactile sensations and the corresponding local signs (*a'* and *b'*) informs the individual that sensation *a* is, for example, on the right and sensation *b* on the left. Sensory circles, unlike what affirmed by Weber and Müller, contain no such indications. We are inclined to think that these sensations immediately communicate their position because the blind person, who has long moved their body, is able to tell immediately that *a* is to the right and *b* to the left, but this is an acquisition that they now take for granted, since it has become habitual

[114] Wundt (1896, p. 125; Engl. tr. p. 106).
[115] Wundt (1896, p. 126; Engl. tr. p. 107).
[116] Wundt (1896, p. 128; Engl. tr. p. 108).
[117] Wundt (1896, p. 122; Engl. tr. p. 103).
[118] Wundt (1896, pp. 128–30; Engl. tr. pp. 109–10).

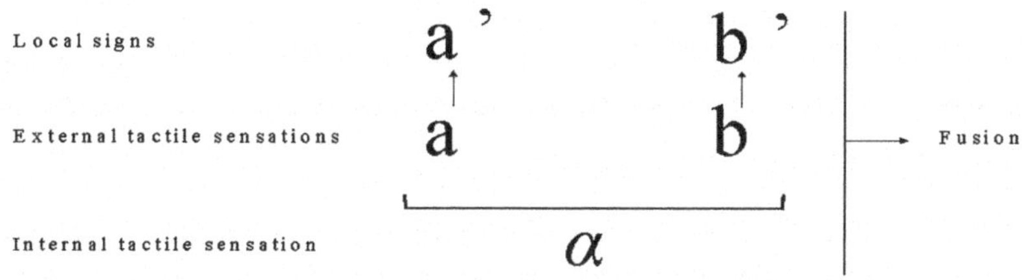

Fig. 10.1 Spatial perception in blind persons according to Wundt's theory

through experience (they have built up a spatial representation of their body and are thus able to readily locate sensations). However, to know originally that *a* is to the right of *b* they had to travel the distance between these points through movement, which is to say, they had to receive concomitant internal tactile sensations. The fusion of external tactile sensations and their local signs with internal tactile sensations is what produces the representation of space for Wundt. Once this fusion has occurred, each element is capable of determining location on its own. This is why, when their body is at rest, the blind person possesses a representation of the location of the contact or can perceive two impressions as separate in space. In this case, the visual image of the place touched does not arise, but rather the spatial location is presented to them as the representation of a movement in the form of a memory image (*Erinnerungsbild*).[119] The opposite can also occur: we can have internal tactile sensations without external tactile sensations, for instance, when we move our arms with our eyes closed and still have a clear representation of their position.[120]

Setting aside the perception of space in the blind, Wundt contends that, in individuals with sight, the localization of tactile impressions and their spatial ordering is founded on the act of arranging tactile sensations in the visual field. It is worth lingering for a moment, however, on what he means by "sense of sight". The following phrase is found at the end of the section *Spatial Ideas*, stating that "another fact against the nativistic and in favor of the genetic theory is the development in the child of the synergy of ocular movements under the influence of external stimuli and the organization of space-perceptions which apparently accompanies it".[121] Wundt understands sight as a complex system, apt for receiving light stimuli and emitting movements in reaction to them. Indeed, he finds the main mistake of some classic empiricist positions to be limitation of the study of vision to the *optical domain*

[119] Wundt (1896, pp. 130–1; Engl. tr. p. 111).

[120] *Ibidem*.

[121] "Nicht minder spricht für die genetische und gegen die nativistische Theorie die Tatsache, dass sich beim menschlichen Kinde die Synergie der Augenbewegungen unter dem Einfluss der Lichtreize entwickelt, und dass damit die Ausbildung der räumlichen Wahrnehmungen Hand in Hand zu gehen scheint", Wundt (1896, p. 166; Engl. tr. p. 141). Interestingly enough, although his psychology is adult-centric and shows little interest in evolutive psychology, Wundt resorts here to the development in the child of the synergy of ocular movements in order to corroborate his position.

alone, without extending the analysis to the *sensory-motor sphere*.[122] As a matter of fact, in Wundt's theory sight is considered in analogy with touch. It is a sort of optical touch: the entire retinal surface behaves like a single tactile region.[123] The difference is that in touch there are many parts of the skin endowed with the capacity to make refined distinctions, while in vision there is only one region: the central visual point, i.e., the retina. The smallest visual angle ranges from 4 to 6 thousandths of a mm, which is roughly the diameter of a retinal cone.[124] So for Wundt, two light impressions must always fall on two different points of the retina to be spatially distinct. Visual acuity is thus determined by the higher density of retinal elements (*Dichtigkeit der Aneinanderlagerung der Netzhautelemente*) with respect to the density of tactile sensory circles.[125] Likewise, as in the case of touch, representation of the distance between two luminous points depends on the *motor energy* or *effort of movement* (*Bewegungsanstrengung*) employed by the eye in passing through this distance.[126]

A certain theoretical tension animates Wundt's theory. He certainly repeatedly affirms the centrality of vision in space perception, going so far as to state that "[…] in normal persons with vision, visual space-perception determines tactual, not the reverse" and that "[…] the fact of general development [phylogenesis], that touch is the more primitive sense, cannot be applied to the development of the individual [ontogenesis]".[127] Yet, the primacy of sight is not as definitive in Wundt as it may seem. Firstly, as mentioned above, Wundt mainly considers the formation of space perception in unimpaired adults, thereby prioritizing a specific kind of psychic fusion, i.e., arrangements of tactile sensations in the visual field. Secondly, he largely conceives the sense of sight in analogy with the tactile domain. If anything, one could argue that the muscular sense is primary in Wundt's analysis.[128] Perhaps better, would be to say that

"Wundt's hypothesis boils down to the idea that our representation of space is in fact the result of the synthesis by association of motion sensations and local signs. Hence, in his psychology, tactile representations and representations of movement are privileged over all other sensory representations, and tactile localization is given pride of place in his own theory of local signs".[129]

---

122 Wundt (1896, p. 165; Engl. tr. p. 140).

123 Wundt (1896, p. 139; Engl. tr. p. 117).

124 Wundt (1896, p. 139; Engl. tr. p. 118).

125 Wundt (1896, p. 149; Engl. tr. p. 127). This is an idea we have already found in Müller and Weber, and that Wundt borrows from them.

126 Wundt (1896, p. 143; Engl. tr. p. 121).

127 "[…] dass beim sehenden Menschen die räumlichen Wahrnehmungen des Gesichtssinns für die des Tastsinns bestimmend sind, nicht umgekehrt. Die Tatsache der generellen Entwicklung, dass der Tastsinn der früher ausgebildete Sinn ist, lässt sich also hier nicht auf die individuelle Entwicklung übertragen", Wundt (1896, p. 165; Engl. tr. pp. 140–1).

128 On Wundt's conception of the muscular sense, see Ross & Bischof (1981); Forest (2004, pp. 7, 21, 24); Smith (2011); Smith (2019, ch. 9).

129 Fisette (2014, p. 56).

10.5 De-Subjectivizing the *A Priori*: Stumpf's Phenomenology of Space

The dichotomy between *morphé* and *hyle*—which led Lotze and Wundt to attribute the ability to spatially arrange sensations to the mind—is challenged and definitively overcome by German psychologist Carl Stumpf (1848–1936). Indeed, in his thought *morphé* is no longer assigned to consciousness or to the anatomical configuration of the sense organs (as in Müller and Weber), but instead attributed only to the *sensory domain.*[130] More precisely, the sensory domain presents itself to the perceiver as an indissoluble mixture of *morphé* and *hyle*[131]—an already structured content upon which there is no need to superimpose form or order, above all a spatial one. In this respect, Stumpf's position can be regarded as a type of *nativism*,[132] since it does not maintain that a spatial arrangement must be added to the sensory sphere through a psychic operation (as in empiricist, constructivist, and mentalist accounts).[133] Yet in Stumpf we are confronted with a form of nativism that cannot be equated to the one characterizing the German physiological debate in the first half of the nineteenth century. Stumpf's nativism is in fact *sui generis*; while embracing different suggestions and positions, it results in a novel understanding of space perception. A student of Brentano in Wurzburg[134] and of Lotze in Göttingen,[135] Stumpf owed his two mentors more than one insight. Yet, though Stumpf's 1873 *Über den psychologischen Ursprung der Raumvorstellung* (*On the Psychological Origin of the Concept of Space*) is dedicated to Lotze, it shows a much tighter connection to Brentano's theory of space, and also embraces the phenomenological approach introduced by the latter.

To understand Stumpf's theory of space[136] a useful point of departure is his theory of sensible appearances (*Erscheinungen*). Stumpf conceives the sensible as independent from the psychological realm, since it represents a pre-predicative domain manifesting laws and structures of its own. At the same time, sensory contents are not considered, as in Herbart's or Wundt's empiricism, punctiform and atomistic elements which combine together according to associative principles, but

[130] Stumpf (1906, p. 4). For an overview on Carl Stumpf's psychology, see, among others, Lewin (1937); Spiegelberg (1960); Ash (1980); Sprung (2006); Martinelli (2009); Bonacchi & Boudewijnse (2011); Fisette & Martinelli (2015).

[131] I borrow this terminology from Pradelle (2015, pp. 231–3), who in turn draws from a Husserlian vocabulary.

[132] A better label might be "phenomenology", which Stumpf himself preferred to "nativism". Yet, it is not erroneous to define Stumpf's space theory as nativistic.

[133] He sharply distinguished phenomena (*Erscheinungen*) from psychic functions (*psychische Funktionen*) (Stumpf [1906, pp. 4–5]; Stumpf [1924, p. 40]. See also Pradelle [2015, p. 232]).

[134] On their relationship, see Stumpf (2015).

[135] See Milkov (2015); Milkov (2023).

[136] On Stumpf's theory of space see, among others, Woodward (1978); Costa (1996); Popescu (2003); Centi (2011); Pradelle (2015); Milkov (2015); Textor (2017).

rather as structured wholes.[137] Following in the wake of Brentano's, Stumpf's phenomenology is conceived as a science of the structural forms of sensory objects,[138] since it holds these structural forms to be immanent to sensory materials and not added later by consciousness. Among these structural forms is of course space, which proves to be an intrinsic and *a priori* property of the percept.[139] In this respect, as scholars have noted, Stumpf's attempt to overcome empiricism by moving toward apriorism should not be misread as a disguised adhesion to Kant's agenda.[140] For Stumpf, the notion of *a priori* pertains directly to the percept and is not a formal condition of knowledge. In other words, he denies the linkage between apriorism and subjectivity that Kant posited by suggesting that the formal properties no longer belong to the subject and its pure cognitive forms but to the sensory domain. This de-subjectivation of the *a priori*[141] entails a dislocation of the viewpoint from the subject to the percept. Stumpf's position is thus a form of intuitionism, since it advocates for an appraisal of what is immediately and actually given in perception.

Attention to the "given" is particularly important to Stumpf's understanding of space perception. When Stumpf describes space, he intentionally takes into consideration only the pre-scientific and pre-cognitive notion of it, namely space as it is immediately displayed in intuition.[142] Empiricist theories tend to work with a more abstract, artificial, and complex concept, such that spatial determinations of distance, place, position, size, direction, and one's body location, get treated as independent properties, i.e., as stemming from different psychic activities. Yet, if considered as a content of intuition, Stumpf claims, space is all of these things all at once.[143] Furthermore, space is not, properly speaking, detachable from sensory qualities.[144] In our intuition of space, a quality of color is not separable from the property of extension—an idea already present in Berkeley, and much later in Brentano[145], which would come to have a major influence on Husserl's phenomenology.[146] Color and extension always manifest themselves together as a unit, making the very idea of a color without extension purely nonsensical.

Since space for Stumpf is perceptually given as a whole and is also inseparable from qualitative sensations, it is clear that its apprehension is not achieved by one faculty alone but rather co-perceived by various senses.[147] However, co-perception

[137] Stumpf (1906, p. 4).

[138] Stumpf (1924, pp. 34–40).

[139] This idea was developed at length in Stumpf (1873).

[140] Pradelle (2015, pp. 234–7).

[141] Pradelle (2015, p. 237).

[142] Stumpf (1873, p. 2).

[143] *Ibidem.*

[144] Stumpf (1873, p. 6).

[145] Berkeley (1709). On Brentano's psychology, see Antonelli & Boccaccini (2021).

[146] On Stumpf's influence on Husserl, see Costa (1996); Popescu (2003); Fisette (2009); Ierna (2009); Dewalque (2012).

[147] Stumpf (1873, p. 6).

does not mean that spatial representation arises solely when multiple senses are simultaneously stimulated—albeit this is what usually happens in ordinary perception. Rather, co-perception means that space is a cross-sensory notion, indissolubly linked to sensations grasped by different senses. It is a sort of a *common sensible*, not because it derives from the association of various sensory data (Stumpf is very much critic of associationism), but because it is a kind of *aisthesis* that can be apprehended by multiple sensory modalities—as already affirmed by Aristotle and Locke, and reinstated by Müller.[148] Here is what Stumpf claims in *Raumvorstellung*:

> [...] space is not perceived by a sense dedicated specifically to it, but co-perceived (*mit wahrgenommen*) by different senses; through the eye we perceive not only colors, [...] but also location, size, etc. Normally, one determines the unity or diversity of the senses psychologically according to the diversity of sensory contents (*Sinnesinhalte*) (it is impossible to classify color and sound under one category, whereas it is possible with individual colors, red, green). It seems, however, that space can no more be brought under one class with the quality of color than the latter can with sound. So how is it possible to attribute completely different contents to one and the same sense, and to attribute one and the same content to two different senses?[149]

Although each sensory modality displays its own specific qualitative content—as sight, for instance, cannot perceive sounds or smells—various senses have access to spatial properties. This is because each sensory quality (color, sound, cold, etc.) is indissolubly and immediately linked to spatial configurations. That being said, for Stumpf, as for many others before him, it is mainly the sense of touch and the sense of sight that provide us a privileged access to space.[150] The idea of co-perception does not prevent Stumpf from acknowledging the primacy of some senses over others. Yet if Bain insisted on the centrality of the muscular sense, Müller and Weber on touch, and Wundt on the *nexus* touch-movement, Stumpf's theory posits a certain "equality" between touch and sight. In this sense, the "age of touch" comes to an end with this thought. His position however is still relevant for our analysis because it not only clarifies the end point of a longstanding debate on the apriority of space—which would afterwards take on different connotations and acquire a quite different physiognomy, especially with the advent of *Gestaltpsychologie*—but also

---

[148] On Aristotle's theory of *common sensibles* and its role in Müller's reflection, see footnote n. 23 of this chapter. As for Locke, both sight and touch offer equal access to the property of "form". In general, all spatial determinations—extension, shape, size, etc.—are for him cognitively given by both touch and sight (Locke [1786, vol. 1, pp. 90, 127]).

[149] "[...] Der Raum wird nicht durch einen ihm eigens gewidmeten Sinn erfasst, sondern bei einzelnen Sinnen mit wahrgenommen; wir erfassen durch das Auge nicht bloß Farben, obgleich sie den eigentümlichen Sinnesinhalt bilden, sondern mit der Farbe auch Ort, Größe usw. Nun bestimmt man gewöhnlich Einheit oder Verschiedenheit der Sinne psychologisch nach der Verschiedenheit der Sinnesinhalte (es ist unmöglich, Farbe und Ton usw. Unter eine Gattung zu bringen, während es bei den einzelnen Farben, Rot, Grün, möglich ist). Es scheint aber, dass Raum ebensowenig mit der Farbenqualität unter eine Gattung zu bringen ist, wie diese mit dem Ton. Wie kommt es also, dass man einem und demselben Sinn ganz verschiedene Inhalte, und dass man einen und denselben Inhalt zwei verschiedenen Sinnen zuschreibt?", Stumpf (1873, p. 6).

[150] Stumpf (1873, p. 2).

contributes an interesting viewpoint on tactile perception of space, adding a further layer to the nativistic appraisal of the notion.

As I have discussed above, despite defining spatiality as an innate and congenital notion based on the anatomical and physiological properties of some sense organs, Müller's and Weber's positions were fairly cautious attributing apriority only to ideas of extension and surface, i.e., to bidimensional space. Depth perception was in fact treated differently, as derived from the combination of tactile and visual data with successive movements. The idea that depth is acquired through experience and not given immediately in sensory perception was, as a matter of fact, widely accepted and often represented a point of agreement between nativism and empiricism. Stumpf's position, however, proves to be different and in a quite interesting way: since the idea of space presumed by phenomenology is the one given by intuition, and since intuitive space is a rather multifaceted notion, our spatial intuition of an object is not limited to only two dimensions. Rather, the percept presents immediately itself to the perceiver in a three-dimensional form. Depth constitutes an integrative part of this percept, not a later acquisition. Yet, though it is quite "intuitive" to consider sight as a sense of depth perception (*Tiefenvorstellung*), it seems difficult to determine how a tactile experience could provide the subject with an immediate intuition of depth.

In this regard, Stumpf remarks that, as soon as we touch a surface, we immediately perceive the surface as either plane or curved. To imagine it otherwise would be impossible, as "to imagine a triangle that is neither right-angled, acute-angled, nor obtuse-angled".[151] Now, each of those surfaces implies a third dimension, since "they describe something about the surface that relates to depth", in particular the presence or absence of outward/inward deflections (*Vorhandensein oder Fehlen von Ausbiegungen*).[152] One, Stumpf continues, would "[...] mistakenly believe that only the curved surface implies depth, and that the flat surface negates it".[153] This is not the case, however. "Every negative concept", upon closer inspection, "contains everything that the positive one contains, and only adds the negation. So whichever of the two is the positive concept, if one of them contains the idea of the third dimension, the other must also contain it."[154] Depth is, in other words, directly implied—and hence immediately given—by any surface whatsoever. A plane surface, not carrying any hint of outward or inward deflections, tells the perceiver that

[151] "Es ist ebenso unmöglich, eine Fläche vorzustellen, die keins von beiden wäre, wie eine Linie, die weder gerade noch ungerade, ein Dreieck, das weder recht- noch spitz- noch stumpfwinklig, eine Farbe, die weder gelb noch rot, usw. Wäre", Stumpf (1873, p. 176). This is the classic example proposed by Berkeley against Locke's theory of abstract ideas (Berkeley [1874, p. 182]).

[152] "Ebenheit und Krümmung aber involvieren die dritte Dimension. Sie sagen etwas von der Fläche aus, was Bezug hat auf die Tiefe; das Vorhandensein oder Fehlen von Ausbiegungen nach der Tiefe hin", Stumpf (1873, p. 177).

[153] "Fälschlich würden wir glauben, dass nur die krumme Fläche Tiefevorstellungen impliziere, und die ebene sie vielmehr negiere", *ibidem*.

[154] "Denn jeder negative Begriff enthält alles, was der positive enthält, und fügt nur eben die Negation hinzu. Welcher von beiden also auch der positive Begriff ist, so muss, wenn einer von beiden die Vorstellung der dritten Dimension enthält, auch der andere sie enthalten", *ibidem*.

they are confronted with something *that is not* three-dimensional. This kind of information is indeed two-fold: it amounts to saying that the surface is bidimensional, but also points, albeit in a negative way, to the idea of a third dimension. On the other hand, by touching a curved surface, the presence of inward or outward deflections positively communicates that the surface is not only high and long, but also deep.

Stumpf of course does not go so far as to affirm that exact appreciation of these three dimensions is given by the perception of a surface. Further exploration is needed to determine the precise height, length, and width of a touched object. Yet, the idea of an *a priori* intuition of the third dimension is something that was lacking in the debate on spatial perception of the period which complements, in a quite novel fashion, what had already been stated by Müller's and Weber's nativism.

10.6 Conclusions

The question that informed this investigation on space perception in nineteenth-century Germany is how the sense of touch interplays with the notion of *a priori*. This is a question of crucial importance. Not only does it make it possible to trace the evolution of a debate which, beginning with the so-called Molyneux problem, has occupied the minds of a large number of philosophers and scientists, but it also allows us to look at this very debate from a different perspective.

There is no doubt that Kant's introduction of the concept of the *a priori* in the late eighteenth century could only redefine the boundaries of the subsequent discussion of the perception of space and the related question of the innate or acquired origin of knowledge. Yet, the way in which this redefinition was achieved presents elements of unquestionable novelty. This chapter attempted to examine one specific (though not secondary) element, namely the relationship that the *a priori* has with touch in the debate on spatial perception. The idea behind our analysis is that touch (sometimes understood as a sense organ, sometimes as a sensory modality) is not simply one of several keys to spatiality. Rather, it plays a privileged and paradigmatic role in the perception of space. Studies on this topic have often tended to either equate the roles of sight and touch in the nineteenth-century German debate, or to focus specifically on the role of sight, probably because the Helmholtz-Hering debate was considered exemplary in this regard. However, if we look at the various and meticulous analyses of spatial perception proposed at the time, the picture seems to take on a different aspect (see Table 10.1).

In Müller's and Weber's reflections, space is regarded as content of an *a priori* perception, stemming directly from inborn, innate, and congenital properties of the sense organs. Although they were united in affirming that both the tactile and the optical organ were able to immediately provide the subject with the spatial, bidimensional configuration of an object, they also interpreted the activity of these sense organs as a form of touch. In criticizing such previous accounts of the problem, Lotze emphasized how they tended to depict sense organs, such as sight and

Table 10.1 Table summarizing the main positions discussed in this chapter

Author	Underlying perspective concerning space	Sensations and spatial localization	Role of touch in spatial perception	Relationship between touch and sight	*A priori* elements in space perception
Johannes Peter Müller, physiologist	Nativism	Immediate localization of tactile and visual sensations (only in 2D)	Touch and sight both yield 2D space; touch primary for resistance, exteriority, solidity	Isomorphism between tactile and visual sensations (nerve correspondence)	Innate spatial disposition of nerves grants sensations a 2D spatial quality
Ernst Heinrich Weber, physiologist	Nativism	Space arises from awareness of sensory circles and their distances (body mapping)	Primacy of touch: (1) Unit area as standard of spatial estimation; (2) Sensory circles tied to tactile surfaces	Sight analogized to touch; its spatial power depends on tactile properties	An "intellectual instinct" compels spatial organization
Hermann Lotze, philosopher	Constructivism / mentalism	Sensations lack spatial quality (intensive, not extensive). Space re-created in mind via *local signs*.	Tactile space perceived through local signs (similar to Weber)	Touch and sight both provide local signs; mind arranges sensations spatially. Yet, local signs = coenesthesia (internal touch).	Space is originally *a priori* in mind; local signs explain assignment of place to sensations
Wilhelm Wundt, psychologist	Constructivism / genetic theory	Spatial ideas arise from fixed order of parts; not in sensations but formed by mind via *fusion* of contents	Space representation stems from motion sensations + local signs; tactile/motor representations central	Space perception = fusion of tactile, visual, kinesthetic sensations; vision awakened by tactile local signs. Two readings: Circularity touch–sight, or primacy of touch/motion	No *a priori* spatial elements
Carl Stumpf, psychologist	Phenomenology / nativism	Space is given in sensation as intuition, immediately 3D (depth intrinsic)	Touch not privileged (co-perception of space) but crucial for depth perception	Space co-perceived by multiple senses (sight, touch, etc.); relative equality between touch and sight	Structural forms (incl. space) are intrinsic, *a priori* properties of percepts

touch, as "groping hands" aimed at grasping qualitative sensations and conveying them to the brain in association with their spatial properties. Lotze's theory of local signs was certainly critical of this very position. Nonetheless it shares some of its crucial ideas. First and foremost, the idea of the apriority of space. For Lotze, however, the *a priori* is connected to a special disposition of the mind (and not to inborn physiological mechanisms), and thus overcomes the previous correlation between sense organs and apriorism. Yet, despite this major change of focal point, Lotze ends up describing local signs as sensations resembling the feelings of fatigue and vigor that accompany the state of one's own body. Local signs thus are interpreted in tight connection with the phenomenon of *coenesthesia*, which was traditionally regarded as a form of internal tact or common sensibility.

The use of touch as a *paradigm* through which to explain the origin of space perception is not absent even in empiricist accounts as Wundt's. In Wundt's analysis of space, the idea that visual space perception determines tactile space in sighted persons, and not the reverse, still does not deprive touch of its leading role. He positions a combination of tactile (/semi-tactile) and motor sensations, fused together by the mind, at the core of space perception. It is not the centrality of touch which is definitively left out of Wundt's reflection, but the presence of *a priori* elements in representations of space: everything is acquired through experience and subsequently processed by psychic functions. In this sense, empiricism, at its peak, elides the apriority of space, though Wundt still acknowledges touch as a fundamental component of space perception. It is interesting to note, as a sort of chiasmatic movement, the return of nativism in the late nineteenth-century German debate in the form of Stumpf's phenomenology. If in Stumpf the apriority of space is reaffirmed, touch's role is finally overcome, as touch becomes a mere component of the multi-sensorial perception of space.

The historical parable traced here ends with a misplacement of the paradigmatic role of touch in spatial perception. It is undeniable though that touch has been an indispensable *fulcrum* in the development of this very debate. Indeed, both nativism and empiricism (/constructivism) have dealt with touch in spatial perception in different ways, emphasizing its fundamental importance in such perception. Similarly, different and often conflicting positions have addressed the peculiar link between touch and the *a priori*. The picture that emerges is an extremely varied one, which cautions against a uniform description of the relationship between touch and spatiality in nineteenth-century German discussions. Yet, it can be said that a certain "age of touch" has dominated this debate and ultimately defined its overall physiognomy.

References

Antonelli, M., & Boccaccini, F. (2021). *Franz Brentano. Mente, coscienza, realtà*. Carocci.
Ash, M. G. (1980). Experimental Psychology in Germany before 1944: Aspects of an Academic Identity Problem. *Psychological Research, 42*, 75–86. https://doi.org/10.1007/BF00308693
Bain, A. (1855). *The Senses and the Intellect*. W. Parker and son.
Berkeley, G. (1709). *An Essay towards a New Theory of Vision*. Aaron Rhames.

Berkeley, G. (1874). *A Treatise Concerning the Principles of Human Knowledge*. J. B. Lippincott & Co. (Original work published 1710)
Bonacchi, S., & Boudewijnse, G.-H. (2011). *Carl Stumpf–From Philosophical Reflection to Interdisciplinary Scientific Investigation*. Krammer.
Bonnier, P. (1905). L'Aschématie. *Revue Neurologique, 13*, 605–609. https://doi.org/10.1016/j.yebeh.2009.09.020
Boring, E. G. (1929). *A History of Experimental Psychology*. The Century Co.
Boring, E. G. (1942). *Sensation and Perception in the History of Experimental Psychology*. Appleton/Century/Crofts, Inc..
Brentano, F. (1895). *Meine letzten Wünsche für Österreich*. Cotta.
Bridges, J. W. (1912). Doctrine of Specific Nerve Energies. *The Journal of Philosophy. Psychology and Scientific Methods, 9*(3), 57–65. https://doi.org/10.2307/2013786
Bück-Rich, U. (1970). Ernst Heinrich Weber und der Anfang einer Physiologie der Hautsinne. In *Zürcher Medizingeschichtliche Abhandlungen No. 70*. Juris Druck.
Cassirer, E. (1912). Hermann Cohen and the Renewal of Kantian Philosophy (L. Patton Trans). *Angelaki, 10*(1), 95–108. https://doi.org/10.1080/09697250500225768
Centi, B. (2011). Stumpf and Lotze on Space, Reality, Relation. In S. Bonacchi & G.-H. Boudewijnse (Eds.), *Carl Stumpf—From Philosophical Reflection to Interdisciplinary Scientific Investigation* (pp. 69–81). Krammer.
Chottin, M. (2014). *Le Partage de l'empirisme: une histoire du problème de Molyneux aux XVIIe et XVIIIe siècles*. Champion.
Costa, V. (1996). Alle origini della fenomenologia: Husserl e Stumpf sul problema dello spazio. *Rivista di Storia della Filosofia, 1*, 165–185.
De Vignemont, F., & Massin, O. (2015). Touch. In M. Matthen (Ed.), *The Oxford Handbook of the Philosophy of Perception* (pp. 294–313). Oxford University Press.
Degenaar, M. (1996). *Molyneux's Problem. Three Centuries of Discussion on the Perception of Forms*. Kluwer.
Dewalque, A. (2012). Intentionnalité *cum fundamento in re*: La constitution des champs sensoriels chez Stumpf et Husserl. *Bulletin d'Analyse Phénoménologique, 1*, 70–96.
Diderot, D. (2011). Letter on the Blind for the Use of Those Who Can See. In K. E. Tunstall (Ed.), *Blindness and Enlightenment* (pp. 177–228). Continuum. (Original work published 1749)
Ferretti, G., & Glenney, B. (2020). *Molyneux's Question and the History of Philosophy*. Routledge.
Fisette, D. (2009). Stumpf and Husserl on Phenomenology and Descriptive Psychology. *Gestalt Theory, 32(2), 175–190.*
Fisette, D. (2014). La théorie des signes locaux de Hermann Lotze et la controverse empirisme-nativisme au XIXe siècle. In *Lotze et son héritage* (pp. 45–71). Peter Lang.
Fisette, D. (2006). *Carl Stumpf. Renaissance de la philosophie*. Vrin.
Fisette, D., & Martinelli, R. (Eds.). (2015). *Philosophy from an Empirical Standpoint: Essays on Carl Stumpf*. Rodopi.
Forest, D. (2004). Le concept de proprioception dans l'histoire de la sensibilité interne. *Revue d'Histoire Des Sciences, 57*(1), 5–31.
Fries, J. F. (1839). *Über den optischen Mittelpunkt im menschlichen Auge*. Cröker.
Gibson, J. J. (1966). *The Senses Considered as Perceptual Systems*. Houghton Mifflin Co.
Gundlach, H. (1992). *Entstehung und Gegenstand der Psychophysik*. Springer.
Gutiérrez, J. (2021). Hermann Von Helmholtz, Ewald Hering and Color Vision: A Controversy over Styles of Reasoning? *Manuscrito, 44*(1), 37–97. https://doi.org/10.1590/0100-6045.2021.v44n1.jg
Hamilton, W. (1861). *Lectures on Metaphysics and Logic* (Vol. 2). Blackwood and sons.
Hatfield, G. (1990). *The Natural and the Normative: Theories of Spatial Perception from Kant to Helmholtz*. MIT Press.
von Helmholtz, H. (1867). *Handbuch der physiologischen Optik*. Leopold Voss.
von Helmholtz, H. (1893). Recent Progress of the Theory of Vision. In E. Atkinson (Ed.), *Popular Lectures on Scientific Subjects*. Longmans, Green, and Co.

Herbart, J. F. (1816). *Lehrbuch zur Psychologie*. August Wilhelm Unzer.
Hoffmeister, J. (1955). *Wörterbuch der Philosophischen Begriffe*. Felix Meiner.
Ierna, C. (2009). Husserl et Stumpf sur la Gestalt et la fusion. *Philosophiques, 36*(2), 489–510. https://doi.org/10.7202/039482ar
Isaac, A. M. C. (2019). Realism without Tears I: Müller's Doctrine of Specific Nerve Energies. *Studies in History and Philosophy of Science, 78*, 83–92. https://doi.org/10.1016/j.shpsa.2019.01.002
James, W. (1887). The Perception of Space. I. *Mind, 12*(45), 1–30. https://doi.org/10.1093/mind/os-12.45.1
Leary, D. E. (1978). The Philosophical Development of the Conception of Psychology in Germany, 1780–1850. *The History of the Behavioral Sciences, 14*(2), 113–121. https://doi.org/10.1002/1520-6696(197804)14:2
Leary, D. E. (1982). Immanuel Kant and the Development of Modern Psychology. In W. R. Woodward & M. G. Ash (Eds.), *The Problematic Science: Psychology in Nineteenth-Century Thought* (pp. 17–42). Praeger.
Leibniz, G. W. (1921). *Nouveaux essais Sur l'entendement humain*. Flammarion. (Original work published 1765).
Lenoir, T. (1993). The Eye as Mathematician: Clinical Practice, Instrumentation, and Helmholtz's Construction of an Empiricist Theory of Vision. In D. Cahan (Ed.), *Hermann von Helmholtz and the Foundations of Nineteenth-Century Science* (pp. 109–153). University of California Press.
Lenoir, T. (2006). Operationalizing Kant: Manifolds, Models, and Mathematics in Helmholtz's Theories of Perception. In M. Friedman & A. Nordmann (Eds.), *The Kantian Legacy in Nineteenth-Century Science* (pp. 141–210). MIT Press.
Lewin, K. (1937). Carl Stumpf. *Psychological Review, 44*, 189–194. 347–422.
Locke, J. (1786). *An Essay Concerning Human Understanding*. (Vol. 1–2). W. Sleater, H. Chamberlaine, and J. Potts. (Original work published 1690).
Locke, J. (1978). *The Correspondence of John Locke*. Clarendon Press.
Lotze, R. H. (1846). Seele und Seelenleben. In R. Wagner (Ed.), *Handwörterbuch der Physiologie* (Vol. 4, pp. 172–190). Vieweg.
Lotze, R. H. (1852). *Medicinische Psychologie oder Physiologie der Seele*. Weidmann.
Lotze, R. H. (1856). *Mikrokosmos. Ideen zur Naturgeschichte und geschichte der Menschheit. Versuch einer Anthropologie* (Vol. 1). Hirzel. Engl. tr. *Microcosmus: An Essay Concerning Man and His Relation to the World* (E. Hamilton & E. E. C. Jones Trans.). Scribner, 1886.
Lotze, R. H. (1873). Mittheilung an C. Stumpf in Betreff der Lehre von den Localzeichen. In C. Stumpf, *Über den psychologischen Ursprung der Raumvorstellung (pp. 315–324). Hirzel.*
Lotze, R. H. (1877). De la formation de la notion d'espace. *Revue Philosophique De La France Et De L'Étranger, 4*, 345–365.
Lotze, R. H. (1879). *System der Philosophie. Zweiter Theil: Drei Bücher der Metaphysik*. Hirzel.
Lotze, R. H. (1881). *Grundzügen der Psychologie. Diktate aus den Vorlesungen* (2nd ed.). Hirzel.
Lotze, R. H. (1937). 21 Briefe von Hermann Lotze an Ernst Friedrich Apelt (1835–1841). *Blatter fur deutsche Philosophie, 10*, 329–331.
Martinelli, R.. (2009). C. Stumpf, *La rinascita della filosofia. Saggi e conferenze. Quodlibet.*
Meischner, W. (1993). Ernst Heinrich Weber und die Psychologie. *Proceedings of the II. Weber-Symposium* (pp. 116–22). Fachbereich Physik der Martin Luther Universität. Halle: Wittenberg und Wilhelm Weber Gesellschaft.
Mill, J. S. (1930). *A System of Logic: Ratiocinative and Inductive. Being a Connected View of the Principles of Evidence, and Methods of Scientific Investigation*. Longman, Greens, and Co. (Original work published 1843).
Milkov, N. (2015). Carl Stumpf's Debt to Hermann Lotze. In D. Fisette & R. Martinelli (Eds.), *Philosophy from an Empirical Standpoint: Essays on Carl Stumpf* (pp. 101–122). Rodopi.
Milkov, N. (2023). *Hermann Lotze's Influence on Twentieth Century Philosophy*. De Gruyter.
Molyneux, W. (1692). *Dioptrica Nova: A Treatise of Dioptricks, in two parts*. Tooke.

Montgomery, E. (1889). Mental Activity. *Mind, 14*, 488–510. https://doi.org/10.1093/mind/xiv.56.488
Moore, G. E. (1957). Nativism. In J. M. Baldwin (Ed.), *Dictionary of Philosophy and Psychology* (Vol. 2). Peter Smith. (Original work published 1902).
Morgan, M. J. (1977). *Molyneux's Question: Vision, Touch and the Philosophy of Perception*. Cambridge University Press.
Müller, J. P. (1826). *Zur vergleichenden Physiologie des Gesichtssinnes des Menschen und der Thiere. Cnobloch.*
Müller, J. P. (1835). *Handbuch der Physiologie des Menschen* (Vol. 1). Verlag von J. Hölscher. Engl. tr. *Elements of Physiology* (W. Baly, Trans.) (Vol. 1). Taylor and Walton, 1838.
Müller, J. P. (1840). *Handbuch der Physiologie des Menschen* (Vol. 2). Verlag von J. Hölscher. Engl. tr. *Elements of Physiology* (W. Baly, Trans.) (Vol. 2). Taylor and Walton, 1842.
Murray, D. J., & Farahmand, B. (1995). The Historical Background of Weber's Research on Touch. In C.-A. Possamaï (Ed.), *Fechner Day 95* (pp. 21–28). International Society for Psychophysics.
Nicolas, S. (2002). La fondation de la psychophysique de Fechner. *L'Annee Psychologique, 102*(2), 255–298. https://doi.org/10.3406/psy.2002.29592
Nicolas, S., & Vobořil, D. (2019). Weber's Compass and Aesthesiometers: History of the Technical Evolution of Devices for Tactile Discrimination. *L'Année Psychologique, 119*(1), 97–170. https://doi.org/10.3917/anpsy1.191.0097
Pastore, N. (1971). *Selective History of Theories of Visual Perception: 1650–1950*. Oxford University Press.
Popescu, V. (2003). Espace et mouvement chez Stumpf et Husserl. Une approche méréologique. *Studia Phaenomenologica, 3*(1), 115–129. https://doi.org/10.7761/sp.3.1-2.115
Pradelle, D. (2013). *Généalogie de la raison. Essai sur l'historicité du sujet transcendantal de Kant à Heidegger*. Puf.
Pradelle, D. (2015). The Autonomy of the Sensible and the Desubjectification of the *A priori* by Stumpf. In D. Fisette & R. Martinelli (Eds.), *Philosophy from an Empirical Standpoint: Essays on Carl Stumpf* (pp. 229–262). Rodopi.
Reid, T. (1994). *An Inquiry into the Human Mind on the Principles of Common Sense* In W. Hamilton (Ed.), *The Works of Thomas Reid* (Vol. 1). Thoemmes Press. (Original work published 1764).
Ribot, T.-A. (1878). Les théories allemandes sur l'espace tactile. *Revue Philosophique, 6*, 130–145.
Rollinger, R. D. (2001). Lotze on the Sensory Representation of Space. In L. Albertazzi (Ed.), *The Dawn of Cognitive Science* (pp. 103–122). Springer Science.
Ross, H. E. (1995). Weber Then and Now. *Perception, 24*, 599–602. https://doi.org/10.1068/p240599
Ross, H. E., & Bischof, K. (1981). Wundt's Views on Sensations of Innervation: A Reevaluation. *Perception, 10*(3), 319–329. https://doi.org/10.1068/p100319
Ross, H. E., & Murray, D. J. (2017). Introduction. In H. E. Ross & D. J. Murray (Eds.), *E. H. Weber on the Tactile Senses* (pp. 1–20). Routledge.
Smith, R. (2011). "The Sixth Sense": Towards a History of Muscular Sensation. *Gesnerus, 68*(1), 218–271. https://doi.org/10.1163/22977953-06802004
Smith, R. (2019). *The Sense of Movement. An Intellectual History. Process Press.*
Spiegelberg, H. (1960). *The Phenomenological Movement: A Historical Introduction*. Nijhoff.
Sprung, H. (2006). *Carl Stumpf–Eine Biografie. Von der Philosophie zur Experimentellen Psychologie*. Profil.
Starobinski, J. (1977). Le concept de cénesthésie et les idées neuropsychologiques de Moritz Schiff. *Gesnerus, 34*, 2–20.
Stumpf, C. (1873). *Uber den Psychologischen Ursprung der Raumvorstellung*. Verlag von S. Hirzel.
Stumpf, C. (1893). Zum Begriff der Lokalzeichen. *Zeitschrift für Psychologie und Physiologie der Sinnesorgane, 4*, 70–73.

Stumpf, C. (1906). Erscheinungen und psychische Funktionen. In *Abhandlungen der Königlich-Preußischen Akademie der Wissenschaften, Philos.-historische Klasse*. Verlag der Königlichen Akademie der Wissenschaften.

Stumpf, C. (1924). Carl Stumpf (Selbstdarstellung). In R. Schmidt (Ed.), *Die Philosophie der Gegenwart in Selbstdarstellungen. Bd. 5* (pp. 1–57). Meiner.

Stumpf, C. (2015). Stumpf's "Einleitung zu Brentanos Briefen an mich", followed by selected letters from Brentano and Stumpf. In D. Fisette & R. Martinelli (Eds.), *Philosophy from an Empirical Standpoint: Essays on Carl Stumpf* (pp. 491–530). Rodopi.

Sully, J. (1878). The Question of Visual Perception in Germany. *Mind, 3*(9), 1–23. https://doi.org/10.1093/mind/os-3.10.167

Textor, M. (2017). "Apprehending a Multitude as Unity": Stumpf on Perceiving Space and Hearing Chords. In S. Lapointe (Ed.), *Philosophy of Mind in the Nineteenth Century: The History of the Philosophy of Mind, Volume 5* (pp. 198–213). Routledge.

Treviranus, G. R. (1837). *Beiträge zur Aufklärung der Erscheinungen und Gesetze des organischen Lebens (Vol. 3)*. Verlag von Johann Georg Heyse.

Turner, R. S. (1994). *In the Eye's Mind: Vision and the Helmholtz-Hering Controversy*. Princeton University Press.

Weber, E. H. (1834). De subtilitate tactu. In *De pulsu, resorptione, auditu et tactu. Annotationes anatomicae* (pp. 44–175). Apud C. F. Koehler. Engl. tr. *E. H. Weber on the Tactile Senses* (pp. 21–136) (H. E. Ross, & D. J. Murray, Trans. and Eds.). Routledge 2017.

Weber, E. H. (1905). *Tastsinn und Gemeingefühl.* Verlag von Wilhelm Engelmann. Engl. Tr. *E. H. Weber on the tactile senses* (pp. 137–246) (H. E. Ross, D. J. Murray, Trans. and Eds.). Routledge 2017. (Original work published 1846).

Woodward, W. R. (1975). Hermann Lotze's Critique of Johannes Müller's Doctrine of Specific Sense Energies. *Medical History, 19*(2), 147–157. https://doi.org/10.1017/S002572730002010X

Woodward, W. R. (1978). From Association to Gestalt: The Fate of Hermann Lotze's Theory of Spatial Perception, 1846–1920. *Isis, 69*(4), 572–582. https://doi.org/10.1086/352115

Wundt, W. (1896). *Grundriss der Psychologie*. Verlag von Wilhelm Engelmann. Engl. tr. *Outlines of Psychology* (C. H. Judd, Trans.). Williams and Norgate, 1897.

Index

A
Adam of Bockenfield, 12, 28, 29, 34–40, 45
Adam of Whitby, 28, 29, 34, 37, 45
Albert the Great, 6, 13, 52, 55, 59, 65–69, 71
Albertini, I.F., 195, 201
Alexander of Aphrodisias, 112
Aristotle, 5, 24, 52, 80, 104, 122, 175, 233
Arnauld, A., 155, 161, 165
Aubry, C. (Vauclius Dathirius Bonglarus), 197
Auriol, P., 86, 91, 92, 96
Auzout, A., 181
Averroes (Ibn Rushd), 6, 13, 27–29, 34, 41, 45, 53, 55, 90, 91, 100, 106, 110, 130, 133
Avicenna (Ibn Sīnā), 6, 10, 13, 54–56, 58, 59, 61, 63, 67–71, 103, 106, 110, 112, 114, 115, 130

B
Babin, F., 16
Bacon, R., 91
Baglivi, G., 184
Bain, A., 250, 251, 257
Baines, T., 181
Baldwin, J.M., 231
Baldwin of Canterbury, 42
Barner, J., 174, 175
Bauhin, C., 178, 179
Bell, C., 9, 11
Bellini, L., 180, 181, 183, 186, 200
Belluto, B., 15, 121–140
Berkeley, G., 7, 11, 18, 209–213, 215, 216, 219, 222, 225, 226, 230, 236, 256, 258
Bernard of Arezzo, 78, 80, 98
Bernard of Gordon, 58
Blaes, G., 202
Boerhaave, H., 200
Bonaventure of Bagnoregio, 42
Bonfigliuoli, S., 181, 194, 197
Bonnier, P., 236
Borelli, G.A., 174, 175, 179–183, 186, 197
Boyle, R., 17
Brentano, F., 250, 255, 256
Brown, T., 210, 225
Buridan, J., 6, 13, 53, 56, 57, 59, 81, 82
Burley, W., 41, 42

C
Capponi, G.B., 193
Cardano, G., 57
Casseri, G.C., 178, 179
Castelli, P., 175
Cavendish, William, 1st Marquess of Newcastle, 154
Chatton, W., 86
Colombo, R., 184
Condillac, É.B., 7, 11, 18, 210, 213–215, 217, 223–225, 248

D
Damasio, A., 16, 146
Democritus, 5, 7, 14, 28, 40, 42, 81, 83, 88, 89, 94–98
Descartes, R., 4–8, 16, 17, 143–169, 222, 230
Despars, J., 18, 58, 69, 71
Destutt de Tracy, A., 210
Diderot, D., 1, 2, 230
Dino del Garbo, 13, 69, 70

C. Beneduce et al. (eds.), *The Sense of Touch*, Studies in the History of Philosophy of Mind 32, https://doi.org/10.1007/978-3-032-16308-0

du Hamel, J-B., 195
du Laurens, A., 178, 185, 186
Duns Scotus, J., 15, 122, 135

E
Elisabeth, Princess of Bohemia, 155
Empedocles, 28
Epicurus, 84, 92, 94
Euclid, 28
Eustachi, B., 181

F
Fabrici, G. (Hieronymus Fabricius of Aquapendente), 177
Fantuzzi, G., 191, 194
Fechner, G.T., 235
Fernel, J., 235
Fieno, G., 98
Finch, J., 181
Forte, A., 186
Fracassati, C., 179–181, 183, 200
Fracastoro, G., 57
Fries, J.F., 244

G
Galen, 6, 10, 14, 15, 95, 104, 106, 109–112, 114–117, 125, 130, 131, 177, 178, 193
Gallego, P., 13, 68, 71
Gassendi, P., 7, 17, 175, 179, 188, 189
Gentile da Foligno, 13, 69, 71
Geoffrey of Aspall, 12, 31, 32, 39, 41, 42, 44, 46
Giles of Rome, 14, 96–99
Glisson, F., 195
Grosseteste, R., 28
Guglielmini, D., 197
Guido Vernani da Rimini, 58

H
Haller, A., 200
Hamilton, W., 57, 230
Hartley, D., 210
Helmholtz, H., 9, 230–234, 239, 241, 242, 250, 259
Herbart, J.F., 233, 239, 243, 250, 255
Hering, E., 232, 259
Hippocrates, 58, 62, 65, 70, 110, 176, 179
Hugh of Saint Victor, 42
Hume, D., 18, 78, 210, 216, 219
Husserl, E., 256

J
James, W., 231, 242
Jessen, J., 177

K
Kant, I., 9, 18, 230, 231, 233, 234, 236, 239, 241, 256, 259

L
Lanzi Paltroni, G.C., 194
Leibniz, G.W., 230
Leucippus, 96, 97
Locke, J., 9, 16, 218, 229, 230, 236–239, 257, 258
Lotze, H., 19, 232, 242–250, 252, 255, 259–261
Lucretius, 4, 5, 14, 15, 92, 94, 104

M
Malebranche, N., 162
Malpighi, M., 6, 10, 17, 173–261
Manzi, G.G., 193
Marsili, L.F., 194, 195
Mastri, B., 15, 121–140
Mazzetti, S., 193–195
Medici, Ferdinando II de', Grand Duke of Tuscany, 181
Mercuriale, G., 177
Merleau-Ponty, M., 18, 226
Mesland, D., 169
Mill, J.S., 230
Mini, P., 191, 193
Molinetti, A., 194
Molyneux, W., 9, 229, 230, 237, 259
More, H., 4, 157, 168, 194
Müller, J.P., 19, 232–243, 245, 247, 248, 251, 252, 254, 255, 257–260
Muscio (Mustio), 63

N
Newton, I., 230
Nicholas of Autrecourt, 13, 77–99
Nicolas de Bar-le-Duc, 96

O
Olivi, P.J., 6, 13, 56

P
Peckham, J., 31
Pedro Hurtado de Mendoza, 127
Peter Lombard, 65, 73
Peter of Abano (Pietro d'Abano), 6, 13, 52, 58, 59, 64–66, 106
Peter of Spain, 13, 64–66, 68, 71
Pighi, G., 194
Plato, 8, 10, 25, 28, 111, 186
Plutarch, 95
Poinsot, J. (John of St. Thomas), 15, 121–140
Pomponazzi, P., 65
Priestley, J., 210
Protagoras, 83, 94, 95
Pseudo-Aristotle, 53, 63

R
Reid, T., 7, 18, 210, 215–219, 223, 225, 238
Rhazes (Abū Bakr al-Rāzī), 62
Ribot, T.A., 231, 232
Richard Rufus of Cornwall, 41
Riolan, J. (the Elder), 14, 105–113, 115–117
Rossetti, D., 186
Ruffo, G., 175

S
Savonarola, M., 60, 61
Sbaraglia, G.G., 17, 191, 193, 194, 199, 200, 202
Scaliger, J.C., 57
Schenck, J.T., 200
Scilla, A., 175, 184
Sextus Empiricus, 14, 95
Shepherd, M., 5, 7, 9, 17, 18, 209–226
Sherrington, C., 144, 145
Simplicius, 112
Soranus of Ephesus, 63
Stancari, A.G., 195
Steno, N., 195
Stewart, D., 210
Stumpf, C., 19, 232, 245, 255–261
Suárez, F., 15, 121–140

T
Thomas Aquinas, 53, 54, 58, 64, 65, 91, 123
Thomas of Cantimpré, 45
Tommaso del Garbo, 66
Torrigiano, P., 58
Treviranus, G.R., 235, 236
Trotula, 63

V
Valsalva, A.M., 191
Vanini, G.C., 57
Vath, J., 13, 59
Vericus (or Wericus or Ulricus), 13, 68, 69, 71
Vesalius, A., 177, 178, 181
Volpari, G.A., 193–195

W
Weber, E.H., 11, 19, 233–242, 244, 245, 247, 248, 252, 254, 255, 257–260
William of Ockham, 86
Wodeham, A., 86
Wundt, W., 19, 232, 242, 248–255, 257, 260, 261

The manufacturer's authorised representative in the EU is Springer Nature Customer Service Centre GmbH, Europaplatz 3, 69115 Heidelberg, Germany. If you have any concerns regarding our products, please contact ProductSafety@springernature.com

Printed and bound by CPI Group (UK) Ltd, Croydon, CR0 4YY
17/09/2026
02218432-0006